S0-ARN-832

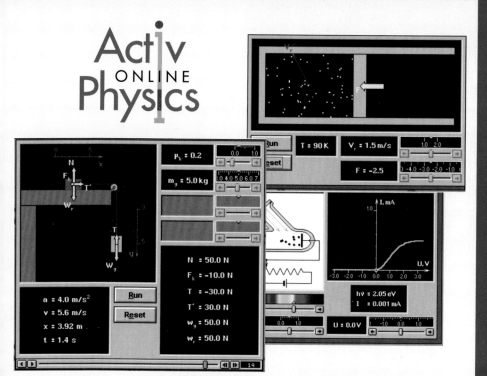

ActivPhysics™ OnLine utilizes visualization, simulation and multiple representations to help you better understand key physical processes, experiment quantitatively, and develop your critical-thinking skills. This library of online interactive simulations are coupled with thought-provoking questions and activities to guide your understanding of physics.

Website: www.aw-bc.com/knight

Minimum System Requirements:
Windows: 250 MHz; OS 98, NT, ME, 2000, XP
Macintosh: 233 MHz; OS 9.2, 10
Both:
- 64 RAM installed
- 1024 x 768 screen resolution
- Browsers: Internet Explorer 5.5, 6.0; Netscape 6.2.3, 7.0
- Plug Ins: Macromedia's Flash 6.2.3

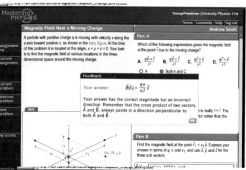

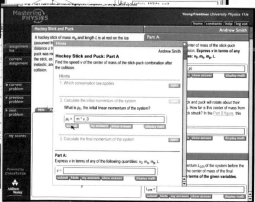

MasteringPhysics™ is the first Socratic tutoring system developed specifically for physics students like you. It is the result of years of detailed studies of how students work physics problems, and where they get stuck and need help. Studies show students who used MasteringPhysics significantly improved their scores on traditional final exams and the Force Concept Inventory (a conceptual test) when compared with traditional hand-graded homework.

With your purchase of a new copy of Knight's *Physics for Scientists and Engineers*, you should have received a Student Access Kit for **MasteringPhysics™** if your professor required it as a component of your course. The kit contains instructions and a code for you to access MasteringPhysics.

If you did not purchase a new textbook and your professor requires you to enroll in the **MasteringPhysics** online homework and tutorial program, you may purchase an online subscription with a major credit card. Go to www.masteringphysics.com and follow the links to purchasing online.

Minimum System Requirements:
Windows: 250 MHz; OS 98, NT, ME, 2000, XP
Macintosh: 233 MHz; OS 9.2, 10
RedHat Linux 8.0
All:
- 64 RAM installed
- 1024 x 768 screen resolution
- Browsers: Internet Explorer 5.0, 5.5, 6.0; Netscape 6.2.3, 7.0; Mozilla 1.2, 1.3

MasteringPhysics™ is powered by MyCyberTutor by Effective Educational Technologies

AW003

Table of Problem-Solving Stra

Note for users of the five-volume edition:
Volume 1 (pp. 1–481) includes chapters 1
Volume 2 (pp. 482–607) includes chapters
Volume 3 (pp. 608–779) includes chapters
Volume 4 (pp. 780–1194) includes chapters
Volume 5 (pp. 1148–1383) includes chapter

Chapters 37–42 are not in the Standard Editi

Physics for Scientists and Engineers

A Strategic Approach

ActivPhysics™ OnLine Activities

 www.aw-bc.com/knight

Physics for Scientists and Engineers
Volume 5
A Strategic Approach

Randall D. Knight

California Polytechnic State University, San Luis Obispo

PEARSON

Addison
Wesley

San Francisco   Boston   New York
Cape Town   Hong Kong   London   Madrid   Mexico City
Montreal   Munich   Paris   Singapore   Sydney   Tokyo   Toronto

Executive Editor:	Adam Black, Ph.D.
Development Editor:	Alice Houston, Ph.D.
Project Manager:	Laura Kenney Editorial & Production Services
Associate Editor:	Liana Allday
Media Producer:	Claire Masson
Marketing Manager:	Christy Lawrence
Market Development:	Susan Winslow
Manufacturing Supervisor:	Vivian McDougal
Art Director:	Blakely Kim
Production Service:	Thompson Steele, Inc.
Text Design:	Mark Ong, Side by Side Studios
Cover Design:	Yvo Riezebos Design
Illustrations:	Precision Graphics
Photo Research:	Cypress Integrated Systems
Cover Printer:	Phoenix Color Corporation
Printer and Binder:	R. R. Donnelley & Sons
Cover Image:	Rainbow/PictureQuest
Credits:	see page x

Library of Congress Cataloging-in-Publication Data
Knight, Randall Dewey.
    Physics for scientists and engineers : a strategic approach / Randall D. Knight.
      p. cm.
    Includes index.
    ISBN 0-8053-8960-1 (extended ed. with MasteringPhysics)
     1. Physics   I. Title.

QC23.2.K65 2004
530--dc22

2003062809

ISBN 0-8053-8976-8 Volume 5 with MasteringPhysics
ISBN 0-8053-9017-0 Volume 5 without MasteringPhysics

1 2 3 4 5 6 7 8 9 10—DOW—06 05 04 03
www.aw-bc.com

Brief Contents

About the Author

Randy Knight has taught introductory physics for over 20 years at Ohio State University and California Polytechnic University, where he is currently Professor of Physics. Professor Knight received a bachelor's degree in physics from Washington University in St. Louis and a Ph.D. in physics from the University of California, Berkeley. He was a post-doctoral fellow at the Harvard-Smithsonian Center for Astrophysics before joining the faculty at Ohio State University. It was at Ohio State that he began to learn about the research in physics education that, many years later, led to this book.

Professor Knight's research interests are in the field of lasers and spectroscopy, and he has published over 25 research papers. He recently led the effort to establish an environmental studies program at Cal Poly, where, in addition to teaching introductory physics, he also teaches classes on energy, oceanography, and environmental issues. When he's not in the classroom or in front of a computer, you can find Randy hiking, sea kayaking, playing the piano, or spending time with his wife Sally and their seven cats.

Credits

Relativity and Quantum Physics, Volume 5

PART VII
Page **1150:** IBM Research, Almaden Research Center.

CHAPTER 36
Page **1151:** John Y. Fowler. Page **1152:** Topham/The Image Works. Page **1172:** Stanford Linear Accelerator Center. Page **1186:** Science Photo Library/Photo Researchers. Page **1187:** Wellcome Dept. of Cognitive Neurology/Science Photo Library/Photo Researchers.

CHAPTER 37
Page **1195:** Richard Megna/Fundamental Photos. Page **1197:** DaimlerChrysler. Page **1201** T: Science Photo Library/Photo Researchers. Page **1201** B: Science Museum/Science and Society Picture Library.
Page **1212:** Gerard Herzberg/Atomic Spectra and Atomic Structure, Prentice-Hall, 1937.

CHAPTER 38
Page **1220:** Courtesy of International Business Machines Corporation. Unauthorized use not permitted. Page **1225** B: DPA/HM/The Image Works. Page **1225** T: Bettman/Corbis. Page **1231:** Dr. Claus Jonsson. Page **1235:** Bettman/Corbis.

CHAPTER 39
Page **1253:** Digital Instruments Inc. Page **1254:** Dr. Claus Jonsson.

CHAPTER 40
Page **1277:** IBM Corporate Archives. Page **1278:** Bettman/Corbis. Page **1310** L: Digital Instruments Inc. Page **1310** R: Prelim Ed., from G. Binnign and H. Rohres, *Surface Science*, 144, p. 321, 1984.

CHAPTER 41
Page **1317:** Ray Nelson/Phototake. Page **1321:** Tom Pantages. Page **1328:** Prelim Ed., from G. Binnign and H. Rohres, *Surface Science*, 144, p. 321, 1984. Page **1335:** Courtesy National Institute of Standards and Technology. Page **1338:** Archivo Iconografico, S.A./Corbis. Page **1343:** Meggers Gallery/American Institute of Physics/Science Photo Library/Photo Researchers.

CHAPTER 42
Page **1352:** Landmann Patrick/Corbis SYGMA. Page **1360:** Bettman/Corbis. Page **1362:** Hulton/Getty Images. Page **1363:** Kevin Fleming/Corbis. Page **1371:** ICRR Institute for Cosmic Ray Research. Page **1375:** Lonnie Duka/Index Stock. Page **1376:** Howard Sochurek/Corbis.

Preface to the Instructor

In 1997 we published *Physics: A Contemporary Perspective*. This was the first comprehensive, calculus-based textbook to make extensive use of results from physics education research. The development and testing that led to this book had been partially funded by the National Science Foundation. In the preface we noted that it was a "work in progress" and that we very much wanted to hear from users—both instructors and students—to help us shape the book into a final form.

And hear from you we did! We received feedback and reviews from roughly 150 professors and, especially important, 4500 of their students. This textbook, the newly titled *Physics for Scientists and Engineers: A Strategic Approach*, is the result of synthesizing that feedback and using it to produce a book that we hope is uniquely tuned to helping today's students succeed. It is the first introductory textbook built from the ground up on research into how students can more effectively learn physics.

Objectives

My primary goals in writing *Physics for Scientists and Engineers: A Strategic Approach* have been:

- To produce a textbook that is more focused and coherent, less encyclopedic.
- To move key results from physics education research into the classroom in a way that allows instructors to use a range of teaching styles.
- To provide a balance of quantitative reasoning and conceptual understanding, with special attention to concepts known to cause student difficulties.
- To develop students' problem-solving skills in a systematic manner.
- To support an active-learning environment.

These goals and the rationale behind them are discussed at length in my small paperback book, *Five Easy Lessons: Strategies for Successful Physics Teaching* (Addison Wesley, 2002). Please request a copy from your local Addison Wesley sales representative if it would be of interest to you (ISBN 0-8053-8702-1).

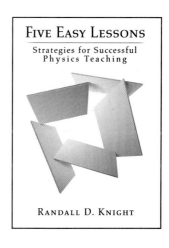

FIVE EASY LESSONS
Strategies for Successful
Physics Teaching

RANDALL D. KNIGHT

Textbook Organization

The 42-chapter extended edition (ISBN 0-8053-8685-8) of *Physics for Scientists and Engineers* is intended for use in a three-semester course. Most of the 36-chapter standard edition (ISBN 0-8053-8982-2), ending with relativity, can be covered in two semesters, but the judicious omission of a few chapters will avoid rushing through the material and give students more time to develop their knowledge and skills.

There's a growing sentiment that quantum physics is quickly becoming the province of engineers, not just scientists, and that even a two–semester course should include a reasonable introduction to quantum ideas. The *Instructor's Guide* outlines a couple of routes through the book that allow most of the quantum physics chapters to be reached in two semesters. I've written the book with the hope that an increasing number of instructors will choose one of these routes.

- **Extended edition,** with modern physics (ISBN 0-8053-8685-8): chapters 1–42.
- **Standard edition** (ISBN 0-8053-8982-2): chapters 1–36.
- **Volume 1** (ISBN 0-8053-8963-6) covers mechanics: chapters 1–15.
- **Volume 2** (ISBN 0-8053-8966-0) covers thermodynamics: chapters 16–19.
- **Volume 3** (ISBN 0-8053-8969-5) covers waves and optics: chapters 20–24.
- **Volume 4** (ISBN 0-8053-8972-5) covers electricity and magnetism, plus relativity: chapters 25–36.
- **Volume 5** (ISBN 0-8053-8975-X) covers relativity and quantum physics: chapters 36–42.
- **Volumes 1–5** boxed set (ISBN 0-8053-8978-4).

The full textbook is divided into seven parts: Part I: *Newton's Laws*, Part II: *Conservation Laws*, Part III: *Applications of Newtonian Mechanics*, Part IV: *Thermodynamics*, Part V: *Waves and Optics*, Part VI: *Electricity and Magnetism*, and Part VII: *Relativity and Quantum Mechanics*. Although I recommend covering the parts in this order (see below), doing so is by no means essential. Each topic is self-contained, and Parts III–VI can be rearranged to suit an instructor's needs. To facilitate a reordering of topics, the full text is available in the five individual volumes listed in the margin.

Organization Rationale: Thermodynamics is placed before waves because it is a continuation of ideas from mechanics. The key idea in thermodynamics is energy, and moving from mechanics into thermodynamics allows the uninterrupted development of this important idea. Further, waves introduce students to functions of two variables, and the mathematics of waves is more akin to electricity and magnetism than to mechanics. Thus moving from waves to fields to quantum physics provides a gradual transition of ideas and skills.

The purpose of placing optics with waves is to provide a coherent presentation of wave physics, one of the two pillars of classical physics. Optics as it is presented in introductory physics makes no use of the properties of electromagnetic fields. There's little reason other than historical tradition to delay optics until after E&M. The documented difficulties that students have with optics are difficulties with waves, not difficulties with electricity and magnetism. However, the optics chapters are easily deferred until the end of Part VI for instructors who prefer that ordering of topics.

More Effective Problem-Solving Instruction

Careful and systematic instruction is provided on all aspects of problem solving. Some of the features that support this approach are described here, and more details are provided in the *Instructor's Guide*.

- An emphasis on using *multiple representations*—descriptions in words, pictures, graphs, and mathematics—to look at a problem from many perspectives.
- The explicit use of *models*, such as the particle model, the wave model, and the field model, to help students recognize and isolate the essential features of a physical process.
- TACTICS BOXES for the development of particular skills, such as drawing a free-body diagram or using Lenz's law. Tactics Box steps are explicitly illustrated in subsequent worked examples, and these are often the starting point of a full problem-solving strategy.

TACTICS BOX 4.3 **Drawing a free-body diagram**

❶ **Identify all forces acting on the object.** This step was described in Tactics Box 4.2.

❷ **Draw a coordinate system.** Use the axes defined in your pictorial representation. If those axes are tilted, for motion along an incline, then the axes of the free-body diagram should be similarly tilted.

❸ **Represent the object as a dot at the origin of the coordinate axes.** This is the particle model.

❹ **Draw vectors representing each of the identified forces.** This was described in Tactics Box 4.1. Be sure to label each force vector.

❺ **Draw and label the *net force* vector $\vec{F}_{\text{net}}$.** Draw this vector beside the diagram, not on the particle. Or, if appropriate, write $\vec{F}_{\text{net}} = \vec{0}$. Then check that $\vec{F}_{\text{net}}$ points in the same direction as the acceleration vector $\vec{a}$ on your motion diagram.

TACTICS BOX 32.2 **Evaluating line integrals**

❶ If $\vec{B}$ is everywhere perpendicular to a line, the line integral of $\vec{B}$ is

$$\int_i^f \vec{B} \cdot d\vec{s} = 0$$

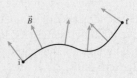

❷ If $\vec{B}$ is everywhere tangent to a line of length L *and* has the same magnitude B at every point, the line integral of $\vec{B}$ is

$$\int_i^f \vec{B} \cdot d\vec{s} = BL$$

- **PROBLEM-SOLVING STRATEGIES** that help students develop confidence and more proficient problem-solving skills through the use of a consistent four-step approach: **MODEL, VISUALIZE, SOLVE, ASSESS**. Strategies are provided for each broad class of problems, such as dynamics problems or problems involving electromagnetic induction. The icon directs students to the specially developed *Skill Builder* tutorial problems in MasteringPhysics™ (see page xi), where they can interactively work through each of these strategies online.

- Worked **EXAMPLES** that illustrate good problem-solving practices through the consistent use of the four-step problem-solving approach and, where appropriate, the Tactics Box steps. The worked examples are often very detailed and carefully lead the student step by step through the *reasoning* behind the solution, not just through the numerical calculations. Steps that are often implicit or omitted in other textbooks, because they seem so obvious to experts, are explicitly discussed since research has shown these are often the points where students become confused.

- **NOTE ▶** Paragraphs within worked examples caution against common mistakes and point out useful tips for tackling problems.

- The *Student Workbook* (see page xi), a unique component of this text, bridges the gap between worked examples and end-of-chapter problems. It provides qualitative problems and exercises that focus on developing the skills and conceptual understanding necessary to solve problems with confidence.

- Approximately 3000 original and diverse *end-of-chapter problems* have been carefully crafted to exercise and test the full range of qualitative and quantitative problem-solving skills. *Exercises*, which are keyed to specific sections, allow students to practice basic skills and computations. *Problems* require a better understanding of the material and often draw upon multiple representations of knowledge. *Challenge Problems* are more likely to use calculus, utilize ideas from more than one chapter, and sometimes lead students to explore topics that weren't explicitly covered in the chapter.

PROBLEM-SOLVING STRATEGY 5.2 **Dynamics problems**

MODEL Make simplifying assumptions.

VISUALIZE

Pictorial representation. Show important points in the motion with a sketch, establish a coordinate system, define symbols, and identify what the problem is trying to find. This is the process of translating words to symbols.

Physical representation. Use a motion diagram to determine the object's acceleration vector $\vec{a}$. Then identify all forces acting on the object and show them on a free-body diagram.

It's OK to go back and forth between these two steps as you visualize the situation.

SOLVE The mathematical representation is based on Newton's second law

$$\vec{F}_{net} = \sum_i \vec{F}_i = m\vec{a}$$

The vector sum of the forces is found directly from the free-body diagram. Depending on the problem, either

- Solve for the acceleration, then use kinematics to find velocities and positions, or
- Use kinematics to determine the acceleration, then solve for unknown forces.

ASSESS Check that your result has the correct units, is reasonable, and answers the question.

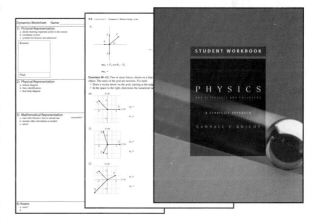

Proven Features to Promote Deeper Understanding

Research has shown that many students taking calculus-based physics arrive with a wealth of misconceptions and subsequently struggle to develop a coherent understanding of the subject. Using a number of unique, reinforcing techniques, this book tackles these issues head-on to enable students to build a solid foundation of understanding.

- A *concrete-to-abstract* approach introduces new concepts through observations about the real world and everyday experience. Step by step, the text then builds up the concepts and principles needed by a theory that will make sense of the observations and make new, testable predictions. This inductive approach better matches how students learn, and it reinforces how physics—and science in general—operates.

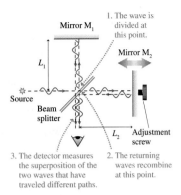

1. The wave is divided at this point.

Mirror M₁

Mirror M₂

L_1

Source

Beam splitter

L_2

Adjustment screw

3. The detector measures the superposition of the two waves that have traveled different paths.

2. The returning waves recombine at this point.

Annotated FIGURE showing the operation of the Michelson interferometer.

■ **STOP TO THINK** questions embedded in each chapter allow students to assess whether they've understood the main idea of a section. The *Stop to Think* questions, which include concept questions, ratio reasoning, and ranking tasks, are primarily derived from physics education research.

■ **NOTE** ▶ paragraphs draw attention to common misconceptions, clarify possible confusions in terminology and notation, and provide important links to previous topics.

■ Unique *annotated figures*, based on research into visual learning modes, make the artwork a teaching tool on a par with the written text. Commentary in blue—the "instructor's voice"—helps students "read" the figure. Students "learn by viewing" how to interpret a graph, how to translate between multiple representations, how to grasp a difficult concept through a visual analogy, and many other important skills.

■ The learning goals and links that begin each chapter outline what the student needs to remember from previous chapters and what to focus on in the chapter ahead.

　▶ **Looking Ahead** lists key concepts and skills the student will learn in the coming chapter.

　◀ **Looking Back** suggests important topics students should review from previous chapters.

■ Unique schematic *Chapter Summaries* help students organize their knowledge in an expert-like hierarchy, from general principles (top) to applications (bottom). Side-by-side pictorial, graphical, textual, and mathematical representations are used to help students with different learning styles and enable them to better translate between these key representations.

■ *Part Overviews and Summaries* provide a global framework for the student's learning. Each part begins with an overview of the chapters ahead. It then concludes with a broad summary to help students draw connections between the concepts presented in that set of chapters. **KNOWLEDGE STRUCTURE** tables in the part summaries, similar to the chapter summaries, help students see a forest rather than dozens of individual trees.

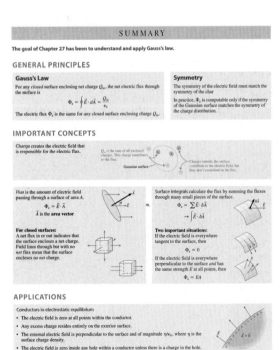

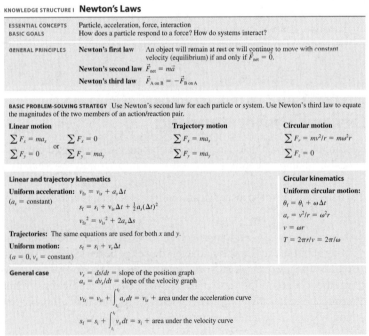

The Student Workbook

A key component of *Physics for Scientists and Engineers: A Strategic Approach* is the accompanying *Student Workbook*. The workbook bridges the gap between textbook and homework problems by providing students the opportunity to learn and practice skills prior to using those skills in quantitative end-of-chapter problems, much as a musician practices technique separately from performance pieces. The workbook exercises, which are keyed to each section of the textbook, focus on developing specific skills, ranging from identifying forces and drawing free-body diagrams to interpreting wave functions.

The workbook exercises, which are generally qualitative and/or graphical, draw heavily upon the physics education research literature. The exercises deal with issues known to cause student difficulties and employ techniques that have proven to be effective at overcoming those difficulties. The workbook exercises can be used in-class as part of an active-learning teaching strategy, in recitation sections, or as assigned homework. More information about effective use of the *Student Workbook* can be found in the *Instructor's Guide*.

Available versions: Extended (ISBN 0-8053-8961-X), Standard (ISBN 0-8053-8984-9), Volume 1 (ISBN 0-8053-8965-2), Volume 2 (ISBN 0-8053-8968-7), Volume 3 (ISBN 0-8053-8971-7), Volume 4 (ISBN 0-8053-8974-1), and Volume 5 (ISBN 0-8053-8977-6).

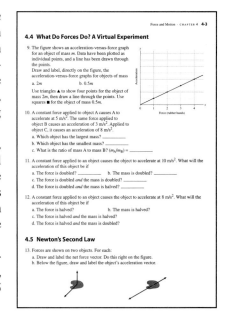

Instructor Supplements

- The **Instructor's Guide for Physics for Scientists and Engineers** (ISBN 0-8053-8985-7) offers detailed comments and suggested teaching ideas for every chapter, an extensive review of what has been learned from physics education research, and guidelines for using active-learning techniques in your classroom.
- The **Instructor's Solutions Manuals**, **Chapters 1–19** (ISBN 0-8053-8986-5), and **Chapters 20–42** (ISBN 0-8053-8989-X), written by Professors Pawan Kahol and Donald Foster, at Wichita State University, provide *complete* solutions to all the end-of-chapter problems. The solutions follow the four-step Model/Visualize/Solve/Assess procedure used in the *Problem-Solving Strategies* and all worked examples. Emphasis is placed on the reasoning behind the solution, rather than just the numerical manipulations. The full text of each solution is available as an editable Word document and as a pdf file on the *Instructor's Supplement CD-ROM* for your own use or for posting on your course website.
- The cross-platform **Instructor's Resource CD-ROMs** (ISBN 0-8053-8996-2) consists of the **Simulation and Image Presentation CD-ROM** and the **Instructor's Supplement CD-ROM**. The *Simulation and Image Presentation CD-ROM* provides a comprehensive library of more than 220 applets from *ActivPhysics OnLine*, as well as all the figures from the textbook (excluding photographs) in JPEG format. In addition, all the tables, chapter summaries, and knowledge structures are provided as JPEGs, and the Tactics Boxes, Problem-Solving Strategies, and key (boxed) equations are provided in editable Word format. The *Instructor's Supplement CD-ROM* provides editable Word versions and pdf files of the *Instructor's Guide* and the *Instructor's Solutions Manuals*. Complete *Student Workbook* solutions are also provided as pdf files.
- **MasteringPhysics**™ (www.masteringphysics.com) is a sophisticated, research-proven online tutorial and homework assignment system that provides students with individualized feedback and hints based on their input. It provides a comprehensive library of conceptual tutorials (including one for each

Problem-Solving Strategy in this textbook), multistep self-tutoring problems, and end-of-chapter problems from *Physics for Scientists and Engineers*. *MasteringPhysics*™ provides instructors with a fast and effective way to assign online homework assignments that comprise a range of problem types. The powerful post-assignment diagnostics allow instructors to assess the progress of their class as a whole or to quickly identify individual students' areas of difficulty.

- **ActivPhysics**™ **OnLine** (www.aw-bc.com/knight) provides a comprehensive library of more than 420 tried and tested *ActivPhysics* applets updated for web delivery using the latest online technologies. In addition, it provides a suite of highly regarded applet-based tutorials developed by education pioneers Professors Alan Van Heuvelen and Paul D'Alessandris. The *ActivPhysics* margin icon directs students to specific exercises that complement the textbook discussion.

  The online exercises are designed to encourage students to confront misconceptions, reason qualitatively about physical processes, experiment quantitatively, and learn to think critically. They cover all topics from mechanics to electricity and magnetism and from optics to modern physics. The highly acclaimed *ActivPhysics OnLine* companion workbooks help students work through complex concepts and understand them more clearly. More than 220 applets from the *ActivPhysics OnLine* library are also available on the *Simulation and Image Presentation CD-ROM*.

- The **Printed Test Bank** (ISBN 0-8053-8994-6) and cross-platform **Computerized Test Bank** (ISBN 0-8053-8995-4), prepared by Professor Benjamin Grinstein, at the University of California, San Diego, contain more than 1500 high-quality problems, with a range of multiple-choice, true/false, short-answer, and regular homework-type questions. In the computerized version, more than half of the questions have numerical values that can be randomly assigned for each student.

- The **Transparency Acetates** (ISBN 0-8053-8993-8) provide more than 200 key figures from *Physics for Scientists and Engineers* for classroom presentation.

Student Supplements

- The **Student Solutions Manuals Chapters 1–19** (ISBN 0-8053-8708-0) and **Chapters 20–42** (ISBN 0-8053-8998-9), written by Professors Pawan Kahol and Donald Foster at Wichita State University, provides *detailed* solutions to more than half of the odd-numbered end-of-chapter problems. The solutions follow the four-step Model/Visualize/Solve/Assess procedure used in the *Problem-Solving Strategies* and all worked examples.

- **MasteringPhysics**™ (www.masteringphysics.com) provides students with individualized online tutoring by responding to their wrong answers and providing hints for solving multistep problems. It gives them immediate and up-to-date assessment of their progress, and shows where they need to practice more.

- **ActivPhysics**™ **OnLine** (www.aw-bc.com/knight) provides students with a suite of highly regarded applet-based tutorials (see above). The accompanying workbooks help students work though complex concepts and understand them more clearly. The *ActivPhysics* margin icon directs students to specific exercises that complement the textbook discussion.

- **ActivPhysics OnLine Workbook Volume 1: Mechanics • Thermal Physics • Oscillations & Waves** (ISBN 0-8053-9060-X)

- **ActivPhysics OnLine Workbook Volume 2: Electricity & Magnetism • Optics • Modern Physics** (ISBN 0-8053-9061-8)

■ The **Addison-Wesley Tutor Center** (www.aw.com/tutorcenter) provides one-on-one tutoring via telephone, fax, email, or interactive website during evening hours and on weekends. Qualified college instructors answer questions and provide instruction for *Mastering Physics*™ and for the examples, exercises, and problems in *Physics for Scientists and Engineers*.

Acknowledgments

I have relied upon conversations with and, especially, the written publications of many members of the physics education community. Those who may recognize their influence include Arnold Arons, Uri Ganiel, Ibrahim Halloun, Richard Hake, David Hestenes, Leonard Jossem, Jill Larkin, Priscilla Laws, John Mallinckrodt, Lillian McDermott, Edward "Joe" Redish, Fred Reif, Rachel Scherr, Bruce Sherwood, David Sokoloff, Ronald Thornton, Sheila Tobias, and Alan Van Heuleven. John Rigden, founder and director of the Introductory University Physics Project, provided the impetus that got me started down this path. Early development of the materials was supported by the National Science Foundation as the *Physics for the Year 2000* project; their support is gratefully acknowledged.

I am grateful to Pawan Kahol and Don Foster for the difficult task of writing the *Instructor's Solutions Manuals*; to Jim Andrews and Susan Cable for writing the workbook answers; to Wayne Anderson, Jim Andrews, Dave Ettestad, Stuart Field, Robert Glosser, and Charlie Hibbard for their contributions to the end-of-chapter problems; and to my colleague Matt Moelter for many valuable contributions and suggestions.

I especially want to thank my editor Adam Black, development editor Alice Houston, editorial assistant Liana Allday, and all the other staff at Addison Wesley for their enthusiasm and hard work on this project. Project manager Laura Kenney, Carolyn Field and the team at Thompson Steele, Inc., copy editor Kevin Gleason, photo researcher Brian Donnelly, and page-layout artist Judy Maenle get much of the credit for making this complex project all come together. In addition to the reviewers and classroom testers listed below, who gave invaluable feedback, I am particularly grateful to Wendell Potter and Susan Cable for their close scrutiny of every word and figure.

Finally, I am endlessly grateful to my wife Sally for her love, encouragement, and patience, and to our many cats for their innate abilities to hold down piles of papers and to type qqqqqqqq whenever it was needed.

Randy Knight, September 2003
rknight@calpoly.edu

Reviewers and Classroom Testers

Gary B. Adams, *Arizona State University*
Wayne R. Anderson, *Sacramento City College*
James H. Andrews, *Youngstown State University*
David Balogh, *Fresno City College*
Dewayne Beery, *Buffalo State College*
Joseph Bellina, *Saint Mary's College*
James R. Benbrook, *University of Houston*
David Besson, *University of Kansas*

Randy Bohn, *University of Toledo*
Art Braundmeier, *University of Southern Illinois, Edwardsville*
Carl Bromberg, *Michigan State University*
Douglas Brown, *Cabrillo College*
Ronald Brown, *California Polytechnic State University, San Luis Obispo*
Mike Broyles, *Collin County Community College*

James Carolan, *University of British Columbia*
Michael Crescimanno, *Youngstown State University*
Wei Cui, *Purdue University*
Robert J. Culbertson, *Arizona State University*
Purna C. Das, *Purdue University North Central*
Dwain Desbien, *Estrella Mountain Community College*
John F. Devlin, *University of Michigan, Dearborn*
Alex Dickison, *Seminole Community College*
Chaden Djalali, *University of South Carolina*
Sandra Doty, *Denison University*
Miles J. Dresser, *Washington State University*
Charlotte Elster, *Ohio University*
Robert J. Endorf, *University of Cincinnati*
Tilahun Eneyew, *Embry-Riddle Aeronautical University*
F. Paul Esposito, *University of Cincinnati*
John Evans, *Lee University*
Michael R. Falvo, *University of North Carolina*
Abbas Faridi, *Orange Coast College*
Stuart Field, *Colorado State University*
Daniel Finley, *University of New Mexico*
Jane D. Flood, *Muhlenberg College*
Thomas Furtak, *Colorado School of Mines*
Richard Gass, *University of Cincinnati*
J. David Gavenda, *University of Texas, Austin*
Stuart Gazes, *University of Chicago*
Katherine M. Gietzen, *Southwest Missouri State University*
Robert Glosser, *University of Texas, Dallas*
William Golightly, *University of California, Berkeley*
Paul Gresser, *University of Maryland*
C. Frank Griffin, *University of Akron*
John B. Gruber, *San Jose State University*
Randy Harris, *University of California, Davis*
Stephen Hass, *University of Southern California*
Nicole Herbots, *Arizona State University*
Scott Hildreth, *Chabot College*
David Hobbs, *South Plains College*
Laurent Hodges, *Iowa State University*
John L. Hubisz, *North Carolina State University*
George Igo, *University of California, Los Angeles*
Bob Jacobsen, *University of California, Berkeley*
Rong-Sheng Jin, *Florida Institute of Technology*
Marty Johnston, *University of St. Thomas*
Stanley T. Jones, *University of Alabama*
Darrell Judge, *University of Southern California*
Pawan Kahol, *Wichita State University*
Teruki Kamon, *Texas A&M University*
Richard Karas, *California State University, San Marcos*
Deborah Katz, *U.S. Naval Academy*
Miron Kaufman, *Cleveland State University*
M. Kotlarchyk, *Rochester Institute of Technology*
Cagliyan Kurdak, *University of Michigan*
Fred Krauss, *Delta College*
H. Sarma Lakkaraju, *San Jose State University*

Darrell R. Lamm, *Georgia Institute of Technology*
Robert LaMontagne, *Providence College*
Alessandra Lanzara, *University of California, Berkeley*
Sen-Ben Liao, *Massachusetts Institute of Technology*
Dean Livelybrooks, *University of Oregon*
Chun-Min Lo, *University of South Florida*
Richard McCorkle, *University of Rhode Island*
James McGuire, *Tulane University*
Theresa Moreau, *Amherst College*
Gary Morris, *Rice University*
Michael A. Morrison, *University of Oklahoma*
Richard Mowat, *North Carolina State University*
Taha Mzoughi, *Mississippi State University*
Vaman M. Naik, *University of Michigan, Dearborn*
Craig Ogilvie, *Iowa State University*
Martin Okafor, *Georgia Perimeter College*
Benedict Y. Oh, *University of Wisconsin*
Georgia Papaefthymiou, *Villanova University*
Peggy Perozzo, *Mary Baldwin College*
Brian K. Pickett, *Purdue University, Calumet*
Joe Pifer, *Rutgers University*
Dale Pleticha, *Gordon College*
Robert Pompi, *SUNY-Binghamton*
David Potter, *Austin Community College*
Chandra Prayaga, *University of West Florida*
Didarul Qadir, *Central Michigan University*
Michael Read, *College of the Siskiyous*
Michael Rodman, *Spokane Falls Community College*
Sharon Rosell, *Central Washington University*
Anthony Russo, *Okaloosa-Walton Community College*
Otto F. Sankey, *Arizona State University*
Rachel E. Scherr, *University of Maryland*
Bruce Schumm, *University of California, Santa Cruz*
Douglas Sherman, *San Jose State University*
Elizabeth H. Simmons, *Boston University*
Alan Slavin, *Trent College*
William Smith, *Boise State University*
Paul Sokol, *Pennsylvania State University*
Chris Sorensen, *Kansas State University*
Anna and Ivan Stern, *AW Tutor Center*
Michael Strauss, *University of Oklahoma*
Arthur Viescas, *Pennsylvania State University*
Chris Vuille, *Embry-Riddle Aeronautical University*
Ernst D. Von Meerwall, *University of Akron*
Robert Webb, *Texas A&M University*
Zodiac Webster, *California State University, San Bernardino*
Robert Weidman, *Michigan Technical University*
Jeff Allen Winger, *Mississippi State University*
Ronald Zammit, *California Polytechnic State University, San Luis Obispo*
Darin T. Zimmerman, *Pennsylvania State University, Altoona*

Preface to the Student

From Me to You

The most incomprehensible thing about the universe is that it is comprehensible.
   —Albert Einstein

The day I went into physics class it was death.
   —Sylvia Plath, *The Bell Jar*

Let's have a little chat before we start. A rather one-sided chat, admittedly, because you can't respond, but that's OK. I've heard from many of your fellow students over the years, so I have a pretty good idea of what's on your mind.

What's your reaction to taking physics? Fear and loathing? Uncertainty? Excitement? All of the above? Let's face it, physics has a bit of an image problem on campus. You've probably heard that it's difficult, maybe downright impossible unless you're an Einstein. Things that you've heard, your experiences in other science courses, and many other factors all color your *expectations* about what this course is going to be like.

It's true that there are many new ideas to be learned in physics and that the course, like college courses in general, is going be to much faster paced than science courses you had in high school. I think it's fair to say that it will be an *intense* course. But we can avoid many potential problems and difficulties if we can establish, here at the beginning, what this course is about and what is expected of you—and of me!

Just what is physics, anyway? Physics is a way of thinking about the physical aspects of nature. Physics is not better than art or biology or poetry or religion, which are also ways to think about nature; it's simply different. One of the things this course will emphasize is that physics is a human endeavor. The information content of this book was not found in a cave or conveyed to us by aliens; it was discovered by real people engaged in a struggle with real issues. I hope to convey to you something of the history and the process by which we have come to accept the principles that form the foundation of today's science and engineering.

You might be surprised to hear that physics is not about "facts." Oh, not that facts are unimportant, but physics is far more focused on discovering *relationships* that exist between facts and *patterns* that exist in nature than on learning facts for their own sake. As a consequence, there's not a lot of memorization when you study physics. Some—there are still definitions and equations to learn—but less than in many other courses. Our emphasis, instead, will be on thinking and reasoning. This is important to factor into your expectations for the course.

Perhaps most important of all, *physics is not math!* Physics is much broader. We're going to look for patterns and relationships in nature, develop the logic that relates different ideas, and search for the reasons *why* things happen as they do. In doing so, we're going to stress qualitative reasoning, pictorial and graphical reasoning, and reasoning by analogy. And yes, we will use math, but it's just one tool among many.

It will save you much frustration if you're aware of this physics–math distinction up front. Many of you, I know, want to find a formula and plug numbers into it—that is, to do a math problem. Maybe that's what you learned in high school science courses, but it is *not* what this course expects of you. We'll certainly do

(a) X-ray diffraction pattern

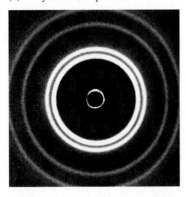

(b) Electron diffraction pattern

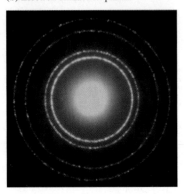

many calculations, but the specific numbers are usually the last and least important step in the analysis.

Physics is about recognizing patterns. The top photograph is an x-ray diffraction pattern that shows how a collimated beam of x rays spreads out after passing through a crystal. The bottom photograph shows what happens when a collimated beam of electrons is shot through the same crystal. What does the obvious similarity in these two photographs tell us about the nature of light and about the nature of matter?

As you study, you'll sometimes be baffled, puzzled, and confused. That's perfectly normal and to be expected. Making mistakes is OK too *if* you're willing to learn from the experience. No one is born knowing how to do physics any more than he or she is born knowing how to play the piano or shoot basketballs. The ability to do physics comes from practice, repetition, and struggling with the ideas until you "own" them and can apply them yourself in new situations. There's no way to make learning effortless, at least for anything worth learning, so expect to have some difficult moments ahead.

But also expect to have some moments of excitement at the joy of discovery. There will be instants at which the pieces suddenly click into place and you *know* that you understand a difficult idea. There will be times when you'll surprise yourself by successfully working a difficult problem that you didn't think you could solve. My hope, as an author, is that the excitement and sense of adventure will far outweigh the difficulties and frustrations.

Many of you, I suspect, would like to know the "best" way to study for this course. There is no best way. People are too different, and what works for one student works less effectively for another. But I do want to stress that *reading the text* is vitally important. Class time will be used to clarify difficulties and to develop tools for using the knowledge, but your instructor will *not* use class time simply to repeat information in the text. The basic knowledge for this course is written down within these pages, and the *number one expectation* is that you will read carefully and thoroughly to find and learn that knowledge.

Despite there being no best way to study, I will suggest *one* way that is successful for many students. It consists of the following four steps:

1. **Read each chapter *before* it is discussed in class.** I cannot stress too highly how important this step is. Class attendance is largely ineffective if you have not prepared. When you first read a chapter, focus on learning new vocabulary, definitions, and notation. There's a list of terms and notations at the end of each chapter. Learn them! You won't understand what's being discussed or how the ideas are being used if you don't know what the terms and symbols mean.

2. **Participate actively in class.** Take notes, ask and answer questions, take part in discussion groups. There is ample scientific evidence that *active participation* is far more effective for learning science than is passive listening.

3. **After class, go back for a *careful* rereading of the chapter.** In your second reading, pay closer attention to the details and the worked examples. Look for the *logic* behind each example (and I've tried to help make this clear), not just at what formula is being used. Do the *Student Workbook* exercises for each section as you finish your reading of it.

4. **Finally, apply what you have learned to the homework problems at the end of each chapter.** I strongly encourage you to form a study group with two or three classmates. There's good evidence that students who study regularly with a group do better than the rugged individualists who try to go it alone.

Did someone mention a workbook? The companion *Student Workbook* is a vital part of this course. It contains questions and exercises that ask you to reason *qualitatively*, to use graphical information, and to give explanations. It is through these exercises that you will learn what the concepts mean and will practice the reasoning skills appropriate to the chapter. You will then have acquired the baseline knowledge that you need *before* turning to the end-of-chapter homework problems. In sports or in music, you would never think of performing before you practice, so why would you want to do so in physics? The workbook is where you practice and work on basic skills.

Many of you, I know, would like to go straight to the homework problems and then thumb through the text looking for a formula that seems like it will work. That approach will not succeed in this course, and it's guaranteed to make you frustrated and discouraged. Very few homework problems are "plug and chug" problems where you simply put numbers into a formula. To work the homework problems successfully, you need a better study strategy—either that outlined above or your own—that helps you learn the concepts and the relationships between the ideas. Many of the chapters in this book have Problem-Solving Strategies to help you develop effective problem-solving skills.

A traditional guideline in college is to study two hours outside of class for every hour spent in class, and this text is designed with that expectation. Of course, two hours is an average. Some chapters are fairly straightforward and will go quickly. Others likely will require much more than two study hours per class hour.

Now that you know more about what is expected of you, what can you expect of me? That's a little trickier, because the book is already written! Nonetheless, it was prepared on the basis of what I think my students throughout the years have expected—and wanted—from their physics textbook.

You should know that these course materials—the text and the workbook—are based upon extensive research about how students learn physics and the challenges they face. The effectiveness of many of the exercises has been demonstrated through extensive class testing. I've written the book in an informal style that I hope you will find appealing and that will encourage you to do the reading. And finally, I have endeavored to make clear not only that physics, as a technical body of knowledge, is relevant to your profession but also that physics is an exciting adventure of the human mind.

I hope you'll enjoy the time we're going to spend together.

Detailed Contents

Volume 1 contains chapters 1–15; Volume 2 contains chapters 16–19; Volume 3 contains chapters 20–24; Volume 4 contains chapters 25–36; Volume 5 contains chapters 36–42.

If the fusion reactor in the core of the Death Star generates 10^{15} W of power, how much mass does the Death Star lose each year as it converts mass into energy?

Relativity and Quantum Physics

Contemporary Physics

Our journey into physics is nearing its end. We began roughly 350 years ago with Newton's discovery of the laws of motion. The conclusion of Part VI brought us to the end of the 19th century, just over 100 years ago. Along the way you've learned about the motion of particles, the conservation of energy, the physics of waves, and the electromagnetic interactions that hold atoms together and generate light waves. We can begin the last phase of our journey with confidence.

Newton's mechanics and Maxwell's electromagnetism were the twin pillars of science at the end of the 19th century and the basis for much of engineering and applied science in the 20th century. Despite the successes of these theories, a series of discoveries starting around 1900 and continuing into the first few decades of the 20th century profoundly altered our understanding of the universe at the most fundamental level. These discoveries forced scientists to reconsider the very nature of space and time and to develop new models of light and matter.

The discoveries and new ideas of the early 20th century led to two new theories: relativity and quantum physics. These two theories form the basis for physics as it is practiced today and are already having a significant impact on 21st-century engineering. We will end our journey into physics with a look at these contemporary topics and some of their applications.

Relativity

The idea of measuring distance with a meter stick and time with a clock or stopwatch has been with us since Chapter 1. The basic notions of space and time seem so self-evident that no one had seriously questioned them. No one, that is, until an unknown young scientist named Albert Einstein began to ponder these issues in the years right around 1900.

It wasn't space and time that first troubled Einstein. Instead, he was bothered by what he saw as paradoxes and difficulties in Maxwell's theory of electromagnetism. Einstein was able to show that electromagnetism is a self-consistent theory only if the speed of electromagnetic waves—the speed of light—is the same in all inertial reference frames, no matter how the reference frames might be moving with respect to each other or to the source of the wave. But this strange behavior of a traveling wave can be true only if *space and time are different* for two experimenters moving relative to each other.

We'll need to explore what it means for space and time to be different for different experimenters. In doing so we'll discover some of the well-known puzzles of relativity, such as length contraction, the twin paradox, and a cosmic speed limit for particles. Our exploration of these fascinating ideas will end with what is perhaps the most famous equation in physics: Einstein's $E = mc^2$.

Quantum Physics

We ended Part V with experimental evidence that light sometimes acts like a particle and that electrons can exhibit wave-like behavior. These were observations only; we did not offer any explanation at the time. We now want to return to that thread of thought and see how it leads to the new ideas of quantum physics.

We'll begin by looking at *evidence* about the atomic world. What do we know about electrons and atoms, and how do we know it? Atoms were first thought to be tiny, indivisible pieces of matter that moved in accordance with Newton's laws of motion. But it became increasingly clear toward the end of the 19th century that atoms *can* be divided into smaller pieces. Furthermore, the classical physics of Newton and Maxwell was unable to explain the behavior of atoms or the light they emit.

We will focus our attention on a key experiment called the *photoelectric effect*. The experiment is straightforward and the data are unambiguous, but difficulties will arise when we try to explain the outcome. It was, once again, Albert Einstein who offered a fresh and original interpretation of the photoelectric effect in terms of the *quantization* of energy—in particular, the idea that the energy of a light wave is bundled into small, discrete packages that we now call *photons*.

We will then look at Niels Bohr's efforts to apply the ideas of energy quantization to atoms. Bohr's model of the atom was the first to explain the discrete spectra emitted by atoms, but he was unable to extend his model beyond the simplest hydrogen atom. Bohr was on the right track, but his ideas were still too classical. The missing ingredient in Bohr's model was de Broglie's hypothesis that matter has wave-like properties—a hypothesis we looked at in Chapter 24.

The complete theory of quantum physics was developed by Erwin Schrödinger soon after he learned of de Broglie's hypothesis. Schrödinger's new theory describes atomic particles in terms of an entirely new concept called a *wave function*. One of our most important tasks in Part VII will be to answer the questions:

■ What is a wave function?
■ How is the wave function interpreted and connected to experimental measurements?
■ What new law of nature governs the behavior of the wave function?

Once we've answered these questions, we'll begin to see how Schrödinger's *quantum mechanics*—the quantum-physics analysis of motion—successfully explains the behavior of electrons and atoms.

We will concentrate on quantum mechanics in one dimension. This restriction will allow us to focus on physical phenomena without becoming sidetracked by the mathematics of quantum mechanics in three dimensions. Although one-dimensional models aren't perfect, they will be adequate for understanding the essential features of scanning tunneling microscopes, various kinds of semiconductor devices, radioactive decay, and other applications.

Quantum physics will give you an entirely new perspective on the nature of matter and light. The quantum world with its wave functions can seem strange and mysterious, yet quantum mechanics gives the most definitive and accurate predictions of any physical theory ever devised.

Atoms and Nuclei

You learned in Chapter 18 how macroscopic measurements of density, pressure, and temperature allow us to learn about the sizes and speeds of atoms. You've also learned how atoms can be polarized (Chapter 25), how electrons and ions can transfer charge as an electric current (Chapter 28), and how atoms act as tiny magnets (Chapter 32). And in Chapter 24 you saw the rather strange evidence that matter on the atomic scale sometimes acts like a wave.

But when you get right down to it, what *is* an atom? What makes an atom of carbon different from an atom of gold? And what's inside an atom? As important as these questions are, there's an even deeper question: How do we *know* about atoms? Atomic phenomena lie beyond the immediate realm of our senses, so learning to interpret macroscopic data in terms of atomic- and nuclear-level events is an essential aspect of learning about atoms and nuclei.

An understanding of atoms and nuclei depends on both classical electromagnetism and quantum mechanics. Although we can give only an introduction to atomic and nuclear physics, you will learn where the electron shell model of chemistry comes from, how atoms emit and absorb light, what's inside the nucleus, and why some nuclei undergo radioactive decay.

A scanning tunneling microscope image of individual iron atoms on a copper surface. The atoms form the Japanese character for "atom."

36 Relativity

These are the fundamental tools with which we learn about space and time.

▶ **Looking Ahead**

The goal of Chapter 36 is to understand how Einstein's theory of relativity changes our concepts of space and time. In this chapter you will learn to:

- Use the principle of relativity.
- Understand how time dilation and length contraction change our concepts of space and time.
- Use the Lorentz transformations of positions and velocities.
- Calculate relativistic momentum and energy.
- Understand how mass and energy are equivalent.

◀ **Looking Back**

The material in this chapter depends on an understanding of relative motion in Newtonian mechanics. Please review:

- Section 6.4 Inertial reference frames and the Galilean transformations.

Space and time seem like straightforward ideas. You can measure lengths with a ruler or meter stick. You can time events with a stop watch. Nothing could be simpler.

So it seemed to everyone until 1905, when an unknown young scientist had the nerve to suggest that this simple view of space and time was in conflict with other principles of physics. In the century since, Einstein's theory of relativity has radically altered our understanding of some of the most fundamental ideas in physics.

Relativity, despite its esoteric reputation, has very real implications for modern technology. Global positioning system (GPS) satellites depend on relativity, as do the navigation systems used by airliners. Nuclear reactors make tangible use of Einstein's famous equation $E = mc^2$ to generate 20% of the electricity used in the United States. The annihilation of matter in positron-emission tomography (PET scanners) has given neuroscientists an unprecedented ability to monitor activity within the brain.

The theory of relativity is fascinating, perplexing, and challenging. It is also vital to our contemporary understanding of the universe in which we live.

Albert Einstein (1879–1955) was one of the most influential thinkers in history.

36.1 Relativity: What's It All About?

What do you think of when you hear the phrase "theory of relativity"? A white-haired Einstein? $E = mc^2$? Black holes? Time travel? Perhaps you've heard that the theory of relativity is so complicated and abstract that only a handful of people in the whole world really understand it.

There is, without doubt, a certain mystique associated with relativity, an aura of the strange and exotic. The good news is that understanding the ideas of relativity is well within your grasp. Einstein's *special theory of relativity,* the portion of relativity we'll study, is not mathematically difficult at all. The challenge is conceptual because relativity questions deeply held assumptions about the nature of space and time. In fact, that's what relativity is all about—space and time.

In one sense, relativity is not a new idea at all. Certain ideas about relativity are part of Newtonian mechanics. You had an introduction to these ideas in Chapter 6, where you learned about reference frames and the Galilean transformations. Einstein, however, thought that relativity should apply to *all* the laws of physics, not just mechanics. The difficulty, as you'll see, is that some aspects of relativity appear to be incompatible with the laws of electromagnetism, particularly the laws governing the propagation of light waves.

Lesser scientists might have concluded that relativity simply doesn't apply to electromagnetism. Einstein's genius was to see that the incompatibility arises from *assumptions* about space and time, assumptions no one had ever questioned because they seem so obviously true. Rather than abandon the ideas of relativity, Einstein changed our understanding of space and time.

Fortunately, you need not be a genius to follow a path that someone else has blazed. However, we will have to exercise the utmost care with regard to logic and precision. We will need to state very precisely just how it is that we know things about the physical world, then ruthlessly follow the logical consequences. The challenge is to stay on this path, not to let our prior assumptions—assumptions that are deeply ingrained in all of us—lead us astray.

What's Special About Special Relativity?

Einstein's first paper on relativity, in 1905, dealt exclusively with inertial reference frames, reference frames that move relative to each other with constant velocity. Ten years later, Einstein published a more encompassing theory of relativity that considers accelerated motion and its connection to gravity. The second theory, because it's more general in scope, is called *general relativity.* General relativity is the theory that describes black holes, curved spacetime, and the evolution of the universe. It is a fascinating theory but, alas, very mathematical and outside the scope of this textbook. If you're interested, many popular science books provide a nontechnical introduction to general relativity.

Motion at constant velocity is a "special case" of motion; namely, motion for which the acceleration is zero. Hence Einstein's first theory of relativity has come to be known as **special relativity.** It is special in the sense of being a restricted, special case of his more general theory, not special in the everyday sense of meaning distinctive or exceptional. Special relativity, with its conclusions about time dilation and length contraction, is what we will study.

36.2 Galilean Relativity

A firm grasp of Galilean relativity is necessary if we are to appreciate and understand what is new in Einstein's theory. Thus we begin with the ideas of relativity that are embodied in Newtonian mechanics.

Reference Frames

Suppose you're passing me as we both drive in the same direction along a freeway. My car's speedometer reads 55 mph while your speedometer shows 60 mph. Is 60 mph your "true" speed? That is certainly your speed relative to someone standing beside the road, but your speed relative to me is only 5 mph. Your speed is 120 mph relative to a driver approaching from the other direction at 60 mph.

An object does not have a "true" speed or velocity. The very definition of velocity, $v = \Delta x/\Delta t$, assumes the existence of a coordinate system in which, during some time interval Δt, the displacement Δx is measured. The best we can manage is to specify an object's velocity relative to, or with respect to, the coordinate system in which it is measured.

Let's define a **reference frame** to be a coordinate system in which experimenters equipped with meter sticks, stopwatches, and any other needed equipment make position and time measurements on moving objects. Three ideas are implicit in our definition of a reference frame:

- A reference frame extends infinitely far in all directions.
- The experimenters are at rest in the reference frame.
- The number of experimenters and the quality of their equipment are sufficient to measure positions and velocities to any level of accuracy needed.

The first two bullets are especially important. It is often convenient to say "the laboratory reference frame" or "the reference frame of the rocket." These are shorthand expressions for "a reference frame, infinite in all directions, in which the laboratory (or the rocket) and a set of experimenters happen to be at rest."

NOTE ▶ A reference frame is not the same thing as a "point of view." That is, each person or each experimenter does not have his or her own private reference frame. **All experimenters at rest relative to each other share the same reference frame.** ◀

Figure 36.1 shows two reference frames called S and S′. The coordinate axes in S are x, y, z and those in S′ are x', y', z'. Reference frame S′ moves with velocity v relative to S or, equivalently, S moves with velocity $-v$ relative to S′. There's no implication that either reference frame is "at rest." Notice that the zero of time, when experimenters start their stopwatches, is the instant that the origins of S and S′ coincide.

We will restrict our attention to *inertial reference frames,* implying that the relative velocity v is constant. You should recall from Chapter 6 that an **inertial reference frame** is a reference frame in which Newton's first law, the law of inertia, is valid. In particular, an inertial reference frame is one in which an isolated particle, one on which there are no forces, either remains at rest or moves in a straight line at constant speed.

Any reference frame that moves at constant velocity with respect to an inertial reference frame is itself an inertial reference frame. Conversely, a reference frame that accelerates with respect to an inertial reference frame is *not* an inertial reference frame. Our restriction to reference frames moving with respect to each other at constant velocity—with no acceleration—is the "special" part of special relativity.

NOTE ▶ An inertial reference frame is an idealization. A true inertial reference would need to be floating in deep space, far from any gravitational influence. In practice, an earthbound laboratory is a good approximation of an inertial reference frame because the accelerations associated with the earth's rotation and motion around the sun are too small to influence most experiments. ◀

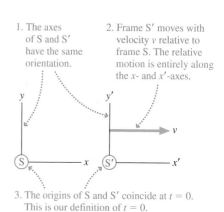

1. The axes of S and S′ have the same orientation.

2. Frame S′ moves with velocity v relative to frame S. The relative motion is entirely along the x- and x'-axes.

3. The origins of S and S′ coincide at $t = 0$. This is our definition of $t = 0$.

FIGURE 36.1 The standard reference frames S and S′.

Which of these is an inertial reference frame (or a very good approximation)?

    a. Your bedroom
    b. A car rolling down a steep hill
    c. A train coasting along a level track
    d. A rocket being launched
    e. A roller coaster going over the top of a hill
    f. A sky diver falling at terminal speed

The Galilean Transformations

Suppose a firecracker explodes at time t. The experimenters in reference frame S determine that the explosion happened at position x. Similarly, the experimenters in S′ find that the firecracker exploded at x' in their reference frame. What is the relationship between x and x'?

Figure 36.2 shows the explosion and the two reference frames. You can see from the figure that $x = x' + vt$, thus

$$
\begin{array}{lcl}
x = x' + vt & & x' = x - vt \\
y = y' & \text{or} & y' = y \\
z = z' & & z' = z
\end{array}
\qquad (36.1)
$$

These equations, which you met in Chapter 6, are the *Galilean transformations of position*. If you know a position measured by the experimenters in one inertial reference frame, you can calculate the position that would be measured by experimenters in any other inertial reference frame.

Suppose the experimenters in both reference frames now track the motion of the object in Figure 36.3 by measuring its position at many instants of time. The experimenters in S find that the object's velocity is $\vec{u}$. During the *same time interval* Δt, the experimenters in S′ measure the velocity to be $\vec{u}'$.

NOTE ▶ In this chapter, we will use v to represent the velocity of one reference frame relative to another. We will use $\vec{u}$ and $\vec{u}'$ to represent the velocities of objects with respect to reference frames S and S′. This notation differs from the notation of Chapter 6, where we used V to represent the relative velocity. ◀

We can find the relationship between $\vec{u}$ and $\vec{u}'$ by taking the time derivatives of Equation 36.1 and using the definition $u_x = dx/dt$:

$$
u_x = \frac{dx}{dt} = \frac{dx'}{dt} + v = u_x' + v
$$

$$
u_y = \frac{dy}{dt} = \frac{dy'}{dt} = u_y'
$$

The equation for u_z is similar. The net result is

$$
\begin{array}{lcl}
u_x = u_x' + v & & u_x' = u_x - v \\
u_y = u_y' & \text{or} & u_y' = u_y \\
u_z = u_z' & & u_z' = u_z
\end{array}
\qquad (36.2)
$$

Equations 36.2 are the *Galilean transformations of velocity*. If you know the velocity of a particle as measured by the experimenters in one inertial reference frame, you can use Equations 36.2 to find the velocity that would be measured by experimenters in any other inertial reference frame.

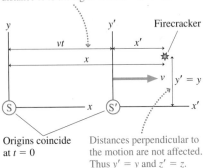

At time t, the origin of S′ has moved distance vt to the right. Thus $x = x' + vt$.

Origins coincide at $t = 0$

Distances perpendicular to the motion are not affected. Thus $y' = y$ and $z' = z$.

FIGURE 36.2 The position of an exploding firecracker is measured in reference frames S and S′.

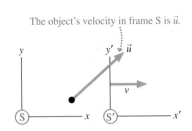

The object's velocity in frame S is $\vec{u}$.

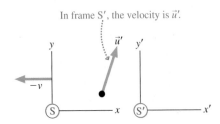

In frame S′, the velocity is $\vec{u}'$.

FIGURE 36.3 The velocity of a moving object is measured in reference frames S and S′.

EXAMPLE 36.1 The speed of sound

An airplane is flying at speed 200 m/s with respect to the ground. Sound wave 1 is approaching the plane from the front, sound wave 2 is catching up from behind. Both waves travel at 340 m/s relative to the ground. What is the speed of each wave relative to the plane?

MODEL Assume that the earth (frame S) and the airplane (frame S′) are inertial reference frames. Frame S′, in which the airplane is at rest, moves with velocity $v = 200$ m/s relative to frame S.

VISUALIZE Figure 36.4 shows the airplane and the sound waves.

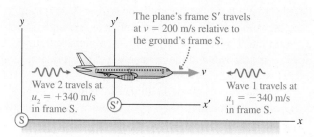

The plane's frame S′ travels at $v = 200$ m/s relative to the ground's frame S.

Wave 2 travels at $u_2 = +340$ m/s in frame S.

Wave 1 travels at $u_1 = -340$ m/s in frame S.

FIGURE 36.4 Experimenters in the plane measure different speeds for the sound waves than do experimenters on the ground.

SOLVE The speed of a mechanical wave, such as a sound wave or a wave on a string, is its speed *relative to its medium*. Thus the *speed of sound* is the speed of a sound wave through a reference frame in which the air is at rest. This is reference frame S, where wave 1 travels with velocity $u_1 = -340$ m/s and wave 2 travels with velocity $u_2 = +340$ m/s. Notice that the Galilean transformations use *velocities*, with appropriate signs, not just speeds.

The airplane travels to the right with reference frame S′ at velocity v. We can use the Galilean transformations of velocity to find the velocities of the two sound waves in frame S′:

$$u_1' = u_1 - v = -340 \text{ m/s} - 200 \text{ m/s} = -540 \text{ m/s}$$

$$u_2' = u_2 - v = 340 \text{ m/s} - 200 \text{ m/s} = 140 \text{ m/s}$$

ASSESS This isn't surprising. If you're driving 50 mph, a car coming the other way at 55 mph is approaching you at 105 mph. A car coming up behind you at 55 mph seems to be gaining on you at the rate of only 5 mph. Wave speeds behave the same. Notice that a mechanical wave would appear to be stationary to a person moving at the wave speed. To a surfer, the crest of the ocean wave remains at rest under his or her feet.

STOP TO THINK 36.2 Ocean waves are approaching the beach at 10 m/s. A boat heading out to sea travels at 6 m/s. How fast are the waves moving in the boat's reference frame?

a. 16 m/s   b. 10 m/s   c. 6 m/s   d. 4 m/s

The Galilean Principle of Relativity

Experimenters in reference frames S and S′ measure different values for position and velocity. What about the force on and the acceleration of the particle in Figure 36.5? The strength of a force can be measured with a spring scale. The experimenters in reference frames S and S′ both see the *same reading* on the scale (we'll assume the scale has a bright digital display easily seen by all experimenters), leading them to conclude that the force is the same in both frames. That is, $F' = F$.

We can compare the accelerations measured in the two reference frames by taking the time derivative of the velocity transformation equation $u' = u - v$. (We'll assume, for simplicity, that the velocities and accelerations are all in the x-direction.) The relative velocity v between the two reference frames is *constant*, thus

$$a' = \frac{du'}{dt} = \frac{du}{dt} = a \qquad (36.3)$$

Experimenters in reference frames S and S′ measure different values for an object's position and velocity, but they *agree* on its acceleration.

If $F = ma$ in reference frame S, then $F' = ma'$ in reference frame S′. Stated another way, if Newton's second law is valid in one inertial reference frame, then it is valid in all inertial reference frames. Because other laws of mechanics, such

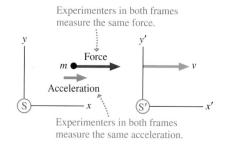

Experimenters in both frames measure the same force.

Force

Acceleration

Experimenters in both frames measure the same acceleration.

FIGURE 36.5 Experimenters in both reference frames test Newton's second law by measuring the force on a particle and its acceleration.

as the conservation laws, follow from Newton's laws of motion, we can state this conclusion as the *Galilean principle of relativity:*

> **Galilean principle of relativity** The laws of mechanics are the same in all inertial reference frames.

The Galilean principle of relativity is easy to state, but to understand it we must understand what is and is not "the same." To take a specific example, consider the law of conservation of momentum. Figure 36.6a shows two particles about to collide. Their total momentum in frame S, where particle 2 is at rest, is $P_i = 9 \text{ kg m/s}$. This is an isolated system, hence the law of conservation of momentum tells us that the momentum after the collision will be $P_f = 9 \text{ kg m/s}$.

Figure 36.6b has used the velocity transformation to look at the same particles in frame S′ in which particle 1 is at rest. The initial momentum in S′ is $P_i' = -18 \text{ kg m/s}$. Thus it is not the *value* of the momentum that is the same in all inertial reference frames. Instead, the Galilean principle of relativity tells us that the *law* of momentum conservation is the same in all inertial reference frames. If $P_f = P_i$ in frame S, then it must be true that $P_f' = P_i'$ in frame S′. Consequently, we can conclude that P_f' will be -18 kg m/s after the collision in S′.

Using Galilean Relativity

The principle of relativity is concerned with the laws of mechanics, not with the values that are needed to satisfy the laws. If momentum is conserved in one inertial reference frame, it is conserved in all inertial reference frames. Even so, a problem may be easier to solve in one reference frame than in others.

Elastic collisions provide a good example of using reference frames. You learned in Chapter 10 how to calculate the outcome of a perfectly elastic collision between two particles in the reference frame in which particle 2 is initially at rest. We can use that information together with the Galilean transformations to solve elastic-collision problems in any inertial reference frame.

(a) Collision seen in frame S

(b) Collision seen in frame S′

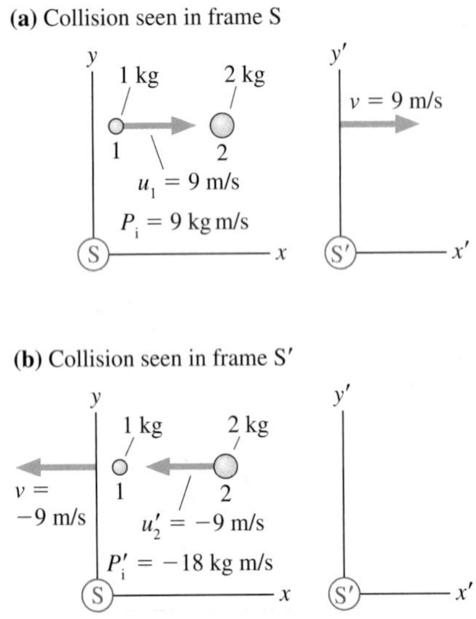

FIGURE 36.6 Total momentum measured in two reference frames.

> **TACTICS BOX 36.1 Analyzing elastic collisions**
>
> ❶ Use the Galilean transformations to transform the initial velocities of particles 1 and 2 from frame S to a reference frame S′ in which particle 2 is at rest.
> ❷ The outcome of the collision in S′ is given by
>
> $$u_{1f}' = \frac{m_1 - m_2}{m_1 + m_2} u_{1i}'$$
>
> $$u_{2f}' = \frac{2m_1}{m_1 + m_2} u_{1i}'$$
>
> ❸ Transform the two final velocities from frame S′ back to frame S.

EXAMPLE 36.2 An elastic collision
A 300 g ball moving to the right at 2 m/s has a perfectly elastic collision with a 100 g ball moving to the left at 4 m/s. What are the direction and speed of each ball after the collision?

MODEL The velocities are measured in the laboratory frame, which we call frame S.

VISUALIZE Figure 36.7a shows both the balls and reference frame S′ in which ball 2 is at rest.

(a)

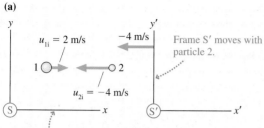

The collision takes place in frame S.

(b)

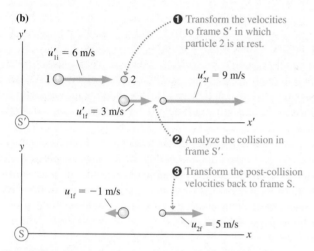

❶ Transform the velocities to frame S' in which particle 2 is at rest.

❷ Analyze the collision in frame S'.

❸ Transform the post-collision velocities back to frame S.

FIGURE 36.7 Using reference frames to solve an elastic-collision problem.

SOLVE The three steps of Tactics Box 36.1 are illustrated in Figure 36.7b. We're given u_{1i} and u_{2i}. The Galilean transformations of these velocities to frame S', using $v = -4$ m/s, are

$$u_{1i}' = u_{1i} - v = (2 \text{ m/s}) - (-4 \text{ m/s}) = 6 \text{ m/s}$$
$$u_{2i}' = u_{2i} - v = (-4 \text{ m/s}) - (-4 \text{ m/s}) = 0 \text{ m/s}$$

The 100 g ball is at rest in frame S', which is what we wanted. The velocities after the collision are

$$u_{1f}' = \frac{m_1 - m_2}{m_1 + m_2} u_{1i}' = 3 \text{ m/s}$$

$$u_{2f}' = \frac{2m_1}{m_1 + m_2} u_{1i}' = 9 \text{ m/s}$$

We've finished the collision analysis, but we're not done because these are the post-collision velocities in frame S'. Another application of the Galilean transformations tells us that the post-collision velocities in frame S are

$$u_{1f} = u_{1f}' + v = (3 \text{ m/s}) + (-4 \text{ m/s}) = -1 \text{ m/s}$$
$$u_{2f} = u_{2f}' + v = (9 \text{ m/s}) + (-4 \text{ m/s}) = 5 \text{ m/s}$$

Thus the 300 g ball rebounds to the left at a speed of 1 m/s and the 100 g ball is knocked to the right at a speed of 5 m/s.

ASSESS You can easily verify that momentum is conserved: $P_f = P_i = 0.20$ kg m/s. The calculations in this example were easy. The important point of this example, and one worth careful thought, is the *logic* of what we did and why we did it.

36.3 Einstein's Principle of Relativity

The 19th century was an era of optics and electromagnetism. Thomas Young demonstrated in 1801 that light is a wave, and by midcentury scientists had devised techniques for measuring the speed of light. Faraday discovered electromagnetic induction in 1831, setting in motion a train of events leading to Maxwell's conclusion, in 1864, that light is an electromagnetic wave.

If light is a wave, what is the medium in which it travels? This was perhaps *the* most important scientific question of the second half of the 19th century. The medium in which light waves were assumed to travel was called the **ether.** Experiments to measure the speed of light were assumed to be measuring its speed through the ether. But just what *is* the ether? What are its properties? Can we collect a jar full of ether to study? Despite the significance of these questions, experimental efforts to detect the ether or measure its properties kept coming up empty handed.

Maxwell's theory of electromagnetism didn't help the situation. The crowning success of Maxwell's theory was his prediction that light waves travel with speed

$$c = \frac{1}{\sqrt{\epsilon_0 \mu_0}} = 3.00 \times 10^8 \text{ m/s}$$

This is a very specific prediction with no wiggle room. The difficulty with such a specific prediction was the implication that Maxwell's laws of electromagnetism

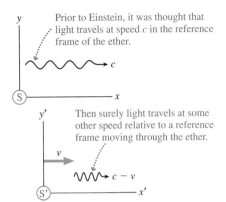

Prior to Einstein, it was thought that light travels at speed c in the reference frame of the ether.

Then surely light travels at some other speed relative to a reference frame moving through the ether.

FIGURE 36.8 It seems as if the speed of light should differ from c in a reference frame moving through the ether.

are valid *only* in the reference frame of the ether. After all, as Figure 36.8 shows, the light speed should certainly be larger or smaller than c in a reference frame moving through the ether, just as the sound speed is different to someone moving through the air.

As the 19th century closed, it appeared that Maxwell's theory did not obey the classical principle of relativity. There was just one reference frame, the reference frame of the ether, in which the laws of electromagnetism seemed to be true. And to make matters worse, the fact that no one had been able to detect the ether meant that no one could identify the one reference frame in which Maxwell's equations "worked."

It was in this muddled state of affairs that a young Albert Einstein made his mark on the world. Even as a teenager, Einstein had wondered how a light wave would look to someone "surfing" the wave, traveling alongside the wave at the wave speed. You can do that with a water wave or a sound wave, but light waves seemed to present a logical difficulty. An electromagnetic wave sustains itself by virtue of the fact that a changing magnetic field induces an electric field and a changing electric field induces a magnetic field. But to someone moving with the wave, *the fields would not change*. How could there be an electromagnetic wave under these circumstances?

Several years of thinking about the connection between electromagnetism and reference frames led Einstein to the conclusion that *all* the laws of physics, not just the laws of mechanics, should obey the principle of relativity. In other words, the principle of relativity is a fundamental statement about the nature of the physical universe. Thus we can remove the restriction in the Galilean principle of relativity and state a much more general principle:

> **Principle of relativity** All the laws of physics are the same in all inertial reference frames.

All of the results of Einstein's theory of relativity flow from this one simple statement.

The Constancy of the Speed of Light

If Maxwell's equations of electromagnetism are laws of physics, and there's every reason to think they are, then, according to the principle of relativity, Maxwell's equations must be true in *every* inertial reference frame. On the surface this seems to be an innocuous statement, equivalent to saying that the law of conservation of momentum is true in every inertial reference frame. But follow the logic:

1. Maxwell's equations are true in all inertial reference frames.
2. Maxwell's equations predict that electromagnetic waves, including light, travel at speed $c = 3.00 \times 10^8$ m/s.
3. Therefore, **light travels at speed c in all inertial reference frames.**

Figure 36.9 shows the implications of this conclusion. *All* experimenters, regardless of how they move with respect to each other, find that *all* light waves, regardless of the source, travel in their reference frame with the *same* speed c. If Cathy's velocity toward Bill and away from Amy is $v = 0.9c$, Cathy finds, by making measurements in her reference frame, that the light from Bill approaches her at speed c, not at $c + v = 1.9c$. And the light from Amy, which left Amy at speed c, catches up from behind at speed c *relative to Cathy*, not the $c - v = 0.1c$ you would have expected.

Although this prediction goes against all shreds of common sense, the experimental evidence for it is strong. Laboratory experiments are difficult because

This light wave leaves Amy at speed c relative to Amy. It approaches Cathy at speed c relative to Cathy.

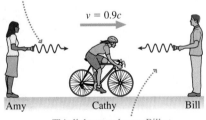

$v = 0.9c$

Amy   Cathy   Bill

This light wave leaves Bill at speed c relative to Bill. It approaches Cathy at speed c relative to Cathy.

FIGURE 36.9 Light travels at speed c in all inertial reference frames, regardless of how the reference frames are moving with respect to the light source.

even the highest laboratory speed is insignificant in comparison to c. In the 1930s, however, the physicists R. J. Kennedy and E. M. Thorndike realized that they could use the earth itself as a laboratory. The earth's speed as it circles the sun is about 30,000 m/s. The *relative* velocity of the earth in January differs by 60,000 m/s from its velocity in July, when the earth is moving in the opposite direction. Kennedy and Thorndike were able to use a very sensitive and stable interferometer to show that the numerical values of the speed of light in January and July differ by less than 2 m/s.

More recent experiments have used unstable elementary particles, called π mesons, that decay into high-energy photons of light. The π mesons, created in a particle accelerator, move through the laboratory at 99.975% the speed of light, or $v = 0.99975c$ as they emit photons at the speed c in the π meson's reference frame. As Figure 36.10 shows, you would expect the photons to travel through the laboratory with speed $c + v = 1.99975c$. Instead, the measured speed of the photons in the laboratory was, within experimental error, 3.00×10^8 m/s.

In summary, *every* experiment designed to compare the speed of light in different reference frames has found that light travels at 3.00×10^8 m/s in every inertial reference frame, regardless of how the reference frames are moving with respect to each other.

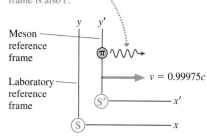

FIGURE 36.10 Experiments find that the photons travel through the laboratory with speed c, not the speed $1.99975c$ that you might expect.

How Can This Be?

You're in good company if you find this impossible to believe. Suppose I shot a ball forward at 50 m/s while driving past you at 30 m/s. You would certainly see the ball traveling at 80 m/s relative to you and the ground. What we're saying with regard to light is equivalent to saying that the ball travels at 50 m/s relative to my car and *at the same time* travels at 50 m/s relative to the ground, even though the car is moving across the ground at 30 m/s. It seems logically impossible.

You might think that this is merely a matter of semantics. If we can just get our definitions and use of words straight, then the mystery and confusion will disappear. Or perhaps the difficulty is a confusion between what we "see" versus what "really happens." In other words, a better analysis, one that focuses on what really happens, would find that light "really" travels at different speeds in different reference frames.

Alas, what "really happens" is that light travels at 3.00×10^8 m/s in every inertial reference frame, regardless of how the reference frames are moving with respect to each other. It's not a trick. There remains only one way to escape the logical contradictions.

The definition of velocity is $u = \Delta x/\Delta t$, the ratio of a distance traveled to the time interval in which the travel occurs. Suppose you and I both make measurements on an object as it moves, but you happen to be moving relative to me. Perhaps I'm standing on the corner, you're driving past in your car, and we're both trying to measure the velocity of a bicycle. Further, suppose we have agreed in advance to measure the bicycle as it moves from the tree to the lamppost in Figure 36.11 on the next page. Your $\Delta x'$ differs from my Δx because of your motion relative to me, causing you to calculate a bicycle velocity u' in your reference frame that differs from its velocity u in my reference frame. This is just the Galilean transformations showing up again.

Now let's repeat the measurements, but this time let's measure the velocity of a light wave as it travels from the tree to the lamppost. Once again, your $\Delta x'$ differs from my Δx, although the difference will be pretty small unless your car is moving at well above the legal speed limit. The obvious conclusion is that your light speed u' differs from my light speed u. But it doesn't. The experiments show that, for a light wave, we'll get the *same* values: $u' = u$.

The only way this can be true is if your Δt is not the same as my Δt. If the time it takes the light to move from the tree to the lamppost in your reference frame, a

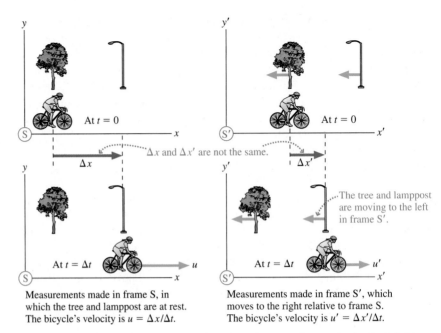

Measurements made in frame S, in which the tree and lamppost are at rest. The bicycle's velocity is $u = \Delta x/\Delta t$.

Measurements made in frame S', which moves to the right relative to frame S. The bicycle's velocity is $u' = \Delta x'/\Delta t'$.

FIGURE 36.11 Measuring the velocity of an object by appealing to the basic definition $u = \Delta x/\Delta t$.

time we'll now call $\Delta t'$, differs from the time Δt it takes the light to move from the tree to the lamppost in my reference frame, then we might find that $\Delta x'/\Delta t' = \Delta x/\Delta t$. That is, $u' = u$ even though you are moving with respect to me.

We've assumed, since the beginning of this textbook, that time is simply time. It flows along like a river, and all experimenters in all reference frames simply use it. For example, suppose the tree and the lamppost both have big clocks that we both can see. Shouldn't we be able to agree on the time interval Δt the light needs to move from the tree to the lamppost?

Perhaps not. It's demonstrably true that $\Delta x' \neq \Delta x$. It's experimentally verified that $u' = u$ for light waves. Something must be wrong with *assumptions* that we've made about the nature of time. The principle of relativity has painted us into a corner, and our only way out is to reexamine our understanding of time.

36.4 Events and Measurements

To question some of our most basic assumptions about space and time requires extreme care. We need to be certain that no assumptions slip into our analysis unnoticed. Our goal is to describe the motion of a particle in a clear and precise way, making the barest minimum of assumptions.

Events

The fundamental entity of relativity is called an **event**. An event is a physical activity that takes place at a definite point in space and at a definite instant of time. A firecracker exploding is an event. A collision between two particles is an event. A light wave hitting a detector is an event.

Events can be observed and measured by experimenters in different reference frames. An exploding firecracker is as clear to you as you drive by in your car as it is to me standing on the street corner. We can quantify where and when an event occurs with four numbers: the coordinates (x, y, z) and the instant of time t. These four numbers, illustrated in Figure 36.12, are called the **spacetime coordinates** of the event.

An event has spacetime coordinates (x, y, z, t) in frame S and different spacetime coordinates (x', y', z', t') in frame S'.

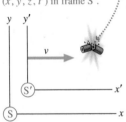

FIGURE 36.12 The location and time of an event are described by its spacetime coordinates.

The spatial coordinates of an event measured in reference frames S and S′ may differ. It now appears that the instant of time recorded in S and S′ may also differ. Thus the spacetime coordinates of an event measured by experimenters in frame S are (x, y, z, t) and the spacetime coordinates of the *same event* measured by experimenters in frame S′ are (x', y', z', t').

The motion of a particle can be described as a sequence of two or more events. We introduced this idea in the previous section when we agreed to measure the velocity of a bicycle and then of a light wave by comparing the object passing the tree (first event) to the object passing the lamppost (second event).

Measurements

Events are what "really happen," but how do we learn about an event? That is, how do the experimenters in a reference frame determine the spacetime coordinates of an event? This is a problem of *measurement.*

We defined a reference frame to be a coordinate system in which experimenters can make position and time measurements. That's a good start, but now we need to be more precise as to *how* the measurements are made. Imagine that a reference frame is filled with a cubic lattice of meter sticks, as shown in Figure 36.13. At every intersection is a clock, and all the clocks in a reference frame are *synchronized*. We'll return in a moment to consider how to synchronize the clocks, but assume for the moment it can be done.

Now, with our meter sticks and clocks in place, we can use a two-part measurement scheme:

- The (x, y, z) coordinates of an event are determined by the intersection of meter sticks closest to the event.
- The event's time t is the time displayed on the clock nearest the event.

You can imagine, if you wish, that each event is accompanied by a flash of light to illuminate the face of the nearest clock and make its reading known.

Several important issues need to be noted:

1. The clocks and meter sticks in each reference frame are imaginary, so they have no difficulty passing through each other.
2. Measurements of position and time made in one reference frame must use only the clocks and meter sticks in that reference frame.
3. There's nothing special about the sticks being 1 m long and the clocks 1 m apart. The lattice spacing can be altered to achieve whatever level of measurement accuracy is desired.
4. We'll assume that the experimenters in each reference frame have assistants sitting beside every clock to record the position and time of nearby events.
5. Perhaps most important, t is the time at which the event *actually happens,* not the time at which an experimenter sees the event or at which information about the event reaches an experimenter.
6. All experimenters in one reference frame agree on the spacetime coordinates of an event. In other words, **an event has a unique set of spacetime coordinates in each reference frame.**

The spacetime coordinates of this event are measured by the nearest meter stick intersection and the nearest clock.

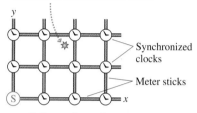

Reference frame S

Reference frame S′ has its own meter sticks and its own clocks.

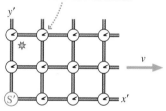

Reference frame S′

FIGURE 36.13 The spacetime coordinates of an event are measured by a lattice of meter sticks and clocks.

STOP TO THINK 36.3   A carpenter is working on a house two blocks away. You notice a slight delay between seeing the carpenter's hammer hit the nail and hearing the blow. At what time does the event "hammer hits nail" occur?

a. At the instant you hear the blow.
b. At the instant you see the hammer hit.
c. Very slightly before you see the hammer hit.
d. Very slightly after you see the hammer hit.

Clock Synchronization

It's important that all the clocks in a reference frame be **synchronized,** meaning that all clocks in the reference frame have the same reading at any one instant of time. We would not be able to use a sequence of events to track the motion of a particle if the clocks differed in their readings. Thus we need a method of synchronization. One idea that comes to mind is to designate the clock at the origin as the *master clock*. We could then carry this clock around to every clock in the lattice, adjust that clock to match the master clock, and finally return the master clock to the origin.

This would be a perfectly good method of clock synchronization in Newtonian mechanics, where time flows along smoothly, the same for everyone. But we've been driven to reexamine the nature of time by the possibility that time is different in reference frames moving relative to each other. Because the master clock would *move,* we cannot assume that the master clock keeps time in the same way as the stationary clocks.

We need a synchronization method that does not require moving the clocks. Fortunately, such a method is easy to devise. Each clock is resting at the intersection of meter sticks, so by looking at the meter sticks, the assistant knows, or can calculate, exactly how far each clock is from the origin. Once the distance is known, the assistant can calculate exactly how long a light wave will take to travel from the origin to each clock. For example, light will take 1.00 μs to travel to a clock 300 m from the origin.

> **NOTE** ▶ It's handy for many relativity problems to know that the speed of light is $c = 300$ m/μs. ◀

To synchronize the clocks, the assistants begin by setting each clock to display the light travel time from the origin, but they don't start the clocks. Next, as Figure 36.14 shows, a light flashes at the origin and, simultaneously, the clock at the origin starts running from $t = 0$ s. The light wave spreads out in all directions at speed c. A photodetector on each clock recognizes the arrival of the light wave and, without delay, starts the clock. The clock had been preset with the light travel time, so each clock as it starts reads exactly the same as the clock at the origin. Thus all the clocks will be synchronized after the light wave has passed by.

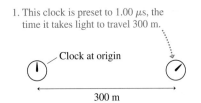

1. This clock is preset to 1.00 μs, the time it takes light to travel 300 m.

Clock at origin

300 m

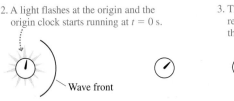

2. A light flashes at the origin and the origin clock starts running at $t = 0$ s.

Wave front

3. The clock starts when the light wave reaches it. It is now synchronized with the origin clock.

FIGURE 36.14 Synchronizing the clocks.

Events and Observations

We noted above that t is the time the event *actually happens*. This is an important point, one that bears further discussion. Light waves take time to travel. Messages, whether they're transmitted by light pulses, telephone, or courier on horseback, take time to be delivered. An experimenter *observes* an event, such as an exploding firecracker, only *at a later time* when light waves reach his or her eyes. But our interest is in the event itself, not the experimenter's observation of the event. The time at which the experimenter sees the event or receives information about the event is not when the event actually occurred.

Suppose at $t = 0$ s a firecracker explodes at $x = 300$ m. The flash of light from the firecracker will reach an experimenter at the origin at $t_1 = 1.0\ \mu$s. The sound of the explosion will reach a sightless experimenter at $t_2 = 0.88$ s. Neither of these is the time t_{event} of the explosion, although the experimenter can work backward from these times, using known wave speeds, to determine t_{event}. In this example, the spacetime coordinates of the event—the explosion—are (300 m, 0 m, 0 m, 0 s).

EXAMPLE 36.3 Finding the time of an event

Experimenter A in reference frame S stands at the origin looking in the positive x-direction. Experimenter B stands at $x = 900$ m looking in the negative x-direction. A firecracker explodes somewhere between them. Experimenter B sees the light flash at $t = 3.00\ \mu$s. Experimenter A sees the light flash at $t = 4.00\ \mu$s. What are the spacetime coordinates of the explosion?

MODEL Experimenters A and B are in the same reference frame and have synchronized clocks.

VISUALIZE Figure 36.15 shows the two experimenters and the explosion at unknown position x.

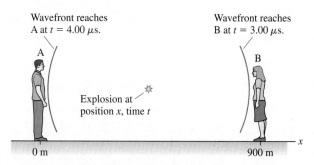

Wavefront reaches A at $t = 4.00\ \mu$s.

Wavefront reaches B at $t = 3.00\ \mu$s.

A

B

Explosion at position x, time t

0 m

900 m

FIGURE 36.15 The light wave reaches the experimenters at different times. Neither of these is the time at which the event actually happened.

SOLVE The two experimenters observe light flashes at two different instants, but there's only one event. Light travels 300 m/μs, so the additional 1.00 μs needed for the light to reach experimenter A implies that distance $(x - 0$ m$)$ is 300 m longer than distance $(900$ m $- x)$. That is,

$$(x - 0\text{ m}) = (900\text{ m} - x) + 300\text{ m}$$

This is easily solved to give $x = 600$ m as the position coordinate of the explosion. The light takes 1.00 μs to travel 300 m to experimenter B, 2.00 μs to travel 600 m to experimenter A. The light is received at 3.00 μs and 4.00 μs, respectively, hence it was emitted by the explosion at $t = 2.00\ \mu$s. The spacetime coordinates of the explosion are (600 m, 0 m, 0 m, 2.00 μs).

ASSESS Although the experimenters *see* the explosion at different times, they agree that the explosion actually *happened* at $t = 2.00\ \mu$s.

Simultaneity

Two events 1 and 2 that take place at different positions x_1 and x_2 but at the *same time $t_1 = t_2$*, as measured in some reference frame, are said to be **simultaneous** in that reference frame. Simultaneity is determined by when the events actually happen, not when they are seen or observed. In general, simultaneous events are *not* seen at the same time because of the difference in light travel times from the events to an experimenter.

EXAMPLE 36.4 Are the explosions simultaneous?

An experimenter in reference frame S stands at the origin looking in the positive x-direction. At $t = 3.0\ \mu$s she sees firecracker 1 explode at $x = 600$ m. A short time later, at $t = 5.0\ \mu$s, she sees firecracker 2 explode at $x = 1200$ m. Are the two explosions simultaneous? If not, which firecracker exploded first?

MODEL Light from both explosions travels toward the experimenter at 300 m/μs.

SOLVE The experimenter *sees* two different explosions, but perceptions of the events are not the events themselves. When did the explosions *actually* occur? Using the fact that light travels 300 m/μs, it's easy to see that firecracker 1 exploded at $t_1 = 1.0\ \mu$s and firecracker 2 also exploded at $t_2 = 1.0\ \mu$s. The events *are* simultaneous.

STOP TO THINK 36.4   A tree and a pole are 3000 m apart. Each is suddenly hit by a bolt of lightning. Mark, who is standing at rest midway between the two, sees the two lightning bolts at the same instant of time. Nancy is at rest under the tree. Define event 1 to be "lightning strikes tree" and event 2 to be "lightning strikes pole." For Nancy, does event 1 occur before, after, or at the same time as event 2?

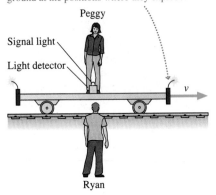

The firecrackers will make burn marks on the ground at the positions where they explode.

FIGURE 36.16 A railroad car traveling to the right with velocity *v*.

36.5 The Relativity of Simultaneity

We've now established a means for measuring the time of an event in a reference frame, so let's begin to investigate the nature of time. The following "thought experiment" is very similar to one suggested by Einstein.

Figure 36.16 shows a long railroad car traveling to the right with a velocity *v* that may be an appreciable fraction of the speed of light. A firecracker is tied to each end of the car, right above the ground. Each firecracker is powerful enough that, when it explodes, it will make a burn mark on the ground at the position of the explosion.

Ryan is standing on the ground, watching the railroad car go by. Peggy is standing in the exact center of the car with a special box at her feet. This box has two light detectors, one facing each way, and a signal light on top. The box works as follows:

1. If a flash of light is received at the right detector before a flash is received at the left detector, then the light on top of the box will turn green.
2. If a flash of light is received at the left detector before a flash is received at the right detector, or if two flashes arrive simultaneously, the light on top will turn red.

The firecrackers explode as the railroad car passes Ryan, and he sees the two light flashes from the explosions simultaneously. He then measures the distances to the two burn marks and finds that he was standing exactly halfway between the marks. Because light travels equal distances in equal times, Ryan concludes that the two explosions were simultaneous in his reference frame, the reference frame of the ground. Further, because he was midway between the two ends of the car, he was directly opposite Peggy when the explosions occurred.

Figure 36.17a shows the sequence of events in Ryan's reference frame. Light travels at speed *c* in all inertial reference frames, so, although the firecrackers were moving, the light waves are spheres centered on the burn marks. Ryan determines that the light wave coming from the right reaches Peggy and the box before the light wave coming from the left. Thus, according to Ryan, the signal light on top of the box turns green.

How do things look in Peggy's reference frame, a reference frame moving to the right at velocity *v* relative to the ground? As Figure 36.17b shows, Peggy sees Ryan moving to the left with speed *v*. Light travels at speed *c* in all inertial reference frames, so the light waves are spheres centered on the ends of the car. If the explosions are simultaneous, as Ryan has determined, the two light waves reach her and the box simultaneously. Thus, according to Peggy, the signal light on top of the box turns red!

Now the light on top must be either green or red. *It can't be both!* Later, after the railroad car has stopped, Ryan and Peggy can place the box in front of them. Either it has a red light or a green light. Ryan can't see one color while Peggy sees the other. Hence we have a paradox. It's impossible for Peggy and Ryan both to be right. But who is wrong, and why?

(a) The events in Ryan's frame

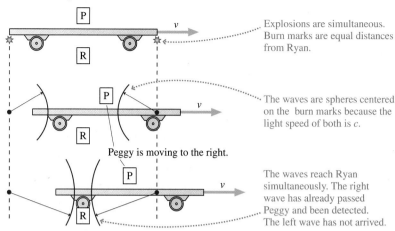

Explosions are simultaneous. Burn marks are equal distances from Ryan.

The waves are spheres centered on the burn marks because the light speed of both is c.

Peggy is moving to the right.

The waves reach Ryan simultaneously. The right wave has already passed Peggy and been detected. The left wave has not arrived.

(b) The events in Peggy's frame

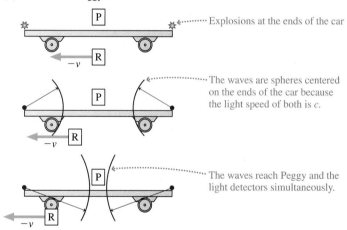

Explosions at the ends of the car

The waves are spheres centered on the ends of the car because the light speed of both is c.

The waves reach Peggy and the light detectors simultaneously.

FIGURE 36.17 Exploding firecrackers seen in two different reference frames.

What do we know with absolute certainty?

1. Ryan detected the flashes simultaneously.
2. Ryan was halfway between the firecrackers when they exploded.
3. The light from the two explosions traveled toward Ryan at equal speeds.

The conclusion that the explosions were simultaneous in Ryan's reference frame is unassailable. The light is green.

Peggy, however, made an assumption. It's a perfectly ordinary assumption, one that seems sufficiently obvious that you probably didn't notice, but an assumption nonetheless. Peggy assumed that the explosions were simultaneous.

Didn't Ryan find them to be simultaneous? Indeed, he did. Suppose we call Ryan's reference frame S, the explosion on the right event R, and the explosion on the left event L. Ryan found that $t_R = t_L$. But Peggy has to use a different set of clocks, the clocks in her reference frame S′, to measure the times t'_R and t'_L at which the explosions occurred. The fact that $t_R = t_L$ in frame S does *not* allow us to conclude that $t'_R = t'_L$ in frame S′.

In fact, the right firecracker must explode *before* the left firecracker in frame S′. Figure 36.17b, with its assumption about simultaneity, was incorrect. Figure 36.18 shows the situation in Peggy's reference frame with the right firecracker exploding first. Now the wave from the right reaches Peggy and the box first, as Ryan had concluded, and the light on top turns green.

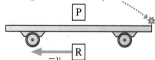

The right firecracker explodes first.

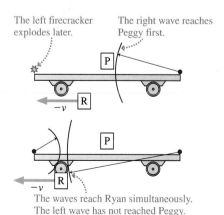

The left firecracker explodes later.

The right wave reaches Peggy first.

The waves reach Ryan simultaneously. The left wave has not reached Peggy.

FIGURE 36.18 The real sequence of events in Peggy's reference frame.

One of the most disconcerting conclusions of relativity is that **two events occurring simultaneously in reference frame S are** *not* **simultaneous in any reference frame S′ that is moving relative to S.** This is called the **relativity of simultaneity.**

The two firecrackers *really* explode at the same instant of time in Ryan's reference frame. And the right firecracker *really* explodes first in Peggy's reference frame. It's not a matter of when they see the flashes. Our conclusion refers to the times at which the explosions actually occur.

The paradox of Peggy and Ryan contains the essence of relativity, and it's worth careful thought. First, review the logic until you're certain that there *is* a paradox, a logical impossibility. Then convince yourself that the only way to resolve the paradox is to abandon the assumption that the explosions are simultaneous in Peggy's reference frame. If you understand the paradox and its resolution, you've made a big step toward understanding what relativity is all about.

STOP TO THINK 36.5   A tree and a pole are 3000 m apart. Each is suddenly hit by a bolt of lightning. Mark, who is standing at rest midway between the two, sees the two lightning bolts at the same instant of time. Nancy is flying her rocket at $v = 0.5c$ in the direction from the tree toward the pole. The lightning hits the tree just as she passes by it. Define event 1 to be "lightning strikes tree" and event 2 to be "lightning strikes pole." For Nancy, does event 1 occur before, after, or at the same time as event 2?

36.6 Time Dilation

17.1   Activ Physics ONLINE

The principle of relativity has driven us to the logical conclusion that time is not the same for two reference frames moving relative to each other. Our analysis thus far has been mostly qualitative. It's time to start developing some quantitative tools that will allow us to compare measurements in one reference frame to measurements in another reference frame.

Figure 36.19a shows a special clock called a **light clock.** The light clock is a box of height h with a light source at the bottom and a mirror at the top. The light source emits a very short pulse of light that travels to the mirror and reflects back to a light detector beside the source. The clock advances one "tick" each time the detector receives a light pulse, and it immediately, with no delay, causes the light source to emit the next light pulse.

Our goal is to compare two measurements of the interval between two ticks of the clock: one taken by an experimenter standing next to the clock and the other by an experimenter moving with respect to the clock. To be specific, Figure 36.19b shows the clock at rest in reference frame S′. We call this the **rest frame** of the clock. Reference frame S′ moves to the right with velocity v relative to reference frame S.

Relativity requires us to measure *events,* so let's define event 1 to be the emission of a light pulse and event 2 to be the detection of that light pulse. Experimenters in both reference frames are able to measure where and when these events occur *in their frame.* In frame S, the time interval $\Delta t = t_2 - t_1$ is one tick of the clock. Similarly, one tick in frame S′ is $\Delta t' = t_2' - t_1'$.

To be sure we have a clear understanding of the relativity result, let's first do a classical analysis. In frame S′, the clock's rest frame, the light travels straight up and down, a total distance $2h$, at speed c. The time interval is $\Delta t' = 2h/c$.

Figure 36.20a shows the operation of the light clock as seen in frame S. The clock is moving to the right at speed v in S, thus the mirror moves distance $\frac{1}{2}v(\Delta t)$

(a) A light clock

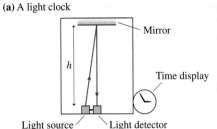

(b) The clock is at rest in frame S′.

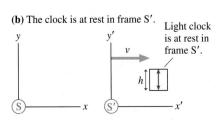

FIGURE 36.19 The ticking of a light clock can be measured by experimenters in two different reference frames.

during the time $\frac{1}{2}(\Delta t)$ in which the light pulse moves from the source to the mirror. The distance traveled by the light during this interval is $\frac{1}{2}u_{\text{light}}(\Delta t)$, where u_{light} is the speed of light in frame S. You can see from the vector addition in Figure 36.20b that the speed of light in frame S′ is $u_{\text{light}} = (c^2 + v^2)^{1/2}$. (Remember, this is a classical analysis in which the speed of light *does* depend on the motion of the reference frame relative to the light source.)

The Pythagorean theorem applied to the right triangle in Figure 36.20a is

$$h^2 + \left(\frac{1}{2}v\Delta t\right)^2 = \left(\frac{1}{2}u_{\text{light}}\Delta t\right)^2 = \left(\frac{1}{2}\sqrt{c^2 + v^2}\,\Delta t\right)^2$$
$$= \left(\frac{1}{2}c\Delta t\right)^2 + \left(\frac{1}{2}v\Delta t\right)^2 \tag{36.4}$$

The term $(\frac{1}{2}v\Delta t)^2$ is common to both sides and cancels. Solving for Δt gives $\Delta t = 2h/c$, identical to $\Delta t'$. In other words, a classical analysis finds that the clock ticks at exactly the same rate in both frame S and frame S′. This shouldn't be surprising. There's only one kind of time in classical physics, measured the same by all experimenters independent of their motion.

The principle of relativity changes only one thing, but that change has profound consequences. According to the principle of relativity, light travels at the same speed in *all* inertial reference frames. In frame S′, the rest frame of the clock, the light simply goes straight up and back. The time of one tick,

$$\Delta t' = \frac{2h}{c} \tag{36.5}$$

is unchanged from the classical analysis.

Figure 36.21 shows the light clock as seen in frame S. The difference from Figure 36.20a is that the light now travels along the hypotenuse at speed c. We can again use the Pythagorean theorem to write

$$h^2 + \left(\frac{1}{2}v\Delta t\right)^2 = \left(\frac{1}{2}c\Delta t\right)^2 \tag{36.6}$$

Solving for Δt gives

$$\Delta t = \frac{2h/c}{\sqrt{1 - v^2/c^2}} = \frac{\Delta t'}{\sqrt{1 - v^2/c^2}} \tag{36.7}$$

The time interval between two ticks in frame S is *not* the same as in frame S′.

It's useful to define $\beta = v/c$, the velocity as a fraction of the speed of light. For example, a reference frame moving with $v = 2.4 \times 10^8$ m/s has $\beta = 0.80$. In terms of β, Equation 36.7 is

$$\Delta t = \frac{\Delta t'}{\sqrt{1 - \beta^2}} \tag{36.8}$$

NOTE ▶ The expression $(1 - v^2/c^2)^{1/2} = (1 - \beta^2)^{1/2}$ occurs frequently in relativity. The value of the expression is 1 when $v = 0$, and it steadily decreases to 0 as $v \rightarrow c$ (or $\beta \rightarrow 1$). The square root is an imaginary number if $v > c$, which would make Δt imaginary in Equation 36.8. Time intervals certainly have to be real numbers, suggesting that $v > c$ is not physically possible. One of the predictions of the theory of relativity, as you've undoubtedly heard, is that nothing can travel faster than the speed of light. Now you can begin to see why. We'll examine this topic more closely in Section 36.9. In the meantime, we'll require v to be less than c. ◀

(a)

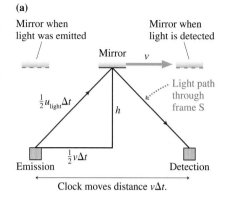

(b)

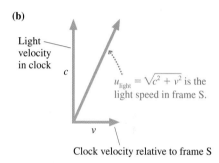

FIGURE 36.20 A classical analysis of the light clock.

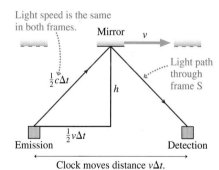

FIGURE 36.21 A light clock analysis in which the speed of light is the same in all reference frames.

Proper Time

Frame S′ has one important distinction. It is the *one and only* inertial reference frame in which the clock is at rest. Consequently, it is the one and only inertial reference frame in which the times of both events—the emission of the light and the detection of the light—are measured by the *same* clock. You can see that the light pulse in Figure 36.19, the rest frame of the clock, starts and ends at the same position and can be measured by one clock. In Figure 36.21, the emission and detection take place at different positions in frame S and must be measured by different clocks.

The time interval between two events that occur at the *same position* is called the **proper time** $\Delta\tau$. Only one inertial reference frame measures the proper time, and it does so with a single clock that is present at both events. An inertial reference frame moving with velocity $v = \beta c$ relative to the proper time frame must use two clocks to measure the time interval because the two events occur at different positions. The time interval in this frame is

$$\Delta t = \frac{\Delta\tau}{\sqrt{1 - \beta^2}} \geq \Delta\tau \qquad \text{(time dilation)} \qquad (36.9)$$

The "stretching out" of the time interval implied by Equation 36.9 is called **time dilation.** Time dilation is sometimes described by saying that "moving clocks run slow." This is not an accurate statement because it implies that some reference frames are "really" moving while others are "really" at rest. The whole point of relativity is that all inertial reference frames are equally valid, that all we know about reference frames is how they move relative to each other. A better description of time dilation is the statement that **the time interval between two ticks is the shortest in the reference frame in which the clock is at rest.** The time interval between two ticks is longer (i.e., the clock "runs slower") when it is measured in any reference frame in which the clock is moving.

NOTE ▶ Equation 36.9 was derived using a light clock because the operation of a light clock is clear and easy to analyze. But the conclusion is really about time itself. *Any* clock, regardless of how it operates, behaves the same. ◀

EXAMPLE 36.5 **From the sun to Saturn**
Saturn is 1.43×10^{12} m from the sun. A rocket travels along a line from the sun to Saturn at a constant speed of $0.9c$ relative to the solar system. How long does the journey take as measured by an experimenter on earth? As measured by an astronaut on the rocket?

MODEL Let the solar system be in reference frame S and the rocket be in reference frame S′ that travels with velocity $v = 0.9c$ relative to S. Relativity problems must be stated in terms of *events*. Let event 1 be "the rocket and the sun coincide" (the experimenter on earth says that the rocket passes the sun; the astronaut on the rocket says that the sun passes the rocket) and event 2 be "the rocket and Saturn coincide."

VISUALIZE Figure 36.22 shows the two events as seen from the two reference frames. Notice that the two events occur at the *same position* in S′, the position of the rocket, and consequently can be measured by *one* clock.

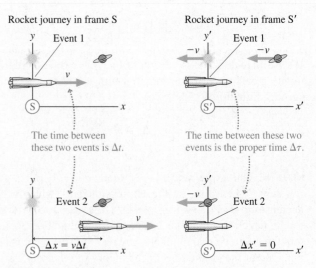

FIGURE 36.22 Pictorial representation of the trip as seen in frames S and S′.

SOLVE The time interval measured in the solar system reference frame, which includes the earth, is simply

$$\Delta t = \frac{\Delta x}{v} = \frac{1.43 \times 10^{12} \text{ m}}{0.9 \times (3.00 \times 10^8 \text{ m/s})} = 5300 \text{ s}$$

Relativity hasn't abandoned the basic definition $v = \Delta x/\Delta t$, although we do have to be sure that Δx and Δt are measured in just one reference frame and refer to the same two events.

How are things in the rocket's reference frame? The two events occur at the *same position* in S′ and can be measured by *one* clock, the clock at the origin. Thus the time measured by the astronauts is the *proper time* $\Delta\tau$ between the two events. We can use Equation 36.9 with $\beta = 0.9$ to find

$$\Delta\tau = \sqrt{1 - \beta^2}\,\Delta t = \sqrt{1 - 0.9^2}\,(5300 \text{ s}) = 2310 \text{ s}$$

ASSESS The time interval measured between these two events by the astronauts is less than half the time interval measured by experimenters on earth. The difference has nothing to do with when earthbound astronomers *see* the rocket pass the sun and Saturn. Δt is the time interval from when the rocket actually passes the sun, as measured by a clock at the sun, until it actually passes Saturn, as measured by a synchronized clock at Saturn. The interval between *seeing* the events from earth, which would have to allow for light travel times, would be something other than 5300 s. Δt and $\Delta\tau$ are different because *time is different* in two reference frames moving relative to each other.

STOP TO THINK 36.6   Molly flies her rocket past Nick at constant velocity v. Molly and Nick both measure the time it takes the rocket, from nose to tail, to pass Nick. Which of the following is true?

a. Both Molly and Nick measure the same amount of time.
b. Molly measures a shorter time interval than Nick.
c. Nick measures a shorter time interval than Molly.

Experimental Evidence

Is there any evidence for the crazy idea that clocks moving relative to each other tell time differently? Indeed, there's plenty. An experiment in 1971 sent an atomic clock around the world on a jet plane while an identical clock remained in the laboratory. This was a difficult experiment because the traveling clock's speed was so small compared to c, but measuring the small differences between the time intervals was just barely within the capabilities of atomic clocks. It was also a more complex experiment than we've analyzed because the clock accelerated as it moved around a circle. Nonetheless, the traveling clock, upon its return, was 200 ns behind the clock that stayed at home, which was exactly as predicted by relativity.

Very detailed studies have been done on unstable particles called *muons* that are created at the top of the atmosphere, at a height of about 60 km, when high-energy cosmic rays collide with air molecules. It is well known, from laboratory studies, that stationary muons decay with a *half-life* of 1.5 μs. That is, half the muons decay within 1.5 μs, half of those remaining decay in the next 1.5 μs, and so on. The decays can be used as a clock.

The muons travel down through the atmosphere at very nearly the speed of light. The time needed to reach the ground, assuming $v \approx c$, is $\Delta t \approx (60,000 \text{ m})/(3 \times 10^8 \text{ m/s}) = 200 \text{ } \mu$s. This is 133 half lives, so the fraction of muons reaching the ground should be $\approx (1/2)^{133} = 10^{-40}$. That is, only 1 out of every 10^{40} muons should reach the ground. In fact, experiments find that about 1 in 10 muons reach the ground, an experimental result that differs by a factor of 10^{39} from our prediction!

The discrepancy is due to time dilation. In Figure 36.23, the two events "muon is created" and "muon hits ground" take place at two different places in the earth's reference frame. However, these two events occur at the *same position* in the muon's reference frame. (The muon is like the rocket in Example 36.5.) Thus

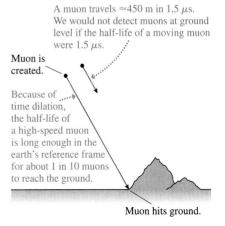

A muon travels ≈450 m in 1.5 μs. We would not detect muons at ground level if the half-life of a moving muon were 1.5 μs.

Muon is created.

Because of time dilation, the half-life of a high-speed muon is long enough in the earth's reference frame for about 1 in 10 muons to reach the ground.

Muon hits ground.

FIGURE 36.23 We wouldn't detect muons at the ground if not for time dilation.

the muon's internal clock measures the proper time. The time-dilated interval $\Delta t = 200~\mu s$ in the earth's reference frame corresponds to a proper time $\Delta \tau \approx 5~\mu s$ in the muon's reference frame. That is, in the muon's reference frame it takes only 5 μs from creation at the top of the atmosphere until the ground runs into it. This is 3.3 half-lives, so the fraction of muons reaching the ground is $(1/2)^{3.3} = 0.1$, or 1 out of 10. We wouldn't detect muons at the ground at all if not for time dilation.

The details are beyond the scope of this textbook, but dozens of high-energy particle accelerators around the world that study quarks and other elementary particles have been designed and built on the basis of Einstein's theory of relativity. The fact that they work exactly as planned is strong testimony to the reality of time dilation.

The Twin Paradox

The most well-known relativity paradox is the twin paradox. George and Helen are twins. On their 25th birthday, Helen departs on a starship voyage to a distant star. Let's imagine, to be specific, that her starship accelerates almost instantly to a speed of $0.95c$ and that she travels to a star that is 9.5 light years (9.5 ly) from earth. Upon arriving, she discovers that the planets circling the star are inhabited by fierce aliens, so she immediately turns around and heads home at $0.95c$.

A **light year,** abbreviated ly, is the distance that light travels in one year. A light year is vastly larger than the diameter of the solar system. The distance between two neighboring stars is typically a few light years. For our purpose, we can write the speed of light as $c = 1$ ly/year. That is, light travels 1 light year per year.

This value for c allows us to determine how long, according to George and his fellow earthlings, it takes Helen to travel out and back. Her total distance is 19 ly and, due to her rapid acceleration and rapid turn around, she travels essentially the entire distance at speed $v = 0.95c = 0.95$ ly/year. Thus the time she's away, as measured by George, is

$$\Delta t_G = \frac{19~\text{ly}}{0.95~\text{ly/year}} = 20~\text{years} \tag{36.10}$$

George will be 45 years old when his sister Helen returns with tales of adventure.

While she's away, George takes a physics class and studies Einstein's theory of relativity. He realizes that time dilation will make Helen's clocks run more slowly than his clocks, which are at rest relative to him. Her heart—a clock—will beat fewer times and the minute hand on her watch will go around fewer times. In other words, she's aging more slowly than he is. Although she is his twin, she will be younger than he is when she returns.

Calculating Helen's age is not hard. We simply have to identify Helen's clock, because it's always with Helen as she travels, as the clock that measures proper time $\Delta \tau$. From Equation 36.9,

$$\Delta t_H = \Delta \tau = \sqrt{1 - \beta^2}\,\Delta t_G = \sqrt{1 - 0.95^2}\,(20~\text{years}) = 6.25~\text{years} \tag{36.11}$$

George will have just celebrated his 45th birthday as he welcomes home his 31-year-and-3-month-old twin sister.

This may be unsettling, because it violates our commonsense notion of time, but it's not a paradox. There's no logical inconsistency in this outcome. So why is it called "the twin paradox"? Read on.

Helen, knowing that she had quite of bit of time to kill on her journey, brought along several physics books to read. As she learns about relativity, she begins to think about George and her friends back on earth. Relative to her, they are all moving away at $0.95c$. Later they'll come rushing toward her at $0.95c$. Time dilation will cause their clocks to run more slowly than her clocks, which are at rest relative to her. In other words, as Figure 36.24 shows, Helen concludes that people on earth are aging more slowly than she is. Alas, she will be much older than they when she returns.

Finally, the big day arrives. Helen lands back on earth and steps out of the starship. George is expecting Helen to be younger than he is. Helen is expecting George to be younger than she is.

Here's the paradox! It's logically impossible for each to be younger than the other at the time when they are reunited. Where, then, is the flaw in our reasoning? It seems to be a symmetrical situation—Helen moves relative to George and George moves relative to Helen—but symmetrical reasoning has led to a conundrum.

But are the situations really symmetrical? George goes about his business day after day without noticing anything unusual. Helen, on the other hand, experiences three distinct periods during which the starship engines fire, she's crushed into her seat, and free dust particles that had been floating inside the starship are no longer, in the starship's reference frame, at rest or traveling in a straight line at constant speed. In other words, George spends the entire time in an inertial reference frame, *but Helen does not*. The situation is *not* symmetrical.

The principle of relativity applies *only* to inertial reference frames. Our discussion of time dilation was for inertial reference frames. Thus George's analysis and calculations are correct. Helen's analysis and calculations are *not* correct because she was trying to apply an inertial reference frame result to a noninertial reference frame.

Helen is younger than George when she returns. This is strange, but not a paradox. It is a consequence of the fact that time flows differently in two reference frames moving relative to each other.

> Helen is moving relative to me at 0.95c. Her clocks are running more slowly than mine, and when she returns she'll be younger than I am.

> 0.95c
>
> 9.5 ly

> George is moving relative to me at 0.95c. His clocks are running more slowly than mine, and when I return he'll be younger than I am.

FIGURE 36.24 The twin paradox.

36.7 Length Contraction

We've seen that relativity requires us to rethink our idea of time. Now let's turn our attention to the concepts of space and distance. Consider the rocket that traveled from the sun to Saturn in Example 36.5. Figure 36.25a shows the rocket moving with velocity v through the solar system reference frame S. Define $L = \Delta x = x_{Saturn} - x_{sun}$ as the distance between the sun and Saturn in frame S or, more generally, the *length* of the spatial interval between two points. The rocket's speed is $v = L/\Delta t$, where Δt is the time measured in frame S for the journey from the sun to Saturn.

Activ
Physics   17.2

(a) Reference frame S: The solar system is stationary.

The rocket moves distance L in time Δt. This is the distance between the sun and Saturn in S.

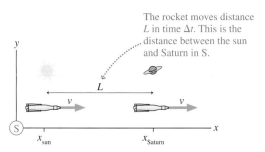

(b) Reference frame S': The rocket is stationary.

Saturn moves distance L' in time $\Delta t' = \Delta \tau$. This is the distance between the sun and Saturn in S'.

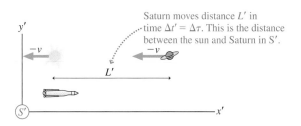

FIGURE 36.25 L and L' are the distances between the sun and Saturn in frames S and S'.

Figure 36.25b shows the situation in reference frame S', where the rocket is at rest. The sun and Saturn move to the left at speed $v = L'/\Delta t'$, where $\Delta t'$ is the time measured in frame S' for Saturn to travel distance L'.

Speed v is the relative speed between S and S′ and is the same for experimenters in both reference frames. That is,

$$v = \frac{L}{\Delta t} = \frac{L'}{\Delta t'} \qquad (36.12)$$

The time interval $\Delta t'$ measured in frame S′ is the proper time $\Delta \tau$ because both events occur at the same position in frame S′ and can be measured by one clock. We can use the time-dilation result, Equation 36.9, to relate $\Delta \tau$ measured by the astronauts to Δt measured by the earthbound scientists. Then Equation 36.12 becomes

$$\frac{L}{\Delta t} = \frac{L'}{\Delta \tau} = \frac{L'}{\sqrt{1 - \beta^2}\,\Delta t} \qquad (36.13)$$

The Δt cancels, and the distance L' in frame S′ is

$$L' = \sqrt{1 - \beta^2}\,L \qquad (36.14)$$

Surprisingly, we find that **the distance between two objects in reference frame S′ is *not the same* as the distance between the same two objects in reference frame S.**

Frame S, in which the distance is L, has one important distinction. It is the *one and only* inertial reference frame in which the objects are at rest. Experimenters in frame S can take all the time they need to measure L because the two objects aren't going anywhere. The distance L between two objects or two points in space measured in the reference frame in which the objects are at rest is called the **proper length** ℓ. Only one inertial reference frame can measure the proper length.

We can use the proper length ℓ to write Equation 36.14 as

$$L' = \sqrt{1 - \beta^2}\,\ell \leq \ell \qquad (36.15)$$

This "shrinking" of the distance between two objects, as measured by an experiment moving with respect to the objects, is called **length contraction.** Although we derived length contraction for the distance between two distinct objects, it applies equally well to the length of any physical object that stretches between two points along the x- and x'-axes. The length of an object is greatest in the reference frame in which the object is at rest. The object's length is less (i.e., the length is contracted) when it is measured in any reference frame moving relative to the object.

The Stanford Linear Accelerator (SLAC) is a 2-mi-long electron accelerator. The accelerator's length is less than 1 m in the reference frame of the electrons.

EXAMPLE 36.6 The distance from the sun to Saturn
In Example 36.5 a rocket traveled along a line from the sun to Saturn at a constant speed of 0.9c relative to the solar system. The Saturn-to-sun distance was given as 1.43×10^{12} m. What is the distance between the sun and Saturn in the rocket's reference frame?

MODEL Saturn and the sun are, at least approximately, at rest in the solar system reference frame S. Thus the given distance is the proper length ℓ.

SOLVE We can use Equation 36.15 to find the distance in the rocket's frame S′:

$$L' = \sqrt{1 - \beta^2}\,\ell = \sqrt{1 - 0.9^2}\,(1.43 \times 10^{12} \text{ m})$$
$$= 0.62 \times 10^{12} \text{ m}$$

ASSESS The sun-to-Saturn distance measured by the astronauts is less than half the distance measured by experimenters on earth. L' and ℓ are different because *space is different* in two reference frames moving relative to each other.

The conclusion that space is different in reference frames moving relative to each other is a direct consequence of the fact that time is different. Experimenters in both reference frames agree on the relative velocity v, leading to Equation 36.12: $v = L/\Delta t = L'/\Delta t'$. We had already learned that $\Delta t' < \Delta t$ because of time dilation. Thus L' *has* to be less than L. That is the only way experimenters in the two reference frames can reconcile their measurements.

To be specific, the earthly experimenters in Examples 36.5 and 36.6 find that the rocket takes 5300 s to travel the 1.43×10^{12} m between the sun and Saturn. The rocket's speed is $v = L/\Delta t = 2.7 \times 10^6$ m/s $= 0.9c$. The astronauts in the rocket find that it takes only 2310 s for Saturn to reach them after the sun has passed by. But there's no conflict, because they also find that the distance is only 0.62×10^{12} m. Thus Saturn's speed toward them is $v = L'/\Delta t' = (0.62 \times 10^{12}$ m$)/(2310$ s$) = 2.7 \times 10^6$ m/s $= 0.9c$.

Another Paradox?

Carmen and Dan are in their physics lab room. They each select a meter stick, lay the two side by side, and agree that the meter sticks are exactly the same length. Then, for an extra-credit project, they go outside and run past each other, in opposite directions, at a relative speed $v = 0.9c$. Figure 36.26 shows their experiment and a portion of their conversation.

Now, Dan's meter stick can't be both longer and shorter than Carmen's meter stick. Is this another paradox? No! Relativity allows us to compare the *same* events as they're measured in two different reference frames. This did lead to a real paradox when Peggy rolled past Ryan on the train. There the signal light on the box turns green (a single event) or it doesn't, and Peggy and Ryan have to agree about it. But the events by which Dan measures the length (in Dan's frame) of Carmen's meter are *not the same events* as those that Carmen uses to measure the length (in Carmen's frame) of Dan's meter stick.

There's no conflict between their measurements. In Dan's reference frame, Carmen's meter stick has been length contracted and is less than 1 m in length. In Carmen's reference frame, Dan's meter stick has been length contracted and is less than 1 m in length. If this weren't the case, if both agreed that one of the meter sticks was shorter than the other, then we could tell which reference frame was "really" moving and which was "really" at rest. But the principle of relativity doesn't allow us to make that distinction. Each is moving relative to the other, so each should make the same measurement for the length of the other's meter stick.

The Spacetime Interval

Forget relativity for a minute and think about ordinary geometry. Figure 36.27 shows two ordinary coordinate systems. They are identical except for the fact that one has been rotated relative to the other. A student using the xy-system would measure coordinates (x_1, y_1) for point 1 and (x_2, y_2) for point 2. A second student, using the $x'y'$-system, would measure (x_1', y_1') and (x_2', y_2').

The students soon find that none of their measurements agree. That is, $x_1 \neq x_1'$ and so on. Even the intervals are different: $\Delta x \neq \Delta x'$ and $\Delta y \neq \Delta y'$. Each is a perfectly valid coordinate system, giving no reason to prefer one over the other, but each yields different measurements.

Is there *anything* on which the two students can agree? Yes, there is. The distance d between points 1 and 2 is independent of the coordinates. We can state this mathematically as

$$d^2 = (\Delta x)^2 + (\Delta y)^2 = (\Delta x')^2 + (\Delta y')^2 \tag{36.16}$$

The quantity $(\Delta x)^2 + (\Delta y)^2$ is called an **invariant** in geometry because it has the same value in any Cartesian coordinate system.

Returning to relativity, is there an invariant in the spacetime coordinates, some quantity that has the *same value* in all inertial reference frames? There is, and to find it let's return to the light clock that we analyzed in Figure 36.21. Figure 36.28 on the next page shows the light clock as seen in reference frames S′ and S″. The speed of light is the same in both frames, even though both are moving with respect to each other and with respect to the clock.

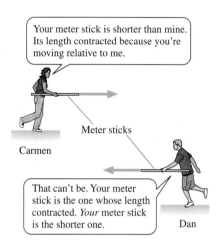

Your meter stick is shorter than mine. Its length contracted because you're moving relative to me.

Meter sticks

Carmen

That can't be. Your meter stick is the one whose length contracted. *Your* meter stick is the shorter one.

Dan

FIGURE 36.26 Carmen and Dan each measure the length of the other's meter stick as they move relative to each other.

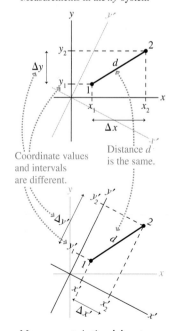

Measurements in the xy-system

Coordinate values and intervals are different.

Distance d is the same.

Measurements in the $x'y'$-system

FIGURE 36.27 Distance d is the same in both coordinate systems.

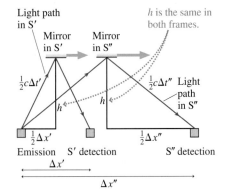

Light path in S'

Mirror in S'

Mirror in S''

h is the same in both frames.

$\frac{1}{2}c\Delta t'$

$\frac{1}{2}c\Delta t''$ Light path in S''

h

h

$\frac{1}{2}\Delta x'$

$\frac{1}{2}\Delta x''$

Emission S' detection

S'' detection

$\Delta x'$

$\Delta x''$

FIGURE 36.28 The light clock seen by experimenters in reference frames S' and S''.

Notice that the clock's height h is common to both reference frames. Thus

$$h^2 = \left(\frac{1}{2}c\Delta t'\right)^2 - \left(\frac{1}{2}\Delta x'\right)^2 = \left(\frac{1}{2}c\Delta t''\right)^2 - \left(\frac{1}{2}\Delta x''\right)^2 \quad (36.17)$$

The factor $\frac{1}{2}$ cancels, allowing us to write

$$c^2(\Delta t')^2 - (\Delta x')^2 = c^2(\Delta t'')^2 - (\Delta x'')^2 \quad (36.18)$$

Let us define the **spacetime interval** s between two events to be

$$s^2 = c^2(\Delta t)^2 - (\Delta x)^2 \quad (36.19)$$

What we've shown in Equation 36.18 is that **the spacetime interval s has the same value in all inertial reference frames.** That is, the spacetime interval between two events is an invariant. It is a value that all experimenters, in all reference frames, can agree upon.

EXAMPLE 36.7 Using the spacetime interval

A firecracker explodes at the origin of an inertial reference frame. Then, 2.0 μs later, a second firecracker explodes 300 m away. Astronauts in a passing rocket measure the distance between the explosions to be 200 m. According to the astronauts, how much time elapses between the two explosions?

MODEL The spacetime coordinates of two events are measured in two different inertial reference frames. Call the reference frame of the ground S and the reference frame of the rocket S'. The spacetime interval between these two events is the same in both reference frames.

SOLVE The spacetime interval (or, rather, its square) in frame S is

$$s^2 = c^2(\Delta t)^2 - (\Delta x)^2 = (600 \text{ m})^2 - (300 \text{ m})^2$$
$$= 270,000 \text{ m}^2$$

where we used $c = 300$ m/μs to determine that $c\Delta t = 600$ m. The spacetime interval has the same value in frame S'. Thus

$$s^2 = 270,000 \text{ m}^2 = c^2(\Delta t')^2 - (\Delta x')^2$$
$$= c^2(\Delta t')^2 - (200 \text{ m})^2$$

This is easily solved to give $\Delta t' = 1.85$ μs.

ASSESS The two events are closer together in both space and time in the rocket's reference frame than in the reference frame of the ground.

Einstein's legacy, according to popular culture, was the discovery that "everything is relative." But it's not so. Time intervals and space intervals may be relative, as were the intervals Δx and Δy in the purely geometric analogy with which we opened this section, but some things are *not* relative. In particular, the spacetime interval s between two events is not relative. It is a well-defined number, agreed to by experimenters in each and every inertial reference frame.

STOP TO THINK 36.7 Beth and Charles are at rest relative to each other. Anjay runs past at velocity v while holding a long pole parallel to his motion. Anjay, Beth, and Charles each measure the length of the pole at the instant Anjay passes Beth. Rank in order, from largest to smallest, the three lengths L_A, L_B, and L_C.

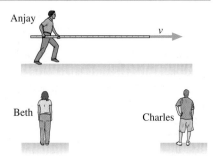

Anjay

v

Beth

Charles

36.8 The Lorentz Transformations

The Galilean transformation $x' = x - vt$ of classical relativity lets us calculate the position x' of an event in frame S' if we know its position x in frame S. Classical relativity, of course, assumes that $t' = t$. Is there a similar transformation in relativity that would allow us to calculate an event's spacetime coordinates (x', t') in frame S' if we know their values (x, t) in frame S? Such a transformation would need to satisfy three conditions. It must

1. Agree with the Galilean transformations in the low-speed limit $v \ll c$.
2. Transform not only spatial coordinates but also time coordinates.
3. Ensure that the speed of light is the same in all reference frames.

We'll continue to use reference frames in the standard orientation of Figure 36.29. The motion is parallel to the x- and x'-axes, and we *define* $t = 0$ and $t' = 0$ as the instant when the origins of S and S' coincide.

The requirement that a new transformation agree with the Galilean transformation when $v \ll c$ suggests that we look for a transformation of the form

$$x' = \gamma(x - vt) \quad \text{and} \quad x = \gamma(x' + vt') \tag{36.20}$$

where γ is a dimensionless function of velocity that satisfies $\gamma \to 1$ as $v \to 0$.

To determine γ, consider the following two events:

Event 1: A flash of light is emitted from the origin of both reference frames $(x = x' = 0)$ at the instant they coincide $(t = t' = 0)$.

Event 2: The light strikes a light detector. The spacetime coordinates of this event are (x, t) in frame S and (x', t') in frame S'.

Light travels at speed c in both reference frames, so the positions of event 2 are $x = ct$ in S and $x' = ct'$ in S'. Substituting these expressions for x and x' into Equation 36.20 gives

$$ct' = \gamma(ct - vt) = \gamma(c - v)t$$
$$ct = \gamma(ct' + vt') = \gamma(c + v)t' \tag{36.21}$$

Solve the first for t', by dividing by c, then substitute this result for t' into the second:

$$ct = \gamma(c + v)\frac{\gamma(c - v)t}{c} = \gamma^2(c^2 - v^2)\frac{t}{c}$$

The t cancels, leading to

$$\gamma^2 = \frac{c^2}{c^2 - v^2} = \frac{1}{1 - v^2/c^2}$$

Thus the γ that "works" in the proposed transformation of Equation 36.20 is

$$\gamma = \frac{1}{\sqrt{1 - v^2/c^2}} = \frac{1}{\sqrt{1 - \beta^2}} \tag{36.22}$$

You can see that $\gamma \to 1$ as $v \to 0$, as expected.

The transformation between t and t' is found by requiring that $x = x$ if you use Equation 36.20 to transform a position from S to S' and then back to S. The details will be left for a homework problem. Another homework problem will let you demonstrate that the y and z measurements made perpendicular to the relative motion are not affected by the motion. We tacitly assumed this condition in our analysis of the light clock.

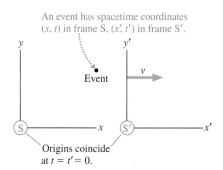

An event has spacetime coordinates (x, t) in frame S, (x', t') in frame S'.

Origins coincide at $t = t' = 0$.

FIGURE 36.29 The spacetime coordinates of an event are measured in inertial reference frames S and S'.

The full set of equations are called the **Lorentz transformations.** They are

$$
\begin{aligned}
x' &= \gamma(x - vt) & x &= \gamma(x' + vt') \\
y' &= y & y &= y' \\
z' &= z & z &= z' \\
t' &= \gamma(t - vx/c^2) & t &= \gamma(t' + vx'/c^2)
\end{aligned}
\qquad (36.23)
$$

The Lorentz transformations transform the spacetime coordinates of *one* event. You should compare these to the Galilean transformation equations in Equation 36.1.

> **NOTE** ▶ These transformations are named after the Dutch physicist H. A. Lorentz, who derived them prior to Einstein. Lorentz was close to discovering special relativity, but he didn't recognize that our concepts of space and time have to be changed before these equations can be properly interpreted. ◀

Using Relativity

Relativity is phrased in terms of *events,* hence relativity problems are solved by interpreting the problem statement in terms of specific events.

PROBLEM-SOLVING STRATEGY 36.1 Relativity

MODEL Frame the problem in terms of events, things that happen at a specific place and time.

VISUALIZE A pictorial representation defines the reference frames.

- Sketch the reference frames, showing their motion relative to each other.
- Show events. Identify objects that are moving with respect to the reference frames.
- Identify any proper time intervals and proper lengths. These are measured in an object's rest frame.

SOLVE The mathematical representation is based on the Lorentz transformations, but not every problem requires the full transformation equations.

- Problems about time intervals can often be solved using time dilation: $\Delta t = \gamma \Delta \tau$.
- Problems about distances can often be solved using length contraction: $L = \ell/\gamma$.

ASSESS Are the results consistent with Galilean relativity when $v \ll c$?

EXAMPLE 36.8 Ryan and Peggy revisited

Peggy is standing in the center of a long, flat railroad car that has firecrackers tied to both ends. The car moves past Ryan, who is standing on the ground, with velocity $v = 0.8c$. Flashes from the exploding firecrackers reach him simultaneously 1.0 μs after the instant that Peggy passes him, and he later finds burn marks on the track 300 m to either side of where he had been standing.

a. According to Ryan, what is the distance between the two explosions and at what times do the explosions occur relative to the time that Peggy passes him?

b. According to Peggy, what is the distance between the two explosions and at what times do the explosions occur relative to the time that Ryan passes her?

MODEL Let the explosion on Ryan's right, the direction in which Peggy is moving, be event R. The explosion on his left is event L.

VISUALIZE Peggy and Ryan are in inertial reference frames. Figure 36.30 shows Peggy's frame S′ moving with $v = 0.8c$ relative to Ryan's frame S. We've defined the reference frames such that Peggy and Ryan are at the origins. The instant they pass, by definition, is $t = t' = 0$ s. The two events are shown in Ryan's reference frame.

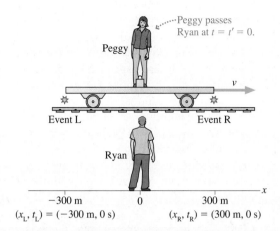

FIGURE 36.30 A pictorial representation of the reference frames and events.

SOLVE

a. The two burn marks tell Ryan that the distance between the explosions was $L = 600$ m. Light travels at $c = 300$ m/μs, and the burn marks are 300 m on either side of him, so Ryan can determine that each explosion took place 1.0 μs before he saw the flash. But this was the instant of time that Peggy passed him, so Ryan concludes that the explosions were simultaneous with each other and with Peggy's passing him. The spacetime coordinates of the two events in frame S are $(x_R, t_R) = (300$ m, 0 μs$)$ and $(x_L, t_L) = (-300$ m, 0 μs$)$.

b. We already know, from our qualitative analysis in Section 36.5, that the explosions are *not* simultaneous in Peggy's reference frame. Event R happens before event L in S', but we don't know how they compare to the time at which Ryan passes Peggy. We can now use the Lorentz transformations to relate the spacetime coordinates of these events as measured by Ryan to the spacetime coordinates as measured by Peggy. Using $v = 0.8c$, we find that γ is

$$\gamma = \frac{1}{\sqrt{1 - v^2/c^2}} = \frac{1}{\sqrt{1 - 0.8^2}} = 1.667$$

For event L, the Lorentz transformations are

$$x_L' = 1.667((-300 \text{ m}) - (0.8c)(0 \text{ } \mu s)) = -500 \text{ m}$$

$$t_L' = 1.667((0 \text{ } \mu s) - (0.8c)(-300 \text{ m})/c^2) = 1.33 \text{ } \mu s$$

And for event R,

$$x_R' = 1.667((300 \text{ m}) - (0.8c)(0 \text{ } \mu s)) = 500 \text{ m}$$

$$t_R' = 1.667((0 \text{ } \mu s) - (0.8c)(300 \text{ m})/c^2) = -1.33 \text{ } \mu s$$

According to Peggy, the two explosions occur 1000 m apart. Furthermore, the first explosion, on the right, occurs 1.33 μs before Ryan passes her at $t' = 0$ s. The second, on the left, occurs 1.33 μs after Ryan goes by.

ASSESS Events that are simultaneous in frame S are *not* simultaneous in frame S'. The results of the Lorentz transformations agree with our earlier qualitative analysis.

A follow-up discussion of Example 36.8 is worthwhile. Because Ryan moves at speed $v = 0.8c = 240$ m/μs relative to Peggy, he moves 320 m during the 1.33 μs between the first explosion and the instant he passes Peggy, then another 320 m before the second explosion. Gathering this information together, Figure 36.31 shows the sequence of events in Peggy's reference frame.

The firecrackers define the ends of the railroad car, so the 1000 m distance between the explosions in Peggy's frame is the car's length L' in frame S'. The car is at rest in frame S', hence length L' is the proper length: $\ell = 1000$ m. Ryan is measuring the length of a moving object, so he should see the car length contracted to

$$L = \sqrt{1 - \beta^2}\ell = \frac{\ell}{\gamma} = \frac{1000 \text{ m}}{1.667} = 600 \text{ m}$$

And, indeed, that is exactly the distance Ryan measured between the burn marks.

Finally, we can calculate the spacetime interval s between the two events. According to Ryan,

$$s^2 = c^2(\Delta t^2) - (\Delta x)^2 = c^2(0 \text{ } \mu s)^2 - (600 \text{ m})^2 = -(600 \text{ m})^2$$

Peggy computes the spacetime interval to be

$$s^2 = c^2(\Delta t')^2 - (\Delta x')^2 = c^2(2.67 \text{ } \mu s)^2 - (1000 \text{ m})^2 = -(600 \text{ m})^2$$

Their calculations of the spacetime interval agree, showing that s really is an invariant, but notice that s itself is an imaginary number.

Length

We've already introduced the idea of length contraction, but we didn't precisely define just what we mean by the *length* of a moving object. The length of an object at rest is clear because we can take all the time we need to measure it with

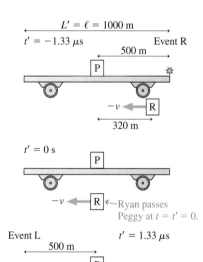

FIGURE 36.31 The sequence of events as seen in Peggy's reference frame.

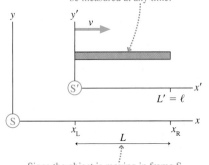

The object is at rest in frame S'. It's length is $L' = \ell$, which can be measured at any time.

$L' = \ell$

Since the object is moving in frame S, simultaneous measurements of its ends must be made in order to find its length L in frame S.

FIGURE 36.32 The length of an object is the distance between *simultaneous* measurements of the positions of the end points.

meter sticks, surveying tools, or whatever we need. But how can we give clear meaning to the length of a moving object?

A reasonable definition of an object's length is the distance $L = \Delta x = x_R - x_L$ between the right and left ends when the positions x_R and x_L are measured *at the same time t*. In other words, length is the distance spanned by the object at *one instant* of time. Measuring an object's length requires *simultaneous* measurements of two positions (i.e., two events are required), hence the result won't be known until the information from two spatially separated measurements can be brought together. That's all right, because relativity deals with what *really happens* rather than with our perceptions of the events.

Figure 36.32 shows an object traveling through reference frame S with velocity v. The object is at rest in reference frame S' that travels with the object at velocity v, hence the length in frame S' is the proper length ℓ. That is, $\Delta x' = x_R' - x_L' = \ell$ in frame S'.

At time t, an experimenter (and his or her assistants) in frame S makes simultaneous measurements of the positions x_R and x_L of the ends of the object. The difference $\Delta x = x_R - x_L = L$ is the length in frame S. The Lorentz transformations of x_R and x_L are

$$x_R' = \gamma(x_R - vt)$$
$$x_L' = \gamma(x_L - vt)$$
(36.24)

where, it is important to note, t is the *same* for both because the measurements are simultaneous.

Subtracting the second equation from the first, we find

$$x_R' - x_L' = \ell = \gamma(x_R - x_L) = \gamma L = \frac{L}{\sqrt{1 - \beta^2}}$$

Solving for L, we find, in agreement with Equation 36.15, that

$$L = \sqrt{1 - \beta^2}\, \ell$$
(36.25)

This analysis has accomplished two things. First, by giving a precise definition of length, we've put our length-contraction result on a firmer footing. Second, we've had good practice at relativistic reasoning using the Lorentz transformation.

NOTE ▶ Length contraction does not tell us how an object would *look*. The visual appearance of an object is determined by light waves that arrive simultaneously at the eye. These waves left points on the object at different times (i.e., *not* simultaneously) because they had to travel different distances to the eye. The analysis needed to determine an object's visual appearance is considerably more complex. Length and length contraction are concerned only with the *actual* length of the object at one instant of time. ◀

The Binomial Approximation

The binomial approximation

If $x \ll 1$, then $(1 + x)^n \approx 1 + nx$

You've met the binomial approximation earlier in this text and in your calculus class. The binomial approximation is useful when we need to calculate a relativistic expression for a nonrelativistic velocity $v \ll c$. Because $v^2/c^2 \ll 1$ in these cases, we can write

$$\sqrt{1 - \beta^2} = (1 - v^2/c^2)^{1/2} \approx 1 - \frac{1}{2}\frac{v^2}{c^2}$$

$$\gamma = \frac{1}{\sqrt{1 - \beta^2}} = (1 - v^2/c^2)^{-1/2} \approx 1 + \frac{1}{2}\frac{v^2}{c^2}$$
(36.26)

The following example illustrates the use of the binomial approximation.

EXAMPLE 36.9 **The shrinking school bus**

An 8.0-m-long school bus drives past at 30 m/s. By how much is its length contracted?

MODEL The school bus is at rest in an inertial reference frame S′ moving at velocity $v = 30$ m/s relative to the ground frame S. The given length, 8.0 m, is the proper length ℓ in frame S′.

SOLVE In frame S, the school bus is length contracted to

$$L = \sqrt{1 - \beta^2}\,\ell$$

The bus's velocity v is much less than c, so we can use the binomial approximation to write

$$L \approx \left(1 - \frac{1}{2}\frac{v^2}{c^2}\right)\ell = \ell - \frac{1}{2}\frac{v^2}{c^2}\ell$$

The *amount* of the length contraction is

$$\ell - L = \frac{1}{2}\frac{v^2}{c^2}\ell = \left(\frac{30 \text{ m/s}}{3.0 \times 10^8 \text{ m/s}}\right)^2 (4.0 \text{ m})$$

$$= 4.0 \times 10^{-14} \text{ m} = 40 \text{ fm}$$

where 1 fm = 1 femtometer = 10^{-15} m.

ASSESS The amount the bus "shrinks" is only slightly larger than the diameter of the nucleus of an atom. It's no wonder that we're not aware of length contraction in our everyday lives. If you had tried to calculate this number exactly, your calculator would have shown $\ell - L = 0$. The difficulty is that the difference between ℓ and L shows up only in the 14th decimal place. A scientific calculator determines numbers to 10 or 12 decimal places, but that isn't sufficient to show the difference. The binomial approximation provides an invaluable tool for finding the very tiny difference between two numbers that are nearly identical.

The Lorentz Velocity Transformations

Figure 36.33 shows an object that is moving in both reference frame S and reference frame S′. Experimenters in frame S determine that the object's velocity is u while experimenters in S′ find it to be u'. For simplicity, we'll assume that the object moves parallel to the x- and x′-axes.

The Galilean velocity transformation $u' = u - v$ was found by taking the time derivative of the position transformation. We can do the same with the Lorentz transformation if we take the derivative with respect to the time in each frame. Velocity u' in frame S′ is

$$u' = \frac{dx'}{dt'} = \frac{d(\gamma(x - vt))}{d(\gamma(t - vx/c^2))} \tag{36.27}$$

where we've used the Lorentz transformations for position x' and time t'.

Carrying out the differentiation gives

$$u' = \frac{\gamma(dx - v\,dt)}{\gamma(dt - v\,dx/c^2)} = \frac{dx/dt - v}{1 - v(dx/dt)/c^2} \tag{36.28}$$

But dx/dt is u, the object's velocity in frame S, leading to

$$u' = \frac{u - v}{1 - uv/c^2} \tag{36.29}$$

You can see that Equation 36.29 reduces to the Galilean transformation $u' = u - v$ when $v \ll c$, as expected.

The reverse transformation, from S′ to S, is found by reversing the sign of v. Altogether,

$$u' = \frac{u - v}{1 - uv/c^2} \quad \text{and} \quad u = \frac{u' + v}{1 + u'v/c^2} \tag{36.30}$$

Equations 36.30 are the Lorentz velocity transformation equations.

NOTE ▶ It is important to distinguish carefully between v, which is the relative velocity of the reference frames in which measurements are carried out, and u and u', which are the velocities of an *object* as measured in two different reference frames. ◀

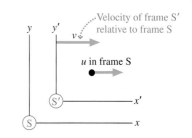

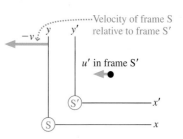

FIGURE 36.33 The velocity of a moving object is measured to be u in frame S and u' in frame S′.

EXAMPLE 36.10 A really fast bullet

A rocket flies past the earth at $0.9c$. As it goes by, the rocket fires a bullet in the forward direction at $0.95c$ with respect to the rocket. What is the bullet's speed with respect to the earth?

MODEL The rocket and the earth are inertial reference frames. Let the earth be frame S and the rocket be frame S'. The velocity of frame S' relative to frame S is $v = 0.9c$. The bullet's velocity in frame S' is $u' = 0.95c$.

SOLVE We can use the Lorentz velocity transformation to find

$$u = \frac{u' + v}{1 + u'v/c^2} = \frac{0.95c + 0.90c}{1 + (0.95c)(0.90c)/c^2} = 0.997c$$

The bullet's speed with respect to the earth is 99.7% of the speed of light.

NOTE ▶ Many relativistic calculations are much easier when velocities are specified as a fraction of c. ◀

ASSESS In Newtonian mechanics, the Galilean transformation of velocity would give $u = 1.85c$. Now, despite the very high speed of the rocket and of the bullet with respect to the rocket, the bullet's speed with respect to the earth remains less than c. This is yet more evidence that objects cannot exceed the speed of light.

Suppose the rocket in Example 36.10 fired a laser beam in the forward direction as it traveled past the earth at velocity v. The laser beam would travel away from the rocket at speed $u' = c$ in the rocket's reference frame S'. What is the laser beam's speed in the earth's frame S? According to the Lorentz velocity transformation, it must be

$$u = \frac{u' + v}{1 + u'v/c^2} = \frac{c + v}{1 + cv/c^2} = \frac{c + v}{1 + v/c} = \frac{c + v}{(c + v)/c} = c \qquad (36.31)$$

Light travels at speed c in both frame S and frame S'. This important consequence of the principle of relativity is "built into" the Lorentz transformations.

36.9 Relativistic Momentum

In Newtonian mechanics, the total momentum of a system is a conserved quantity. Further, as we've seen, the law of conservation of momentum, $P_f = P_i$, is true in all inertial reference frames *if* the particle velocities in different reference frames are related by the Galilean velocity transformations.

The difficulty, of course, is that the Galilean transformations are not consistent with the principle of relativity. It is a reasonable approximation when all velocities are much less than c, but the Galilean transformations fail dramatically as velocities approach c. It's not hard to show that $P_f' \neq P_i'$ if the particle velocities in frame S' are related to the particle velocities in frame S by the Lorentz transformations.

There are two possibilities:

1. The so-called law of conservation of momentum is not really a law of physics. It is approximately true at low velocities but fails as velocities approach the speed of light.
2. The law of conservation of momentum really is a law of physics, but the expression $p = mu$ is not the correct way to calculate momentum when the particle velocity u becomes a significant fraction of c.

Momentum conservation is such a central and important feature of mechanics that it seems unlikely to fail in relativity. How else might the momentum of a particle be defined?

The classical momentum, for one-dimensional motion, is $p = mu = m(\Delta x/\Delta t)$. Δt is the time needed to move distance Δx. That seemed clear enough within a Newtonian framework, but now we've learned that experimenters in different reference frames disagree about the amount of time needed. So whose Δt should we use?

One possibility is to use the time measured *by the particle*. This is the proper time $\Delta \tau$ because the particle is at rest in its own reference frame and needs only one clock. With this in mind, let's redefine the momentum of a particle of mass m moving with velocity $u = \Delta x/\Delta t$ to be

$$p = m\frac{\Delta x}{\Delta \tau} \tag{36.32}$$

We can relate this new expression for p to the familiar Newtonian expression by using the time-dilation result $\Delta \tau = (1 - u^2/c^2)^{1/2}\Delta t$ to relate the proper time interval measured by the particle to the more practical time interval Δt measured by experimenters in frame S. With this substitution, Equation 36.32 becomes

$$p = m\frac{\Delta x}{\Delta \tau} = m\frac{\Delta x}{\sqrt{1 - u^2/c^2}\,\Delta t} = \frac{mu}{\sqrt{1 - u^2/c^2}} \tag{36.33}$$

You can see that Equation 36.33 reduces to the classical expression $p = mu$ when the particle's speed $u \ll c$. That is an important requirement, but whether this is the "correct" expression for p depends on whether the total momentum P is conserved when the velocities of a system of particles are transformed with the Lorentz velocity transformation equations. The proof is rather long and tedious, so we will assert, without actual proof, that the momentum defined in Equation 36.33 does, indeed, transform correctly. **The law of conservation of momentum is still valid in all inertial reference frames *if* the momentum of each particle is calculated with Equation 36.33.**

The factor that multiplies mu in Equation 36.33 looks much like the factor γ in the Lorentz transformation equations for x and t, but there's one very important difference. The v in the Lorentz transformation equations is the velocity of a *reference frame*. The u in Equation 36.33 is the velocity of a particle moving *in a* reference frame.

With this distinction in mind, let's define the quantity

$$\gamma_p = \frac{1}{\sqrt{1 - u^2/c^2}} \tag{36.34}$$

where the subscript p indicates that this is γ for a particle, not for a reference frame. In frame S′, where the particle moves with velocity u', the corresponding expression would be called γ_p'. With this definition of γ_p, the momentum of a particle is

$$p = \gamma_p mu \tag{36.35}$$

EXAMPLE 36.11 Momentum of a subatomic particle

Electrons in a particle accelerator reach a speed of $0.999c$ relative to the laboratory. One collision of an electron with a target produces a muon that moves forward with a speed of $0.95c$ relative to the laboratory. The electron mass is $m_e = 9.11 \times 10^{-31}$ m/s and the muon mass is 1.90×10^{-28} m/s. What is the muon's momentum in the laboratory frame and in the frame of the electron beam?

MODEL Let the laboratory be reference frame S. The reference frame S′ of the electron beam (i.e., a reference frame in which the electrons are at rest) moves in the direction of the electrons at $v = 0.999c$. The muon velocity in frame S is $u = 0.95c$.

SOLVE γ_p for the muon in the laboratory reference frame is

$$\gamma_p = \frac{1}{\sqrt{1 - u^2/c^2}} = \frac{1}{\sqrt{1 - 0.95^2}} = 3.20$$

Thus the muon's momentum in the laboratory is

$$p = \gamma_p mu = (3.20)(1.90 \times 10^{-28}\,\text{kg})(0.95 \times 3.00 \times 10^8\,\text{m/s})$$
$$= 1.73 \times 10^{-19}\,\text{kg\,m/s}$$

The momentum is a factor of 3.2 larger than the Newtonian momentum mu. To find the momentum in the electron-beam reference frame, we must first use the velocity transformation equation to find the muon's velocity in frame S′:

$$u' = \frac{u - v}{1 - uv/c^2} = \frac{0.95c - 0.999c}{1 - (0.95c)(0.999c)/c^2} = -0.962c$$

In the laboratory frame, the faster electrons are overtaking the slower muon. Hence the muon's velocity in the electron-beam frame is negative. γ_p' for the muon in frame S′ is

$$\gamma_p' = \frac{1}{\sqrt{1 - u'^2/c^2}} = \frac{1}{\sqrt{1 - 0.962^2}} = 3.66$$

The muon's momentum in the electron-beam reference frame is

$$p' = \gamma_p' m u'$$

$$= (3.66)(1.90 \times 10^{-28}\ \text{kg})(-0.962 \times 3.00 \times 10^8\ \text{m/s})$$

$$= -2.01 \times 10^{-19}\ \text{kg m/s}$$

ASSESS From the laboratory perspective, the muon moves only slightly slower than the electron beam. But it turns out that the muon moves faster with respect to the electrons, although in the opposite direction, than it does with respect to the laboratory.

The Cosmic Speed Limit

Figure 36.34a is a graph of momentum versus velocity. For a Newtonian particle, with $p = mu$, the momentum is directly proportional to the velocity. The relativistic expression for momentum agrees with the Newtonian value if $u \ll c$, but p approaches ∞ as $u \to c$.

The implications of this graph become clear when we relate momentum to force. Consider a particle subjected to a constant force, such as a rocket that never runs out of fuel. If F is constant, we can see from $F = dp/dt$ that the momentum is $p = Ft$. If Newtonian physics were correct, a particle would go faster and faster as its velocity $u = p/m = (F/m)t$ increased without limit. But the relativistic result, shown in Figure 36.34b, is that the particle's velocity asymptotically approaches the speed of light ($u \to c$) as p approaches ∞. Relativity gives a very different outcome than Newtonian mechanics.

The speed c is a "cosmic speed limit" for material particles. A force cannot accelerate a particle to a speed higher than c because the particle's momentum becomes infinitely large as the speed approaches c. The amount of effort required for each additional increment of velocity becomes larger and larger until no amount of effort can raise the velocity any higher.

Actually, at a more fundamental level, c is a speed limit for *any* kind of **causal influence.** If I throw a rock and break a window, my throw is the *cause* of the breaking window and the rock is the *causal influence.* If I shoot a laser beam at a light detector that is wired to a firecracker, the light wave is the *causal influence* that leads to the explosion. A causal influence can be any kind of particle, wave, or information that travels from A to B and allows A to be the cause of B.

For two unrelated events—a firecracker explodes in Tokyo and a balloon bursts in Paris—the relativity of simultaneity tells us that they may be simultaneous in one reference frame but not in others. Or in one reference frame the firecracker may explode before the balloon bursts but in some other reference frame the balloon may burst first. These possibilities violate our commonsense view of time, but they're not in conflict with the principle of relativity.

For two causally related events—A *causes* B—it would be nonsense for an experimenter in any reference frame to find that B occurs before A. No experimenter in any reference frame, no matter how it is moving, will find that you are born before your mother is born. If A causes B, then it must be the case that $t_A < t_B$ in *all* reference frames.

Suppose there exists some kind of causal influence that *can* travel at speed $u > c$. Figure 36.35 shows a reference frame S in which event A is at the origin ($x_A = 0$). The faster-than-light causal influence—perhaps some yet-to-be-discovered "z ray"—leaves A at $t_A = 0$ and travels to the point at which it will cause event B. It arrives at x_B at time $t_B = x_B/u$.

How do events A and B appear in a reference frame S′ that travels at an ordinary speed $v < c$ relative to frame S? We can use the Lorentz transformations to find out. Because $x_A = 0$ and $t_A = 0$, it's easy to see that $x_A' = 0$ and $t_A' = 0$. That is, the origins of S and S′ overlap at the instant the causal influence leaves

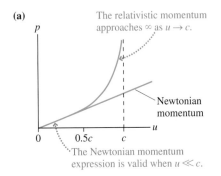

(a)

The relativistic momentum approaches ∞ as $u \to c$.

Newtonian momentum

The Newtonian momentum expression is valid when $u \ll c$.

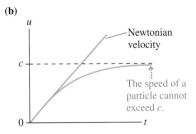

(b)

Newtonian velocity

The speed of a particle cannot exceed c.

FIGURE 36.34 The speed of a particle cannot reach the speed of light.

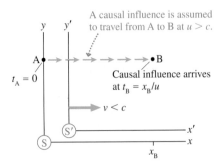

A causal influence is assumed to travel from A to B at $u > c$.

Causal influence arrives at $t_B = x_B/u$

$v < c$

FIGURE 36.35 Assume that a causal influence can travel from A to B at a speed $u > c$.

event A. More interesting is the time at which this influence reaches B in frame S′. The Lorentz time transformation for event B is

$$t'_B = \gamma\left(t_B - \frac{vx_B}{c^2}\right) = \gamma t_B\left(1 - \frac{v(x_B/t_B)}{c^2}\right) = \gamma t_B\left(1 - \frac{vu}{c^2}\right) \quad (36.36)$$

where we first factored out t_B, then made use of the fact that $u = x_B/t_B$ in frame S.

We're assuming $u > c$, so let $u = \alpha c$ where $\alpha > 1$ is a constant. Then $vu/c^2 = \alpha v/c$. Now follow the logic:

1. If $v > c/\alpha$, which is possible because $\alpha > 1$, then $vu/c^2 > 1$.
2. If $vu/c^2 > 1$, then the term $(1 - vu/c^2)$ is negative and $t'_B < 0$.
3. If $t'_B < 0$, then event B happens *before* event A in reference frame S′.

In other words, if a causal influence can travel faster than c, then there exist reference frames in which the effect happens before the cause. We know this can't happen, so our assumption $u > c$ must be wrong. **No causal influence of any kind—particle, wave, or yet-to-be-discovered z rays—can travel faster than c.**

The existence of a cosmic speed limit is one of the most interesting consequences of the theory of relativity. "Warp drive," in which a spaceship suddenly leaps to faster-than-light velocities, is simply incompatible with the theory of relativity. Rapid travel to the stars will remain in the realm of science fiction unless future scientific discoveries find flaws in Einstein's theory and open the doors to yet-undreamed-of theories. While we can't say with certainty that a scientific theory will never be overturned, there is currently not even a hint of evidence that disagrees with the special theory of relativity.

36.10 Relativistic Energy

Energy is our final topic in this chapter on relativity. Space, time, velocity, and momentum are changed by relativity, so it seems inevitable that we'll need a new view of energy.

In Newtonian mechanics, a particle's kinetic energy $K = \frac{1}{2}mu^2$ can be written in terms of its momentum $p = mu$ as $K = p^2/2m$. This suggests that a relativistic expression for energy will likely involve both the square of p and the particle's mass. We also hope that energy will be conserved in relativity, so a reasonable starting point is with the one quantity we've found that is the same in all inertial reference frames: the spacetime interval s.

Let a particle of mass m move through distance Δx during a time interval Δt, as measured in reference frame S. The spacetime interval is

$$s^2 = c^2(\Delta t)^2 - (\Delta x)^2 = \text{invariant}$$

We can turn this into an expression involving momentum if we multiply by $(m/\Delta\tau)^2$, where $\Delta\tau$ is the proper time (i.e., the time measured by the particle). Doing so gives

$$(mc)^2\left(\frac{\Delta t}{\Delta\tau}\right)^2 - \left(\frac{m\Delta x}{\Delta\tau}\right)^2 = (mc)^2\left(\frac{\Delta t}{\Delta\tau}\right)^2 - p^2 = \text{invariant} \quad (36.37)$$

where we used $p = m(\Delta x/\Delta\tau)$ from Equation 36.32.

Now Δt, the time interval in frame S, is related to the proper time by the time-dilation result $\Delta t = \gamma_p \Delta\tau$. With this change, Equation 36.37 becomes

$$(\gamma_p mc)^2 - p^2 = \text{invariant}$$

Finally, for reasons that will be clear in a minute, multiply by c^2, to get

$$(\gamma_p mc^2)^2 - (pc)^2 = \text{invariant} \quad (36.38)$$

To say that the right side is an *invariant* means it has the same value in all inertial reference frames. We can easily determine the constant by evaluating it in the reference frame in which the particle is at rest. In that frame, where $p = 0$ and $\gamma_p = 1$, we find that

$$(\gamma_p mc^2)^2 - (pc)^2 = (mc^2)^2 \tag{36.39}$$

Let's reflect on what this means before taking the next step. The spacetime interval s has the same value in all inertial reference frames. In other words, $c^2(\Delta t)^2 - (\Delta x)^2 = c^2(\Delta t')^2 - (\Delta x')^2$. Equation 36.39 was derived from the definition of the spacetime interval, hence the quantity mc^2 is also an invariant having the same value in all inertial reference frames. In other words, if experimenters in frames S and S′ both make measurements on this particle of mass m, they will find that

$$(\gamma_p mc^2)^2 - (pc)^2 = (\gamma_p' mc^2)^2 - (p'c)^2 \tag{36.40}$$

Experimenters in different reference frames measure different values for the momentum, but experimenters in all reference frames agree that momentum is a conserved quantity. Equations 36.39 and 36.40 suggest that the quantity $\gamma_p mc^2$ is also an important property of the particle, a property that changes along with p in just the right way to satisfy Equation 36.39. But what is this property?

The first clue comes from checking the units. γ_p is dimensionless and c is a velocity, so $\gamma_p mc^2$ has the same units as the classical expression $\frac{1}{2}mv^2$; namely, units of energy. For a second clue, let's examine how $\gamma_p mc^2$ behaves in the low-velocity limit $u \ll c$. We can use the binomial approximation expression for γ_p to find

$$\gamma_p mc^2 = \frac{mc^2}{\sqrt{1 - u^2/c^2}} \approx \left(1 + \frac{1}{2}\frac{u^2}{c^2}\right)mc^2 = mc^2 + \frac{1}{2}mu^2 \tag{36.41}$$

The second term, $\frac{1}{2}mu^2$, is the low-velocity expression for the kinetic energy K. This is an energy associated with motion. But the first term suggests that the concept of energy is more complex than we originally thought. It appears that **there is an inherent energy associated with mass itself.**

With that as a possibility, subject to experimental verification, let's define the **total energy** E of a particle to be

$$E = \gamma_p mc^2 = E_0 + K = \text{rest energy} + \text{kinetic energy} \tag{36.42}$$

This total energy consists of a **rest energy**

$$E_0 = mc^2 \tag{36.43}$$

and a relativistic expression for the *kinetic energy*

$$K = (\gamma_p - 1)mc^2 = (\gamma_p - 1)E_0 \tag{36.44}$$

This expression for the kinetic energy is very nearly $\frac{1}{2}mu^2$ when $u \ll c$ but, as Figure 36.36 shows, differs significantly from the classical value for very high velocities.

Equation 36.43 is, of course, Einstein's famous $E = mc^2$, perhaps the most famous equation in all of physics. Before discussing its significance, we need to tie up some loose ends. First, notice that the right-hand side of Equation 36.39 is the square of the rest energy E_0. Thus we can write a final version of that equation:

$$E^2 - (pc)^2 = E_0^2 \tag{36.45}$$

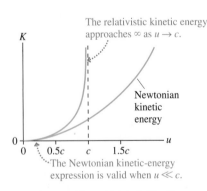

The relativistic kinetic energy approaches ∞ as $u \to c$.

Newtonian kinetic energy

The Newtonian kinetic-energy expression is valid when $u \ll c$.

FIGURE 36.36 The relativistic kinetic energy.

The quantity E_0 is an *invariant* with the same value mc^2 in *all* inertial reference frames.

Second, notice that we can write

$$pc = (\gamma_{\mathrm{p}}mu)c = \frac{u}{c}(\gamma_{\mathrm{p}}mc^2)$$

But $\gamma_{\mathrm{p}}mc^2$ is the total energy E and $u/c = \beta_{\mathrm{p}}$, where the subscript p, as on γ_{p}, implies that we're referring to the motion of a particle within a reference frame, not the motion of two reference frames relative to each other. Thus

$$pc = \beta_{\mathrm{p}}E \qquad (36.46)$$

Figure 36.37 shows the "velocity-energy-momentum triangle," a convenient way to remember the relationships between the three quantities.

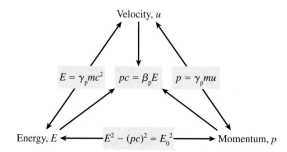

FIGURE 36.37 The velocity-energy-momentum triangle.

EXAMPLE 36.12 Kinetic energy and total energy
Calculate the rest energy and the kinetic energy of (a) a 100 g ball moving with a speed of 100 m/s and (b) an electron with a speed of 0.999c.

MODEL The ball, with $u \ll c$, is a classical particle. We don't need to use the relativistic expression for its kinetic energy. The electron is highly relativistic.

SOLVE
a. For the ball, with $m = 0.10$ kg,

$$E_0 = mc^2 = 9.0 \times 10^{15}\text{ J}$$

$$K = \frac{1}{2}mu^2 = 500\text{ J}$$

b. For the electron, we start by calculating

$$\gamma_{\mathrm{p}} = 1/(1 - u^2/c^2)^{1/2} = 22.4$$

Then, using $m_{\mathrm{e}} = 9.11 \times 10^{-31}$ kg,

$$E_0 = mc^2 = 8.2 \times 10^{-14}\text{ J}$$

$$K = (\gamma_{\mathrm{p}} - 1)E_0 = 170 \times 10^{-14}\text{ J}$$

ASSESS The ball's kinetic energy is a typical kinetic energy. Its rest energy, by contrast, is a staggeringly large number. For a relativistic electron, on the other hand, the kinetic energy is more important than the rest energy.

STOP TO THINK 36.8   An electron moves through the lab at 99% the speed of light. The lab reference frame is S and the electron's reference frame is S'. In which reference frame is the electron's rest mass larger?

a. Frame S, the lab frame
b. Frame S', the electron's frame
c. It is the same in both frames.

FIGURE 36.38 An inelastic collision between two balls of clay does not seem to conserve the total energy E.

The tracks of elementary particles in a bubble chamber show the creation of an electron-positron pair. The negative electron and positive positron spiral in opposite directions in the magnetic field.

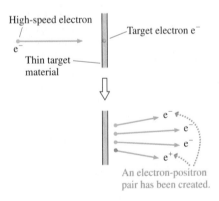

An electron-positron pair has been created.

FIGURE 36.39 An inelastic collision between electrons can create an electron-positron pair.

Mass-Energy Equivalence

Now we're ready to explore the significance of Einstein's famous equation $E = mc^2$. Figure 36.38 shows two balls of clay approaching each other. They have equal masses, equal kinetic energies, and slam together in a perfectly inelastic collision to form one large ball of clay at rest. In Newtonian mechanics, we would say that the initial energy $2K$ is dissipated by being transformed into an equal amount of thermal energy, raising the temperature of the coalesced ball of clay. But Equation 36.42, $E = E_0 + K$, doesn't say anything about thermal energy. The total energy before the collision is $E_i = 2mc^2 + 2K$, with the factor of 2 appearing because there are two masses. It seems like the total energy after the collision, when the clay is at rest, should be $2mc^2$, but this value doesn't conserve total energy.

There's ample experimental evidence that energy is conserved, so there must be a flaw in our reasoning. The statement of energy conservation is

$$E_f = Mc^2 = E_i = 2mc^2 + 2K \qquad (36.47)$$

where M is the mass of clay after the collision. But, remarkably, this requires

$$M = 2m + \frac{2K}{c^2} \qquad (36.48)$$

In other words, **mass is not conserved.** The mass of clay after the collision is larger than the mass of clay before the collision. Total energy can be conserved only if kinetic energy is transformed into an "equivalent" amount of mass.

The mass increase in a collision between two balls of clay is incredibly small, far beyond any scientist's ability to detect. So how do we know if such a crazy idea is true?

Figure 36.39 shows an experiment that has been done countless times in the last 50 years at particle accelerators around the world. An electron that has been accelerated to $u \approx c$ is aimed at a target material. When a high-energy electron collides with an atom in the target, it can easily knock one of the electrons out of the atom. Thus we would expect to see two electrons leaving the target: the incident electron and the ejected electron. Instead, *four* particles emerge from the target: three electrons and a positron. A *positron*, or positive electron, is the antimatter version of an electron, identical to an electron in all respects other than having charge $q = +e$.

In chemical-reaction notation, the collision is

$$e^-(\text{fast}) + e^-(\text{at rest}) \rightarrow e^- + e^- + e^- + e^+$$

An electron and a positron have been *created*, apparently out of nothing. Mass $2m_e$ before the collision has become mass $4m_e$ after the collision. (Notice that charge has been conserved in this collision.)

Although the mass has increased, it wasn't created "out of nothing." This is an inelastic collision, just like the collision of the balls of clay, because the kinetic energy after the collision is less than before. In fact, if you measured the energies before and after the collision, you would find that the decrease in kinetic energy is exactly equal to the energy equivalent of the two particles that have been created: $\Delta K = 2m_e c^2$. The new particles have been created *out of energy!*

Particles can be created from energy and particles can return to energy. Figure 36.40 shows an electron colliding with a positron, its antimatter partner. When a particle and its antiparticle meet, they *annihilate* each other. The mass disappears, and the energy equivalent of the mass is transformed into two high-energy photons of light. Momentum conservation requires two photons, rather than one, and specifies that the two photons have equal energies and be emitted back to back.

If the electron and positron are fairly slow, so that $K \ll mc^2$, then $E_i \approx E_0 = mc^2$. In that case, energy conservation requires

$$E_f = 2E_{photon} = E_i \approx 2m_ec^2 \qquad (36.49)$$

You learned in Chapter 24 that the energy of a photon of light is $E_{photon} = hc/\lambda$, where h is Planck's constant. (Photons and their properties will be discussed again in Chapter 38.) Hence the wavelength of the emitted photons is

$$\lambda = \frac{hc}{m_ec^2} \approx 0.0024 \text{ nm} \qquad (36.50)$$

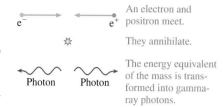

An electron and positron meet.

They annihilate.

The energy equivalent of the mass is transformed into gamma-ray photons.

FIGURE 36.40 The annihilation of an electron-positron pair.

This is an extremely short wavelength, even shorter than the wavelengths of x rays. Photons in this wavelength range are called *gamma rays.* And, indeed, the emission of 0.0024 nm gamma rays is observed in many laboratory experiments in which positrons are able to collide with electrons and thus annihilate. In recent years, with the advent of gamma-ray telescopes on satellites, astronomers have found 0.0024 nm photons coming from many places in the universe, especially galactic centers—evidence that positrons are abundant throughout the universe.

Positron-electron annihilation is also the basis of the medical procedure known as a positron-emission tomography, or PET scans. A patient ingests a very small amount of a radioactive substance that decays by the emission of positrons. This substance is taken up by certain tissues in the body, especially those tissues with a high metabolic rate. As the substance decays, the positrons immediately collide with electrons, annihilate, and create two gamma-ray photons that are emitted back to back. The gamma rays, which easily leave the body, are detected, and their trajectories are traced backward into the body. The overlap of many such trajectories shows quite clearly the tissue where the positron emission is occurring. The results are usually shown as false-color photographs, with redder areas indicating regions of higher positron emission.

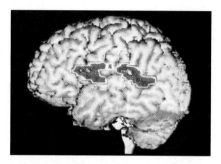

Positron-electron annihilation (a PET scan) provides a noninvasive look into the brain.

Conservation of Energy

The creation and annihilation of particles with mass, processes strictly forbidden in Newtonian mechanics, are vivid proof that neither mass nor the Newtonian definition of energy are conserved. Even so, the *total* energy—the kinetic energy *and* the energy equivalent of mass—remains a conserved quantity.

> **Law of conservation of total energy** The energy $E = \sum E_i$ of an isolated system is conserved, where $E_i = (\gamma_p)_i m_i c^2$ is the total energy of particle i.

Mass and energy are not the same thing, but, as the last few examples have shown, they are *equivalent* in the sense that mass can be transformed into energy and energy can be transformed into mass as long as the total mass is conserved.

Probably the most well-known application of the conservation of total energy is nuclear fission. The uranium isotope ^{236}U, containing 236 protons and neutrons, does not exist in nature. It can be created when a ^{235}U nucleus absorbs a neutron, increasing its atomic mass from 235 to 236. The ^{236}U nucleus quickly fragments into two smaller nuclei and several extra neutrons, a process known as **nuclear fission.** The nucleus can fragment in several ways, but one is

$$n + {}^{235}U \rightarrow {}^{236}U \rightarrow {}^{144}Ba + {}^{89}Kr + 3n$$

Ba and Kr are the atomic symbols for barium and krypton.

This reaction seems like an ordinary chemical reaction—until you check the masses. The masses of atomic isotopes are known with great precision from many decades of measurement in instruments called mass spectrometers. If you add up

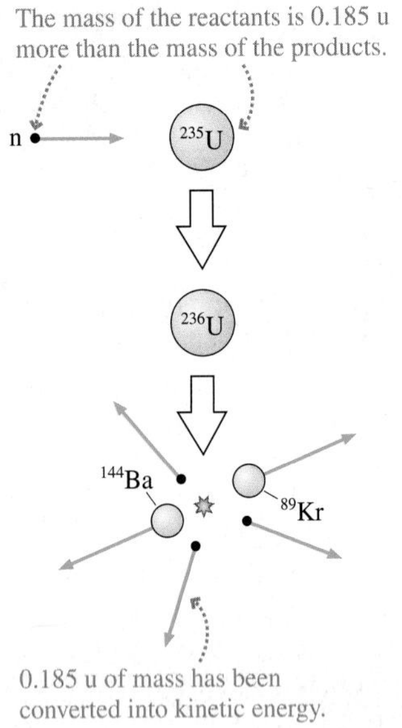

The mass of the reactants is 0.185 u more than the mass of the products.

n •———→ (^{235}U)

⬇

(^{236}U)

⬇

^{144}Ba •   ☆   • ^{89}Kr

0.185 u of mass has been converted into kinetic energy.

FIGURE 36.41 In nuclear fission, the energy equivalent of lost mass is converted into kinetic energy.

the masses on both sides, you find that the mass of the products is 0.185 u smaller than the mass of the initial neutron and ^{235}U, where, you will recall, 1 u = 1.66 × 10^{-27} kg is the unified atomic mass unit. Converting to kilograms gives us the mass loss of 3.07 × 10^{-28} kg.

Mass has been lost, but the energy equivalent of the mass has not. As Figure 36.41 shows, the mass has been converted to kinetic energy, causing the two product nuclei and three neutrons to be ejected at very high speeds. The kinetic energy is easily calculated: $\Delta K = m_{lost}c^2 = 2.8 \times 10^{-11}$ J.

This is a very tiny amount of energy, but it is the energy released from *one* fission. The number of nuclei in a macroscopic sample of uranium is on the order of N_A, Avogadro's number. Hence the energy available if *all* the nuclei fission is enormous. This energy, of course, is the basis for both nuclear power reactors and nuclear weapons.

We started this chapter with an expectation that relativity would challenge our basic notions of space and time. We end by finding that relativity changes our understanding of mass and energy. Most remarkable of all is that each and every one of these new ideas flows from one simple statement: The laws of physics are the same in all inertial reference frames.

SUMMARY

The goal of Chapter 36 has been to understand how Einstein's theory of relativity changes our concepts of space and time.

GENERAL PRINCIPLES

Principle of Relativity All the laws of physics are the same in all inertial reference frames.

- The speed of light c is the same in all inertial reference frames.
- No particle or causal influence can travel at a speed greater than c.

IMPORTANT CONCEPTS

Space

Spatial measurements depend on the motion of the experimenter relative to the events. An object's length is the difference between *simultaneous* measurement of the positions of both ends.

Proper length ℓ is the length of an object measured in a reference frame in which the object is at rest. The L in a frame in which the object moves with velocity v is

$$L = \sqrt{1 - \beta^2}\,\ell \le \ell$$

This is called **length contraction.**

Time

Time measurements depend on the motion of the experimenter relative to the events. Events that are simultaneous in reference frame S are not simultaneous in frame S′ moving relative to S.

Proper time $\Delta\tau$ is the time interval between two events measured in a reference frame in which the events occur at the same position. The time interval Δt in a frame moving with relative velocity v is

$$\Delta t = \Delta\tau / \sqrt{1 - \beta^2} \ge \Delta\tau$$

This is called **time dilation.**

Momentum

The law of conservation of momentum is valid in all inertial reference frames if the momentum of a particle with velocity u is $p = \gamma_{\mathrm{p}} m u$, where

$$\gamma_{\mathrm{p}} = 1 / \sqrt{1 - u^2/c^2}$$

The momentum approaches ∞ as $u \to c$.

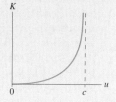

Energy

The law of conservation of energy is valid in all inertial reference frames if the energy of a particle with velocity u is $E = \gamma_{\mathrm{p}} m c^2 = E_0 + K$

Rest energy $E_0 = mc^2$

Kinetic energy $K = (\gamma_{\mathrm{p}} - 1)mc^2$

Invariants

Invariants are quantities that have the same value in all inertial reference frames.

Spacetime interval: $s^2 = (c\Delta t)^2 - (\Delta x)^2$

Particle rest energy: $E_0^2 = (mc^2)^2 = E^2 - (pc)^2$

Mass-energy equivalence

Mass m can be transformed into energy $E = mc^2$.

Energy can be transformed into mass $m = \Delta E/c^2$.

APPLICATIONS

An **event** happens at a specific place in space and time. Spacetime coordinates are (x, t) in frame S and (x', t') in frame S′.

A **reference frame** is a coordinate system with meter sticks and clocks for measuring events. Experimenters at rest relative to each other share the same reference frame.

The **Lorentz transformations** transform spacetime coordinates and velocities between reference frames S and S′.

$$x' = \gamma(x - vt) \qquad x = \gamma(x' + vt')$$
$$y' = y \qquad\qquad y = y'$$
$$z' = z \qquad\qquad z = z'$$
$$t' = \gamma(t - vx/c^2) \qquad t = \gamma(t' + vx'/c^2)$$
$$u' = \frac{u - v}{1 - uv/c^2} \qquad u = \frac{u' + v}{1 + u'v/c^2}$$

where u and u' are the x- and x'-components of velocity.

$$\beta = \frac{v}{c} \quad \text{and} \quad \gamma = 1/\sqrt{1 - v^2/c^2} = 1/\sqrt{1 - \beta^2}$$

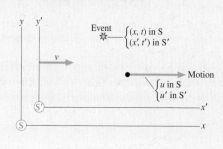

TERMS AND NOTATION

special relativity	simultaneous	invariant
reference frame	relativity of simultaneity	spacetime interval, s
inertial reference frame	light clock	Lorentz transformations
Galilean principle of relativity	rest frame	causal influence
ether	proper time, $\Delta\tau$	total energy, E
principle of relativity	time dilation	rest energy, E_0
event	light year, ly	law of conservation of total energy
spacetime coordinates, (x, y, z, t)	proper length, ℓ	nuclear fission
synchronized	length contraction	

EXERCISES AND PROBLEMS

Exercises

Section 36.2 Galilean Relativity

1. At $t = 1$ s, a firecracker explodes at $x = 10$ m in reference frame S. Four seconds later, a second firecracker explodes at $x = 20$ m. Reference frame S′ moves in the x-direction at a speed of 5 m/s. What are the positions and times of these two events in frame S′?

2. A firecracker explodes in reference frame S at $t = 1$ s. A second firecracker explodes at the same position at $t = 3$ s. In reference frame S′, which moves in the x-direction at speed v, the first explosion is detected at $x′ = 4$ m and the second at $x′ = -4$ m.
    a. What is the speed of frame S′ relative to frame S?
    b. What is the position of the two explosions in frame S?

3. A sprinter crosses the finish line of a race. The roar of the crowd in front approaches her at a speed of 360 m/s. The roar from the crowd behind her approaches at 330 m/s. What are the speed of sound and the speed of the sprinter?

4. A baseball pitcher can throw a ball with a speed of 40 m/s. He is in the back of a pickup truck that is driving away from you. He throws the ball in your direction, and it floats toward you at a lazy 10 m/s. What is the speed of the truck?

5. A boy on a skateboard coasts along at 5 m/s. He has a ball that he can throw at a speed of 10 m/s. What is the ball's speed relative to the ground if he throws the ball
    a. forward?
    b. backward?
    c. to the side?

Section 36.3 Einstein's Principle of Relativity

6. An out-of-control alien spacecraft is diving into a star at a speed of 1×10^8 m/s. At what speed, relative to the spacecraft, is the starlight approaching?

7. A starship blasts past the earth at 2×10^8 m/s. Just after passing the earth, it fires a laser beam out the back of the starship. With what speed does the laser beam approach the earth?

8. A positron moving in the positive x-direction at 2×10^8 m/s collides with an electron at rest. The positron and electron annihilate, producing two gamma-ray photons. Photon 1 travels in the positive x-direction and photon 2 travels in the negative x-direction. What is the speed of each photon?

Section 36.4 Events and Measurements

Section 36.5 The Relativity of Simultaneity

9. Your job is to synchronize the clocks in a reference frame. You are going to do so by flashing a light at the origin at $t = 0$ s. To what time should the clock at $(x, y, z) = (30$ m, 40 m, 0 m$)$ be preset?

10. Bjorn is standing at $x = 600$ m. Firecracker 1 explodes at the origin and firecracker 2 explodes at $x = 900$ m. The flashes from both explosions reach Bjorn's eye at $t = 3$ μs. At what time did each firecracker explode?

11. Bianca is standing at $x = 600$ m. Firecracker 1, at the origin, and firecracker 2, at $x = 900$ m, explode simultaneously. The flash from firecracker 1 reaches Bianca's eye at $t = 3$ μs. At what time does she see the flash from firecracker 2?

12. You are standing at $x = 9$ km. Lightning bolt 1 strikes at $x = 0$ km and lightning bolt 2 strikes at $x = 12$ km. Both flashes reach your eye at the same time. Your assistant is standing at $x = 3$ km. Does your assistant see the flashes at the same time? If not, which does she see first and what is the time difference between the two?

13. You are standing at $x = 9$ km and your assistant is standing at $x = 3$ km. Lightning bolt 1 strikes at $x = 0$ km and lightning bolt 2 strikes at $x = 12$ km. You see the flash from bolt 2 at $t = 10$ μs and the flash from bolt 1 at $t = 50$ μs. According to your assistant, were the lightning strikes simultaneous? If not, which occurred first and what was the time difference between the two?

14. Jose is looking to the east. Lightning bolt 1 strikes a tree 300 m from him. Lightning bolt 2 strikes a barn 900 m from him in the same direction. Jose sees the tree strike 1 μs before he sees the barn strike. According to Jose, were the lightning strikes simultaneous? If not, which occurred first and what was the time difference between the two?

15. You are flying your personal rocketcraft at 0.9c from Star A toward Star B. The distance between the stars, in the stars' reference frame, is 1 ly. Both stars happen to explode simultaneously in your reference frame at the instant you are exactly halfway between them. Do you see the flashes simultaneously? If not, which do you see first and what is the time difference between the two?

Section 36.6 Time Dilation

16. A cosmic ray travels 60 km through the earth's atmosphere in 400 μs, as measured by experimenters on the ground. How long does the journey take according to the cosmic ray?
17. At what speed, as a fraction of c, does a moving clock tick at half the rate of an identical clock at rest?
18. An astronaut travels to a star system 4.5 ly away at a speed of 0.9c. Assume that the time needed to accelerate and decelerate is negligible.
    a. How long does the journey take according to Mission Control on earth?
    b. How long does the journey take according to the astronaut?
    c. How much time elapses between the launch and the arrival of the first radio message from the astronaut saying that she has arrived?
19. A starship voyages to a distant planet 10 ly away. The explorers stay 1 yr, return at the same speed, and arrive back on earth 26 yr after they left. Assume that the time needed to accelerate and decelerate is negligible.
    a. What is the speed of the starship?
    b. How much time has elapsed on the astronauts' chronometers?
20. You fly 5000 km across the United States on an airliner at 250 m/s. You return two days later.
    a. Have you aged more or less than your friends at home?
    b. By how much?
    Hint: Use the binomial approximation.
21. Two clocks are synchronized. One is placed in a race car that drives around a 2.0-km-diameter track at 100 m/s for 24 hours. Afterward, by how much do the two clocks differ?
    Hint: Use the binomial approximation.

Section 36.7 Length Contraction

22. At what speed, as a fraction of c, will a moving rod have a length 60% that of an identical rod at rest?
23. Jill claims that her new rocket is 100 m long. As she flies past your house, you measure the rocket's length and find that it is only 80 m. Should Jill be cited for exceeding the 0.5c speed limit?
24. A muon travels 60 km through the atmosphere at a speed of 0.9997c. According to the muon, how thick is the atmosphere?
25. The Stanford Linear Accelerator (SLAC) accelerates electrons to c = 0.99999997c in a 3.2-km-long tube. If they travel the length of the tube at full speed (they don't, because they are accelerating), how long is the tube in the electrons' reference frame?
26. Our Milky Way galaxy is 100,000 ly in diameter. A spaceship crossing the galaxy measures the galaxy's diameter to be a mere 1 ly.
    a. What is the speed of the spaceship relative to the galaxy?
    b. How long is the crossing time as measured in the galaxy's reference frame?

27. An optical interferometer can detect a displacement of 0.1λ, or $\approx$50 nm. At what speed would a meter stick "shrink" by 50 nm?
    Hint: Use the binomial approximation.

Section 36.8 The Lorentz Transformations

28. An event has spacetime coordinates $(x, t) = (1200$ m, 2.0 μs) in reference frame S. What are the event's spacetime coordinates (a) in reference frame S' that moves in the positive x-direction at 0.8c and (b) in reference frame S'' that moves in the negative x-direction at 0.8c?
29. A rocket travels in the x-direction at speed 0.6c with respect to the earth. An experimenter on the rocket observes a collision between two comets and determines that the spacetime coordinates of the collision are $(x', t') = (3 \times 10^{10}$ m, 200 s). What are the spacetime coordinates of the collision in earth's reference frame?
30. In the earth's reference frame, a tree is at the origin and a pole is at x = 30 km. Lightning strikes both the tree and the pole at t = 10 μs. The lightning strikes are observed by a rocket traveling in the x-direction at 0.5c.
    a. What are the spacetime coordinates for these two events in the rocket's reference frame?
    b. Are the events simultaneous in the rocket's frame? If not, which occurs first?
31. A rocket cruising past earth at 0.8c shoots a bullet out the back door, opposite the rocket's motion, at 0.9c relative to the rocket. What is the bullet's speed relative to the earth?
32. A laboratory experiment shoots an electron to the left at 0.9c. What is the electron's speed relative to a proton moving to the right at 0.9c?
33. A distant quasar is found to be moving away from the earth at 0.8c. A galaxy closer to the earth and along the same line of sight is moving away from us at 0.2c. What is the recessional speed of the quasar as measured by astronomers in the other galaxy?

Section 36.9 Relativistic Momentum

34. A proton is accelerated to 0.999c.
    a. What is the proton's momentum?
    b. By what factor does the proton's momentum exceed its Newtonian momentum?
35. A 1 g particle has momentum 400,000 kg m/s. What is the particle's speed?
36. At what speed is a particle's momentum twice its Newtonian value?
37. What is the speed of a particle whose momentum is mc?

Section 36.10 Relativistic Energy

38. What are the kinetic energy, the rest energy, and the total energy of a 1 g particle with a speed of 0.8c?
39. A quarter-pound hamburger with all the fixings has a mass of 200 g. The food energy of the hamburger (480 food calories) is 2 MJ.
    a. What is the energy equivalent of the mass of the hamburger?
    b. By what factor does the energy equivalent exceed the food energy?

40. How fast must an electron move so that its total energy is 10% more than its rest mass energy?
41. At what speed is a particle's kinetic energy twice its rest energy?
42. At what speed is a particle's total energy twice its rest energy?

Problems

43. A 50 g ball moving to the right at 4.0 m/s overtakes and collides with a 100 g ball moving to the right at 2.0 m/s. The collision is perfectly elastic. Use reference frames and the Chapter 10 result for perfectly elastic collisions to find the speed and direction of each ball after the collision.
44. A 300 g ball moving to the right at 2 m/s has a perfectly elastic collision with a 100 g ball moving to the left at 8 m/s. Use reference frames and the Chapter 10 result for perfectly elastic collisions to find the speed and direction of each ball after the collision.
45. A billiard ball has a perfectly elastic collision with a second billiard ball of equal mass. Afterward, the first ball moves to the left at 2.0 m/s and the second to the right at 4.0 m/s. Use reference frames and the Chapter 10 result for perfectly elastic collisions to find the speed and direction of each ball before the collision.
46. A 9.0 kg artillery shell is moving to the right at 100 m/s when suddenly it explodes into two fragments, one twice as heavy as the other. Measurements reveal that 900 J of energy are released in the explosion and that the heavier fragment was in front of the lighter fragment. Find the velocity of each fragment relative to the ground by analyzing the explosion in the reference frame of (a) the ground and (b) the shell. (c) Is the problem easier to solve in one reference frame?
47. The diameter of the solar system is 10 light hours. A spaceship crosses the solar system in 15 hours, as measured on earth. How long, in hours, does the passage take according to passengers on the spaceship?
   Hint: $c = 1$ light hour per hour.
48. A 30-m-long rocket train car is traveling from Los Angeles to New York at 0.5c when a light at the center of the car flashes. When the light reaches the front of the car, it immediately rings a bell. Light reaching the back of the car immediately sounds a siren.
   a. Are the bell and siren simultaneous events for a passenger seated in the car? If not, which occurs first and by how much time?
   b. Are the bell and siren simultaneous events for a bicyclist waiting to cross the tracks? If not, which occurs first and by how much time?
49. The star Alpha goes supernova. Ten years later and 100 ly away, as measured by astronomers in the galaxy, star Beta explodes.
   a. Is it possible that the explosion of Alpha is in any way responsible for the explosion of Beta? Explain.
   b. An alien spacecraft passing through the galaxy finds that the distance between the two explosions is 120 ly. According to the aliens, what is the time between the explosions?
50. Two events in reference frame S occur 10 μs apart at the same point in space. The distance between the two events is 2400 m in reference frame S′.
   a. What is the time interval between the events in reference frame S′?
   b. What is the velocity of S′ relative to S?

51. a. How fast must a rocket travel on a journey to and from a distant star so that the astronauts age 10 years while the Mission Control workers on earth age 120 years?
   b. As measured by Mission Control, how far away is the distant star?
52. In Section 36.6 we explained that muons can reach the ground because of time dilation. But how do things appear in the muon's reference frame, where the muon's half-life is only 1.5 μs? How can a muon travel the 60 km to reach the earth's surface before decaying? Resolve this apparent paradox. Be as quantitative as you can in your answer.
53. A cube has a density of 2000 kg/m³ while at rest in the laboratory. What is the cube's density as measured by an experimenter in the laboratory as the cube moves through the laboratory at 90% of the speed of light in a direction perpendicular to one of its faces?
54. In an attempt to reduce the extraordinarily long travel times for voyaging to distant stars, some people have suggested traveling at close to the speed of light. Suppose you wish to visit the red giant star Betelgeuse, which is 430 ly away, and that you want your 20,000 kg rocket to move so fast that you age only 20 years during the round trip.
   a. How fast must the rocket travel relative to earth?
   b. How much energy is needed to accelerate the rocket to this speed?
   c. Compare this amount of energy to the total energy used by the United States in the year 2000, which was roughly 1.0×10^{20} J.
55. A rocket traveling at 0.5c sets out for the nearest star, Alpha Centauri, which is 4.25 ly away from earth. It will return to earth immediately after reaching Alpha Centauri. What distance will the rocket travel and how long will the journey last according to (a) stay-at-home earthlings and (b) the rocket crew? (c) Which answers are the correct ones, those in part a or those in part b?
56. The star Delta goes supernova. One year later and 2 ly away, as measured by astronomers in the galaxy, star Epsilon explodes. Let the explosion of Delta be at $x_D = 0$ and $t_D = 0$. The explosions are observed by three spaceships cruising through the galaxy in the direction from Delta to Epsilon at velocities $v_1 = 0.3c$, $v_2 = 0.5c$, and $v_3 = 0.7c$.
   a. What are the times of the two explosions as measured by scientists on each of the three spaceships?
   b. Does one spaceship find that the explosions are simultaneous? If so, which one?
   c. Does one spaceship find that Epsilon explodes before Delta? If so, which one?
   d. Do your answers to parts b and c violate the idea of causality? Explain.
57. Two rockets approach each other. Each is traveling at 0.75c in the earth's reference frame. What is the speed of one rocket relative to the other?
58. A military jet traveling at 1500 m/s has engine trouble and the pilot must bail out. Her ejection seat shoots her forward at 300 m/s relative to the jet. How fast is she traveling relative to the ground according to the relativistic velocity transformation? Would the pilot or anyone else be able to tell the difference between the relativistic and the nonrelativistic result?
59. What is the speed of an electron after being accelerated from rest through a 20×10^6 V potential difference?

60. What is the speed of a proton after being accelerated from rest through a 50×10^6 V potential difference?

61. The half-life of a muon at rest is 1.5 μs. Muons that have been accelerated to a very high speed and are then held in a circular storage ring have a half-life of 7.5 μs.
    a. What is the speed of the muons in the storage ring?
    b. What is the total energy of a muon in the storage ring? The mass of a muon is 207 times the mass of an electron.

62. A solar flare blowing out from the sun at $0.9c$ is overtaking a rocket as it flies away from the sun at $0.8c$. According to the crew on board, with what speed is the flare gaining on the rocket?

63. This chapter has assumed that lengths perpendicular to the direction of motion are not affected by the motion. That is, motion in the x-direction does not cause length contraction along the y- or z-axes. To find out if this is really true, consider two spray-paint nozzles attached to rods perpendicular to the x-axis. It has been confirmed that, when both rods are at rest, both nozzles are exactly 1 m above the base of the rod. One rod is placed in the S reference frame with its base on the x-axis; the other is placed in the S′ reference frame with its base on the x'-axis. The rods then swoop past each other and, as Figure P36.63 shows, each paints a stripe across the other rod.

    We will use proof by contradiction. Assume that objects perpendicular to the motion *are* contracted. An experimenter in frame S finds that the S′ nozzle, as it goes past, is less than 1 m above the x-axis. The principle of relativity says that an experiment carried out in two different inertial reference frames will have the same outcome in both.
    a. Pursue this line of reasoning and show that you end up with a logical contradiction, two mutually incompatible situations.
    b. What can you conclude from this contradiction?

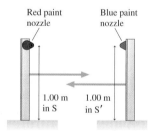

Red paint nozzle   Blue paint nozzle

1.00 m in S   1.00 m in S′

FIGURE P36.63

64. Derive the Lorentz transformations for t and t'.
    Hint: See the comment following Equation 36.22.

65. a. Derive a velocity transformation equation for u_y and u'_y. Assume that the reference frames are in the standard orientation with motion parallel to the x- and x'-axes.
    b. A rocket passes the earth at $0.8c$. As it goes by, it launches a projectile at $0.6c$ perpendicular to the direction of motion. What is the projectile's speed in the earth's reference frame?

66. What is the momentum of a particle with speed $0.95c$ and total energy 2.0×10^{-10} J?

67. What is the momentum of a particle whose total energy is four times its rest energy? Give your answer as a multiple of mc.

68. a. What are the momentum and total energy of a proton with speed $0.99c$?
    b. What is the proton's momentum in a different reference frame in which $E' = 5.0 \times 10^{-10}$ J?

69. At what speed is the kinetic energy of a particle twice its Newtonian value?

70. What is the speed of an electron whose total energy equals the rest mass of a proton?

71. A typical nuclear power plant generates electricity at the rate of 1000 MW. The efficiency of transforming thermal energy into electrical energy is $\frac{1}{3}$ and the plant runs at full capacity for 80% of the year. (Nuclear power plants are down about 20% of the time for maintenance and refueling.)
    a. How much thermal energy does the plant generate in one year?
    b. What mass of uranium is transformed into energy in one year?

72. The sun radiates energy at the rate 3.8×10^{26} W. The source of this energy is fusion, a nuclear reaction in which mass is transformed into energy. The mass of the sun is 2.0×10^{30} kg.
    a. How much mass does the sun lose each year?
    b. What percentage is this of the sun's total mass?
    c. Estimate the lifetime of the sun.

73. The radioactive element radium (Ra) decays by a process known as *alpha decay*, in which the nucleus emits a helium nucleus. (These high-speed helium nuclei were named alpha particles when radioactivity was first discovered, long before the identity of the particles was established.) The reaction is $^{226}\text{Ra} \rightarrow {}^{222}\text{Rn} + {}^4\text{He}$, where Rn is the element radon. The accurately measured atomic masses of the three atoms are 226.015, 222.017, and 4.003. How much energy is released in each decay? (The energy released in radioactive decay is what makes nuclear waste "hot.")

74. The nuclear reaction that powers the sun is the fusion of four protons into a helium nucleus. The process involves several steps, but the net reaction is simply $4\text{p} \rightarrow {}^4\text{He} + \text{energy}$. The mass of a helium nucleus is known to be 6.64×10^{-27} kg.
    a. How much energy is released in each fusion?
    b. What fraction of the initial rest mass energy is this energy?

75. An electron moving to the right at $0.9c$ collides with a positron moving to the left at $0.9c$. The two particles annihilate and produce two gamma-ray photons. What is the wavelength of the photons?

76. Section 36.10 looked at the inelastic collision e^- (fast) + e^- (at rest) $\rightarrow e^- + e^- + e^- + e^+$.
    a. What is the threshold kinetic energy of the fast electron? That is, what minimum kinetic energy must the electron have to allow this process to occur?
    b. What is the speed of an electron with the threshold kinetic energy?

Challenge Problems

77. Two rockets, A and B, approach the earth from opposite directions at speed $0.8c$. The length of each rocket measured in its rest frame is 100 m. What is the length of rocket A as measured by the crew of rocket B?

78. Two rockets are each 1000 m long in their rest frame. Rocket Orion, traveling at $0.8c$ relative to the earth, is overtaking rocket Sirius, which is poking along at a mere $0.6c$. According to the crew on Sirius, how long does Orion take to completely pass? That is, how long is it from the instant the nose of Orion is at the tail of Sirius until the tail of Orion is at the nose of Sirius?

79. Some particle accelerators allow protons (p^+) and antiprotons (p^-) to circulate at equal speeds in opposite directions in a device called a *storage ring*. The particle beams cross each other at various points to cause $p^+ + p^-$ collisions. In one collision, the outcome is $p^+ + p^- \rightarrow e^+ + e^- + \gamma + \gamma$, where γ represents a high-energy gamma-ray photon. The electron and positron are ejected from the collision at $0.9999995c$ and the gamma-ray photon wavelengths are found to be 1.0×10^{-6} nm. What were the proton and antiproton speeds prior to the collision?

80. The rockets of the Goths and the Huns are each 1000 m long in their rest frame. The rockets pass each other, virtually touching, at a relative speed of $0.8c$. The Huns have a laser cannon at the rear of their rocket that shoots a deadly laser beam at right angles to the motion. The captain of the Hun rocket wants to send a threatening message to the Goths by "firing a shot across their bow." He tells his first mate, "The Goths' rocket is length contracted to 600 m. Fire the laser cannon at the instant the nose of our rocket passes the tail of their rocket. The laser beam will cross 400 m in front of them." But things are different in the Goths' reference frame. The Goth captain muses, "The Huns' rocket is length contracted to 600 m, 400 m shorter than our rocket. If they fire the laser cannon as their nose passes the tail of our rocket, the lethal laser blast will go right through our side."

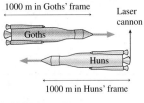

FIGURE CP36.80

The first mate on the Hun rocket fires as ordered. Does the laser beam blast the Goths or not? Resolve this paradox. Show that, when properly analyzed, the Goths and the Huns agree on the outcome. Your analysis should contain both quantitative calculations and written explanation.

81. A very fast pole vaulter lives in the country. One day, while practicing, he notices a 10-m-long barn with the doors open at both ends. He decides to run through the barn at $0.866c$ while carrying his 16-m-long pole. The farmer, who sees him coming, says. "Aha! This guy's pole is length contracted to 8 m. There will be a short interval of time when the pole is entirely inside the barn. If I'm quick, I can simultaneously close both barn doors while the pole vaulter and his pole are inside." The pole vaulter, who sees the farmer beside the barn, thinks to himself, "That farmer is crazy. The barn is length contracted and is only 5 m long. My 16-m-long pole cannot fit into a 5-m-long barn. If the farmer closes the doors just as the tip of my pole reaches the back door, the front door will break off the last 11 m of my pole."

Can the farmer close the doors without breaking the pole? Show that, when properly analyzed, the farmer and the pole vaulter agree on the outcome. Your analysis should contain both quantitative calculations and written explanation. It's obvious that the pole vaulter cannot stop quickly, so you can assume that the doors are paper thin and that the pole breaks through without slowing down.

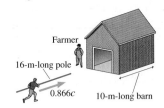

FIGURE CP36.81

STOP TO THINK ANSWERS

Stop to Think 36.1: a, c, and **f.** These move at constant velocity, or very nearly so. The others are accelerating.

Stop to Think 36.2: a. $u' = u - v = -10$ m/s $- 6$ m/s $= -16$ m/s. The *speed* is 16 m/s.

Stop to Think 36.3: c. Even the light has a slight travel time. The event is the hammer hitting the nail, not your seeing the hammer hit the nail.

Stop to Think 36.4: At the same time. Mark is halfway between the tree and the pole, so the fact that he *sees* the lightning bolts at the same time means they *happened* at the same time. It's true that Nancy *sees* event 1 before event 2, but the events actually occurred before she sees them. Mark and Nancy share a reference frame, because they are at rest relative to each other, and all experimenters in a reference frame, after correcting for any signal delays, *agree* on the spacetime coordinates of an event.

Stop to Think 36.5: After. This is the same as the case of Peggy and Ryan. In Mark's reference frame, as in Ryan's, the events are simultaneous. Nancy *sees* event 1 first, but the time when an event is seen is not when the event actually happens. Because all experi-

menters in a reference frame agree on the spacetime coordinates of an event, Nancy's position in her reference frame cannot affect the order of the events. If Nancy had been passing Mark at the instant the lightning strikes occur in Mark's frame, then Nancy would be equivalent to Peggy. Event 2, like the firecracker at the front of Peggy's railroad car, occurs first in Nancy's reference frame.

Stop to Think 36.6: c. Nick measures proper time because Nick's clock is present at both the "nose passes Nick" event and the "tail passes Nick" event. Proper time is the smallest measured time interval between two events.

Stop to Think 36.7: $L_A > L_B = L_C$. Anjay measures the pole's proper length because it is at rest in his reference frame. Proper length is the longest measured length. Beth and Charles may *see* the pole differently, but they share the same reference frame and their *measurements* of the length agree.

Stop to Think 36.8: c. The rest energy E_0 is an invariant, the same in all inertial reference frames. Thus $m = E_0/c^2$ is independent of speed.

37 The End of Classical Physics

Studies of the light emitted by gas discharge tubes helped bring classical physics to an end.

◀ **Looking Ahead**
The goal of Chapter 37 is to understand how scientists discovered the properties of atoms and how these discoveries led to the need for a new theory of light and matter. In this chapter you will learn:

- How the electron was discovered and its charge measured.
- How the nucleus was discovered and its properties identified.
- How to use Rutherford's nuclear model of the atom.
- How atoms emit and absorb light.

◀ **Looking Back**
The material in this chapter depends on many ideas from classical physics. Please review:

- Section 16.2 Atomic masses and the atomic mass number.
- Sections 22.3 and 24.1 Diffraction gratings and spectroscopy.
- Sections 29.2 and 29.6 Electric potential and potential energy.
- Section 32.7 Charged particles in magnetic fields.

Except for relativity, and a brief preview of quantum physics in Chapter 24, everything we have studied until now was known by 1900. Newtonian mechanics, thermodynamics, and Maxwell's theory of electromagnetism form what we call *classical physics*. It is an impressive body of knowledge, with immense explanatory power. Many scientists of the late 1800s felt that they could use these theories to explain just about anything, and some even felt there was nothing significant left to discover.

But within the span of just a few years, right around 1900, investigations into the structure of matter led to many astonishing discoveries that were at odds with classical physics. Discoveries that defied explanation came from investigations as simple as measuring the spectrum of light emitted by gas discharge tubes. It was soon recognized that the laws of classical physics break down when applied to atomic systems. Physicists in the early years of the 20th century had to reexamine their most basic assumptions about the nature of matter and light.

Our goal in this chapter is twofold. The first is to learn how scientists in the 19th and early 20th centuries discovered the properties of atoms. Michael Faraday noted long ago that it is "easy to talk of atoms" but quite another thing to have real knowledge of atoms. We cannot see atoms, so what is the *evidence* that leads us to our current understanding of the atomic theory of matter?

Our second goal is to recognize that many of the newly found atomic properties could not be reconciled with classical physics. Before we launch into quantum physics, it is important to recognize where classical physics failed and why a new theory of light and matter was needed.

37.1 Physics in the 1800s

Scientists in 1800 had three major realms of inquiry. They wanted to understand the nature of matter, of electricity, and of light.

Matter

The idea that matter consists of small, indivisible particles is traceable to Leucippus and his student Democritus, who flourished in ancient Greece around 440–420 BCE. They called these particles *atoms,* Greek for "not divisible." Atomism was not widely accepted, due in no little part to the complete lack of evidence for atoms, but atomic ideas did manage to maintain a minority status throughout the Middle Ages. Then, at about the time of Newton and the beginnings of a mechanistic conception of the world, interest in atoms revived.

Robert Boyle began his study of gases in the 1660s, from which we today know Boyle's law as the fact that pV remains constant for an isothermal process. Newton noted that Boyle's law could be explained if a gas consisted of particles. In 1738, Daniel Bernoulli advanced the idea that gases are composed of small, atom-like particles in random motion. However, the *evidence* for atoms was still far too weak for Bernoulli's ideas to be more than a curiosity.

Things began to change in the early years of the 19th century. The English chemist John Dalton argued that much of what was known about chemical reactions, in particular the law of definite proportions, could be understood if all matter of a particular chemical element consisted of identical, indestructible atoms. The unique feature of Dalton's work, which made it more science than speculation, was his attempt to determine the relative masses of the atoms of different elements. These ideas were further extended by the Italian chemist Amedeo Avogadro, who postulated that atoms could stick together to form more complex entities he called *molecules* and that equal volumes of gases at equal temperatures contain equal numbers of molecules.

The evidence for atoms grew stronger as thermodynamics and the kinetic theory of gases developed in the mid-19th century. Two lines of inquiry led to rough estimates of atomic sizes. First, slight deviations from the ideal-gas law at high pressures could be understood if the atoms were beginning to come into close proximity to one another. Second, the viscosity of a gas, which is readily measured, could be related to the mean free paths of the molecules and the mean free paths, in turn, to the sizes of the atoms. The existence of atoms with diameters of approximately 10^{-10} m was widely accepted by 1890.

Electricity

The generation of static electricity by rubbing amber with fur had been known since antiquity, but the discovery of electric currents in the 17th century raised interesting new questions. For example, is the "electric substance" a continuous fluid or does it consist of granular particles of electricity? There was no direct evidence, but the flow of current suggested a fluid of some sort. This supposition was analogous to the prevailing belief that heat was a fluid called *caloric.*

Eighteenth-century studies of electricity relied on electrostatic generators. These mechanical devices used friction to produce large but mostly uncontrolled potential differences. Volta's invention of the battery in 1800 was a major improvement for investigators because a battery provides a controlled and reproducible potential difference. Further, it was perfect for creating currents in wires. Volta's battery stimulated a wave of research on electricity during the opening decades of the 19th century.

It took only two months from Volta's invention of the battery until the discovery that an electric current through water decomposes the water into hydrogen and oxygen, a process that came to be called **electrolysis.** The basic experiment, as it is done today in chemistry classes, is shown in Figure 37.1. The positive and negative terminals of a battery are connected to pieces of metal called *electrodes*. The negative electrode is called the *cathode* and the positive one is the *anode*. Bubbles of gas appear at the electrodes—hydrogen at one and oxygen at the other—and can be collected in tubes.

The outcome of this experiment does not surprise us today, but at the time of its first performance water had long been regarded as one of the basic elements. The decomposition of water forced scientists to reconsider the basic building blocks of matter. Furthermore, these newly discovered effects suggested a previously unsuspected connection between electricity and matter.

FIGURE 37.1 A current through water decomposes it into hydrogen and oxygen.

Light

The question "What is light?" had long been debated. Newton, as we have noted previously, favored a *corpuscular* theory whereby small particles of light travel in straight lines. His conviction was based largely on the sharp shadows cast by sunlight, in contrast to the diffraction of water waves as they pass barriers. Newton's view was dominant throughout the 18th century.

The situation changed quickly as the 19th century opened. In 1801, the English linguist, physician, and scientist Thomas Young demonstrated the interference of light with his celebrated double-slit experiment, shown in Figure 37.2. The wave model of light was given a more rigorous mathematical foundation in 1818 by the French physicist Augustin Fresnel. Fresnel's theory predicted a number of diffraction effects that had not been previously observed. These predicted effects were initially criticized as being contrary to common sense, but their subsequent experimental verification validated the wave model of light.

But if light is a wave, what is waving? What is the medium? How can light travel through a vacuum? The corpuscular theory had not faced these difficulties, but they were brought to the forefront by the overwhelming evidence that light must be some kind of wave.

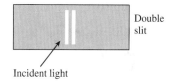

FIGURE 37.2 Young's double-slit experiment showed that light is a wave.

37.2 Faraday

The three lines of inquiry—matter, electricity, and light—came together during the 1820s in the person of Michael Faraday, one of the most remarkable scientific geniuses in history. Faraday conducted three investigations of particular interest to us.

Electrical Conduction in Liquids

Others had already begun to study electrolysis, the conduction of electric current through liquids, but it was Faraday's systematic and careful measurements that revealed the laws governing electrolysis. Faraday showed that electrolysis is most easily understood on the basis of an atomic theory of matter, and he found that there is a *charge* associated with each atom in the solution. Today these charges are called positive and negative *ions*.

In fuel cells, which will power cars in the near future, oxygen and hydrogen are combined to produce water and an electric current. This is the reverse of the electrolysis process shown in Figure 37.1.

Faraday's discoveries implied that

1. Atoms exist.
2. Electric charges are somehow (Faraday did not know how) associated with atoms.
3. There are two different kinds of charge, positive and negative.
4. Electricity is "granular" rather than a continuous fluid. That is, it comes in discrete amounts with a basic unit of charge.

Electrical Conduction in Gases

Faraday also investigated whether electric currents could pass through air. He sealed metal electrodes into a glass tube, lowered the pressure with a primitive vacuum pump, then attached an electrostatic generator. When he started the generator, the tube began to glow with a bright purple color! Faraday's device, called a **gas discharge tube,** is shown in Figure 37.3. The photo that opened this chapter is a modern gas discharge tube.

Faraday's investigations showed that

1. Current flows through a low-pressure gas, creating an electric discharge.
2. The color of the discharge depends on the type of gas in the tube.
3. Regardless of the type of gas, there is a separate, constant glow around the negative electrode (i.e., the cathode). This is called the **cathode glow.**

Today we know the purple color to be characteristic of nitrogen, the primary component of air. You are more likely familiar with the reddish-orange color of the neon discharge tubes used for signs, but neon was not discovered until long after Faraday's time. His investigations, though, showed an unexpected connection between the color of the light and type of atoms in the tube.

Electromagnetic Fields

Perhaps Faraday's most important contributions to physics were in the realms of magnetism and light. You will recall that it was Faraday who introduced the concept of electric and magnetic *fields.* Although these fields were first devised simply as a way of envisioning electric and magnetic processes, Faraday's later studies of electromagnetic induction showed that they have a real existence and real properties. These investigations paved the way for the discovery, about 30 years later, that light is an electromagnetic wave.

Faraday's discoveries were a major step toward providing real *evidence* for the existence of atoms. Altogether, Faraday established that atoms are associated with electricity, he demonstrated that different colors of light are associated with different kinds of atoms, and he prepared the way for showing that light is associated with electricity and magnetism.

Matter, electricity, and light—previously three separate ideas—had been intertwined. Even so, Faraday recognized that he had barely scratched the surface, that far more research was needed before atoms were understood.

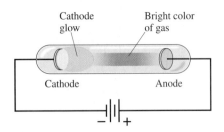

Cathode glow   Bright color of gas

Cathode   Anode

FIGURE 37.3 Faraday's gas discharge tube.

Although we know nothing of what an atom is, yet we cannot resist forming some idea of a small particle which represents it to the mind; and though we are in equal, if not greater, ignorance of electricity, so as to be unable to say whether it is a particular matter or matters, or mere motion of ordinary matter, or some third kind of power or agent, yet there is an immensity of facts which justify us in believing that the atoms of matter are in some way endowed or associated with electric powers to which they owe their most striking qualities, and amongst them their mutual chemical affinity. . . .

Michael Faraday

37.3 Cathode Rays

Faraday's invention of the gas discharge tube had two major repercussions. One set of investigations, to which we will return in Section 37.8, led to the development of spectroscopy and eventually to quantum physics. Another set of investigations led to the discovery of the electron.

An important technological breakthrough came in the 1850s with the development of much-improved vacuum pumps. The German scientist Julius Plücker began a study of Faraday's gas discharge tube using lower gas pressures, and made two important observations:

1. As he reduced the pressure, the colored glow of the gas diminished and the cathode glow became more extended.
2. If the cathode glow extended to the wall of the glass tube, the glass itself emitted a greenish glow at that point.

A few years later, one of Plücker's students found that a solid object sealed inside the tube casts a *shadow* on the glass wall, as shown in Figure 37.4. This discovery suggested that the cathode emits *rays* of some form that travel in straight lines but are easily blocked by solid objects. These rays, which are invisible but cause the glass to glow where they strike it, were quickly dubbed **cathode rays.** This name lives on today in the *cathode-ray tube* that forms the picture tube in televisions and most computer display terminals. But naming the rays did nothing to explain them. What were they?

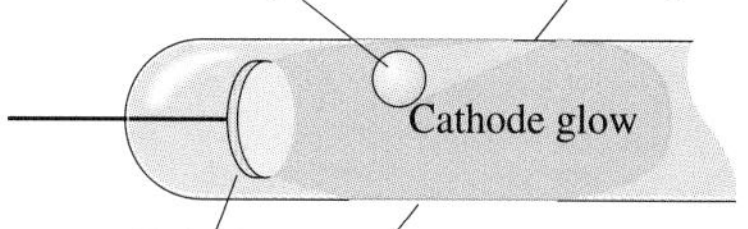

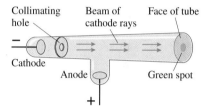

FIGURE 37.4 A solid object in the cathode glow casts a shadow.

Crookes Tubes

The most systematic studies on the new cathode rays were carried out during the 1870s by the English scientist Sir William Crookes. Crookes devised a set of glass tubes, such as the one shown in Figure 37.5, that could be used to make careful studies of cathode rays. His primary innovations were to elongate the tube, use yet lower pressure, and introduce a collimating hole for the rays to pass through. The net result was to generate a well-defined beam of cathode rays that created a small glowing spot where they struck the end of the tube. Today we call his design a **Crookes tube.**

FIGURE 37.5 A Crookes tube.

The work of Crookes and others demonstrated that

1. There is an electric current in a tube in which cathode rays are emitted.
2. The rays are deflected by a magnetic field *as if* they are negative charges.
3. Cathodes made of any metal produce cathode rays. Furthermore, the ray properties are independent of the cathode material.
4. The rays can exert forces on objects and can transfer energy to objects. For example, a thin metal foil in the cathode-ray beam glows red hot.

Crookes' experiments led to more questions than they answered. Were the cathode rays some sort of particles? Or a wave? Were the rays themselves the carriers of the electric current, or were they something else that happened to be emitted whenever there was a current? Item 3 is worthy of note because it suggests that the cathode rays are a *fundamental* entity, not a part of the element from which they are emitted.

Although you can read the final answers in a book today, it is important to realize how difficult these questions were at the time and how experimental evidence was used to answer them. Crookes suggested that molecules in the gas collided with the cathode, somehow acquired a negative charge (i.e., became negative ions), and then "rebounded" with great speed as they were repelled by the negative cathode. These "charged molecules" would travel in a straight line, carry energy and momentum, be deflected by a magnetic field, and cause the tube to glow, or *fluoresce,* where they struck the glass. Crookes' theory predicted, of course, that the negative ions should also be deflected by an electric field. Crookes attempted to demonstrate this deflection by sealing electrodes into the tube and creating an electric field, but his efforts were inconclusive. Other than this troublesome difficulty, Crookes' model seemed to explain the observations.

However, Crookes' theory was immediately attacked. Critics noted that the cathode rays could travel the length of a 90-cm-long tube with no discernible deviation from a straight line. But the mean free path for molecules, due to collisions with other molecules, is only about 6 mm at the pressure in Crookes' tubes.

There was no chance at all that molecules could travel in a straight line for 150 times their mean free path! It was later discovered that the cathode rays could even penetrate very thin ($\approx 2 \mu$m thick) metal foils, something that no atomic-size particle could do. Crookes' theory, seemingly adequate when it was proposed, was wildly inconsistent with subsequent observations.

But if cathode rays were not particles, what were they? An alternative theory was that the cathode rays were electromagnetic waves. After all, light travels in straight lines, casts shadows, carries energy and momentum, and can, under the right circumstances, cause materials to fluoresce. It was known that hot metals emit light—incandescence—so it seemed plausible that the cathode could be emitting waves. A long path through the gas would present no problem, and it was known by 1890 that radio waves could penetrate thin foils. The major obstacle for the wave theory was the deflection of cathode rays by a magnetic field. But the theory of electromagnetic waves was quite new at the time, and many characteristics of these waves were still unknown. Visible light was not deflected by a magnetic field, but it was easy to think that some other form of electromagnetic waves might be so influenced.

The controversy over particles versus waves was intense. British scientists generally favored particles, but their continental counterparts preferred waves. Such controversies are an integral part of science, for they stimulate the best minds to come forward with new ideas and new experiments.

37.4 J. J. Thomson and the Discovery of the Electron

Shortly after Wilhelm Röntgen's 1895 discovery of x rays, the young English physicist J. J. Thomson began using them to study electrical conduction in gases. He found that x rays could discharge an electroscope and concluded that they must be ionizing the air molecules, thereby making the air conductive. That is, the x rays were splitting the molecules into charged fragments—ions!

This simple observation was of profound significance. Until then, the only form of ionization known was the creation of positive and negative ions in solutions where, for example, a molecule such as NaCl splits into two smaller charged pieces. Although the underlying process was not yet understood, the fact that two atoms could acquire charge as a molecule splits apart did not jeopardize the idea that the atoms themselves were indivisible. But after observing that even monatomic gases, such as helium, could be ionized by x rays, Thomson realized that **the atom itself must have charged constituents that could be split apart!** This was the first direct evidence that the atom is a complex structure, not a fundamental, indivisible unit of matter.

Thomson was also conducting experiments to investigate the nature of cathode rays. One of his first goals was to establish, once and for all, that cathode rays are charged particles. Other scientists, using a Crookes tube like the one shown in Figure 37.6a, had measured an electric current in a cathode-ray beam. Although its presence seemed to demonstrate that the rays are charged particles, proponents of the wave model argued that the current might be a separate, independent event that just happened to be following the same straight line as the rays.

Thomson realized that he could use magnetic deflection of the cathode rays to settle the issue. He built a modified tube, shown in Figure 37.6b, in which the collecting electrode was off to the side. Under normal operation, the cathode rays struck the center of the tube face and created a greenish spot on the glass. No current was measured by the electrode under these circumstances. Thomson then placed the tube in a magnetic field that deflected the cathode rays to the side. He could determine their trajectory by the location of the green spot as it moved

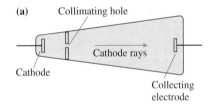

(a) Collimating hole

Cathode rays

Cathode

Collecting electrode

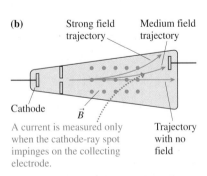

(b) Strong field trajectory   Medium field trajectory

Cathode

$\vec{B}$

A current is measured only when the cathode-ray spot impinges on the collecting electrode.

Trajectory with no field

FIGURE 37.6 Experiments to measure the electric current in a cathode-ray tube.

across the face of the tube. Just at the point when the field was strong enough to deflect the cathode rays onto the electrode, a current was detected! At an even stronger field, when the cathode rays were deflected completely to the other side of the electrode, the current ceased.

This was the first conclusive demonstration that cathode rays really are negatively charged particles. But why were they not deflected by an electric field? Thomson's first efforts to deflect the cathode rays met with the same inconclusive results that others had found, but his experience with the x-ray ionization of gases soon led him to recognize the difficulty. He realized that the rapidly moving cathode-ray particles must be colliding with the few remaining gas molecules in the tube with sufficient energy to *ionize* them by splitting them into charged pieces. The electric field created by these charges neutralized the field of the electrodes, hence there was no deflection.

Fortunately, vacuum technology was getting ever better. By using the most sophisticated techniques of his day, Thomson was able to lower the pressure to a point where ionization of the gas was not a problem. Then, just as he had expected, the cathode rays *were* deflected by an electric field!

Thomson's experiment was a decisive victory for the charged-particle model, but it still did not indicate anything about the nature of the particles. What were they?

Thomson's Crossed-Field Experiment

Thomson could measure the deflection of cathode-ray particles for various strengths of the magnetic field, but magnetic deflection depends both on the particle's charge-to-mass ratio q/m *and* on its speed. Measuring the charge-to-mass ratio, and thus learning something about the particles themselves, requires some means of measuring their velocity. To do so, Thomson devised the experiment for which he is most remembered.

Thomson built a tube containing the parallel-plate electrodes visible in the photo in Figure 37.7a. He then placed the tube between the poles of a magnet. Figure 37.7b shows that the electric and magnetic fields were perpendicular to each other, thus creating what came to be known as a **crossed-field experiment.**

The magnetic field, which is perpendicular to the particle's velocity $\vec{v}$, exerts a magnetic force on the charged particle of magnitude

$$F_{\text{B}} = qvB \qquad (37.1)$$

The magnetic field alone would cause a negatively charged particle to move along a *downward* circular arc. The particle doesn't move in a complete circle because the velocity is large and because the magnetic field is limited in extent. As you learned in Chapter 32, the radius of the arc is

$$r = \frac{mv}{qB} \qquad (37.2)$$

The net result is to *deflect* the beam of particles downward. This deflection is easily measured by observing the green spot where the particles strike the glass at the end of the tube. It is then a straightforward geometry problem to determine the radius of curvature r from the measured deflection.

Thomson's new idea was to create an electric field between the parallel-plate electrodes that would exert an *upward* force on the negative charges, pushing them back toward the center of the tube. The magnitude of the electric force on each particle is

$$F_{\text{E}} = qE \qquad (37.3)$$

Thomson adjusted the electric field strength until the cathode-ray beam, in the presence of both electric and magnetic fields, was exactly in the center of the tube.

J. J. Thomson.

(a)

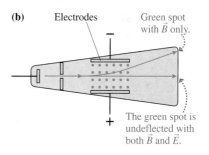
(b) Electrodes — Green spot with $\vec{B}$ only.

The green spot is undeflected with both $\vec{B}$ and $\vec{E}$.

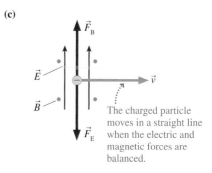
(c)

The charged particle moves in a straight line when the electric and magnetic forces are balanced.

FIGURE 37.7 Thomson's crossed-field experiment to measure the velocity of cathode rays. The photograph shows his original tube and the coils he used to produce the magnetic field.

Zero deflection occurs when the magnetic and electric forces exactly balance each other, as Figure 37.7c shows. The force vectors point in opposite directions, and their magnitudes are equal when

$$F_B = qvB = F_E = qE$$

Notice that the charge q cancels. Once E and B are set, a charged particle can pass undeflected through the crossed fields only if its speed is

$$v = \frac{E}{B} \tag{37.4}$$

By balancing the magnetic force against the electric force, Thomson could determine the velocity of the charged-particle beam. Once he knew v, he could then use Equation 37.2 to find the charge-to-mass ratio:

$$\frac{q}{m} = \frac{v}{rB} \tag{37.5}$$

Thomson found that the charge-to-mass ratio of cathode rays is $q/m \approx 1 \times 10^{11}$ C/kg. This seems not terribly accurate in comparison to a modern value of 1.76×10^{11} C/kg, but keep in mind both the experimental limitations of his day and the fact that, prior to his work, no one had *any* idea of the charge-to-mass ratio.

EXAMPLE 37.1 A crossed-field experiment
An electron is fired between two parallel-plate electrodes that are 5.0 mm apart and 3.0 cm long. A potential difference ΔV between the electrodes establishes an electric field between them. A 3.0-cm-wide, 1.0 mT magnetic field overlaps the electrodes and is perpendicular to the electric field. When $\Delta V = 0$ V, the electron is deflected by 2.0 mm as it passes between the plates. What value of ΔV will allow the electron to pass through the plates without deflection?

MODEL Assume the fields between the electrodes are uniform and that they are zero outside the electrodes.

VISUALIZE Figure 37.8 shows an electron passing through the magnetic field between the plates when $\Delta V = 0$ V. The curvature has been exaggerated to make the geometry clear.

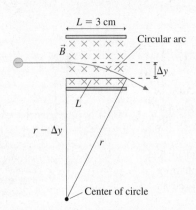

FIGURE 37.8 The electron's trajectory in Example 37.1.

SOLVE We can find the needed electric field, and thus ΔV, if we know the electron's speed. We can find the electron's speed from the radius of curvature of its circular arc in a magnetic field. Figure 37.8 shows a right triangle with hypotenuse r and width L. We can use the Pythagorean theorem to write

$$(r - \Delta y)^2 + L^2 = r^2$$

where Δy is the electron's deflection in the magnetic field. This is easily solved to find the radius of the arc:

$$r = \frac{(\Delta y)^2 + L^2}{2\Delta y} = \frac{(0.0020 \text{ m})^2 + (0.030 \text{ m})^2}{2(0.0020 \text{ m})} = 0.226 \text{ m}$$

The speed of an electron traveling along an arc with this radius is found from Equation 37.2:

$$v = \frac{erB}{m} = 4.0 \times 10^7 \text{ m/s}$$

Thus the electric field that will allow the electron to pass through without deflection is

$$E = vB = 40{,}000 \text{ V/m}$$

The electric field of a parallel-plate capacitor of spacing d is related to the potential difference by $E = \Delta V/d$, so the necessary potential difference is

$$\Delta V = Ed = (40{,}000 \text{ V/m})(0.0050 \text{ m}) = 200 \text{ V}$$

ASSESS A fairly small potential difference is sufficient to counteract the magnetic deflection.

The Electron

Notable as this accomplishment was, Thomson did not stop there. Next he measured q/m for different cathode materials. They were all the same: Whatever the charged particles were, they were identical for all the different elements. Thomson then compared his result to the charge-to-mass ratio of the hydrogen ion, known from electrolysis to have a value of $\approx 1 \times 10^8$ C/kg. This value was roughly 1000 times smaller than for the cathode-ray particles, which could imply that a cathode-ray particle has a much larger charge than a hydrogen ion, or a much smaller mass, or some combination of these.

Electrolysis experiments suggested the existence of a basic unit of charge, so it was tempting to assume that the cathode-ray charge was the same as the charge of a hydrogen ion. However, cathode rays were so different from the hydrogen ion that such an assumption could not be justified without some other evidence. To provide that evidence, Thomson called attention to previous experiments showing that cathode rays can penetrate thin metal foils but atoms cannot. This can be true, Thomson argued, only if cathode-ray particles are vastly smaller and thus much less massive than atoms.

In a paper published in 1897, J. J. Thomson assembled all of the evidence to announce the discovery that cathode rays are negatively charged particles, that they are much less massive ($\approx 0.1\%$) than atoms, and that they are identical when generated by different elements. In other words, Thomson had discovered a **subatomic particle,** one of the constituents of which atoms themselves are constructed. In recognition of the role this particle plays in electricity, it was later named the **electron.**

Thomson's discovery did not immediately convince everyone that the cathode-ray particles were a ubiquitous component of all atoms. But experiments by Thomson and others over the next few years showed that negative particles emitted from hot metal wires (a process discovered by Thomas Edison in his development of the lightbulb) had the same q/m; that one type of radioactive decay (today called *beta radiation*) consisted of particles with the same q/m; and that certain changes in the spectra of atoms when placed in a magnetic field could be understood if the atoms had a charged constituent with the same q/m. By 1900 it was clear to all that electrons were a fundamental building block of atoms. J. J. Thomson was awarded the Nobel prize in 1906.

STOP TO THINK 37.1   J. J. Thomson's conclusion that cathode-ray particles are *fundamental* constituents of atoms was based primarily on which observation?

a. They have a negative charge.
b. They are the same from all cathode materials.
c. Their mass is much less than that of hydrogen.
d. They penetrate very thin metal foils.

37.5 Millikan and the Fundamental Unit of Charge

Thomson measured the electron's charge-to-mass ratio and *surmised* that the mass must be much smaller than that of an atom, but clearly it was desirable to measure the charge q or the mass m directly. Thomson had discovered that air can be ionized by x rays, and he subsequently found that the water vapor in moist air condensed to form small droplets around the ions. By measuring the mass and charge of water droplets, Thomson and his students were able to determine that the unit of charge involved in the ionization of gases was roughly 1×10^{-19} C.

These were not very accurate experiments, but the value obtained was close to the charge of the hydrogen ion as determined (also rather crudely) from electrolysis. Thomson interpreted this information—that the same unit of charge is responsible for conduction in both liquids and gases—as implying the existence of a *fundamental unit of charge*. This fundamental unit of charge is designated *e*. A hydrogen ion has a charge $q_{\text{H ion}} = +e$ and an electron has $q_{\text{elec}} = -e$.

In 1906, the American scientist Robert Millikan began making his own measurements of *e*. He based his technique on his discovery that he could "catch" a charged oil droplet and then accurately measure its motion using the combined influence of gravity and an electric field.

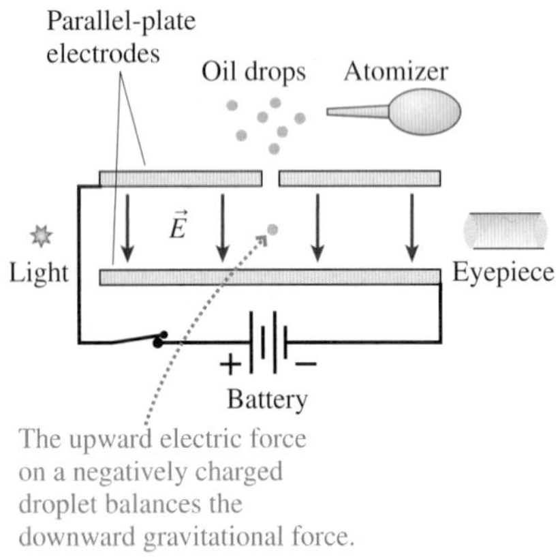

The **Millikan oil-drop experiment,** as we call it today, is illustrated in Figure 37.9. A squeeze-bulb atomizer sprayed out a very fine mist of oil droplets, some of which were charged from friction in the sprayer. These slowly settled toward a horizontal pair of parallel-plate electrodes, where a few droplets passed through a small hole in the top plate. Millikan observed the drops by shining a bright light between the plates and using an eyepiece to see the droplets' reflections. He then established an electric field by applying a voltage to the plates.

A drop will remain suspended between the plates, moving neither up nor down, if the electric field exerts an upward force on a charged drop that exactly balances the downward gravitational force. The forces balance when

$$m_{\text{drop}} g = q_{\text{drop}} E \tag{37.6}$$

and thus the charge on the drop is measured to be

$$q_{\text{drop}} = \frac{m_{\text{drop}} g}{E} \tag{37.7}$$

Notice that *m* and *q* are the mass and charge of the oil droplet, not that of an electron. But because the droplet is charged by acquiring (or losing) electrons, the charge of the droplet should be related to the fundamental unit of charge.

The field strength *E* could be determined accurately from the voltage applied to the plates, so the limiting factor in measuring q_{drop} was Millikan's ability to determine the mass of these small drops. Ideally, the mass could be found by measuring a drop's diameter and using the known density of the oil. However, the drops were too small ($\approx 1 \ \mu\text{m}$) to measure accurately by viewing through the eyepiece.

Instead, Millikan devised an ingenious method to find the size of the droplets. Objects this small are *not* in free fall. The air resistance forces are so large that the drops fall with a very small but constant speed. The motion of a sphere through a viscous medium is a problem that had been solved in the 19th century, and it was known that the sphere's terminal speed depends on its radius and on the viscosity of air. So rather than holding the droplets motionless, Millikan used the electric field to cause them to move slowly up and down through a known distance. He could determine the droplets' velocities by timing them with a stopwatch. Then, using the known viscosity of air, he could calculate their radii, compute their masses, and, finally, arrive at a value for their charge. Although it was a somewhat roundabout procedure, Millikan was able to measure the charge on a droplet with an accuracy of $\pm 0.1\%$ (one part in a thousand).

Millikan measured many hundreds of droplets, some for hours at a time, under a wide variety of conditions. He found that some of his droplets were positively charged and some negatively charged, but **all had charges that were integer multiples of a certain minimum charge value.** Millikan concluded that "the electric charges found on ions all have either exactly the same value or else some

FIGURE 37.9 Millikan's oil-drop apparatus to measure the fundamental unit of charge.

small exact multiple of that value." That value, the fundamental unit of charge that we now call e, is measured to be

$$e = 1.602 \times 10^{-19} \text{ C}$$

We can then combine the measured e with the measured charge-to-mass ratio e/m to find that the mass of the electron is

$$m_{\text{elec}} = 9.11 \times 10^{-31} \text{ kg}$$

Taken together, the experiments of Thomson, Millikan, and others provided overwhelming evidence that electric charge comes in discrete units and that *all* charges found in nature are multiples of a fundamental unit of charge we call e.

EXAMPLE 37.2 Suspending an oil drop

Oil has a density of 860 kg/m³. A 1.0-μm-diameter oil droplet acquires 10 extra electrons as it is sprayed. What potential difference between two parallel plates 1.0 cm apart will cause the droplet to be suspended in air?

MODEL Assume a uniform electric field $E = \Delta V/d$ between the plates.

SOLVE The magnitude of the charge on the drop is $q_{\text{drop}} = 10e$. The mass of the charge is related to its density ρ and volume V by

$$m_{\text{drop}} = \rho V = \frac{4}{3}\pi R^3 \rho = 4.50 \times 10^{-16} \text{ kg}$$

where the droplet's radius is $R = 5.0 \times 10^{-7}$ m. The electric field that will suspend this droplet against the force of gravity is

$$E = \frac{m_{\text{drop}}g}{q_{\text{drop}}} = 2760 \text{ V/m}$$

Establishing this electric field between two plates spaced by $d = 0.010$ m requires a potential difference

$$\Delta V = Ed = 27.6 \text{ V}$$

37.6 Rutherford and the Discovery of the Nucleus

By 1900, it was clear that atoms are not indivisible but, instead, are constructed of charged particles. Atomic sizes were known to be $\approx 10^{-10}$ m, but the electrons common to all atoms are much smaller and much less massive than the smallest atom. How do they "fit" into the larger atom? What is the positive charge of the atom? Where are the charges located inside the atoms?

J. J. Thomson proposed the first model of an atom. Because the electrons are very small and light compared to the whole atom, it seemed reasonable to think that the positively charged part would take up most of the space. Thomson suggested that the atom consists of a spherical "cloud" of positive charge, roughly 10^{-10} m in diameter, in which the smaller negative electrons are embedded. The positive charge exactly balances the negative, so the atom as a whole has no net charge. This model of the atom has often been called the "plum-pudding model" or the "raisin-cake model" for reasons that should be clear from the picture of Figure 37.10.

Thomson proposed that small, negative electrons are embedded in a sphere of positive charge.

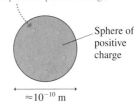

Sphere of positive charge

$\approx 10^{-10}$ m

FIGURE 37.10 Thomson's raisin-cake model of the atom.

Thomson was never able to make any predictions that would enable his model to be tested, and the Thomson atom did not stand the tests of time. His model is of interest today primarily to remind us that our current models of the atom are by no means obvious. Science has many side-steps and dead ends as it progresses.

One of Thomson's students was a New Zealander named Ernest Rutherford. While Rutherford and Thomson were studying the ionizing effects of x rays, in 1896, the French physicist Antoine Henri Becquerel announced the discovery that some new form of "rays" were emitted by crystals of uranium. These rays, like x rays, could expose film, pass through objects, and ionize the air. Yet they were emitted continuously from the uranium without having to "do" anything to it. This was the discovery of **radioactivity,** a topic we'll study in Chapter 42.

With x rays only a year old and cathode rays not yet completely understood, it was not known whether all these various kinds of rays were truly different or merely variations of a single type. Rutherford immediately began a study of these new rays. He quickly discovered that at least two *different* rays are emitted by a uranium crystal. The first, which he called **alpha rays,** were easily absorbed by a piece of paper. The second, **beta rays,** could penetrate through at least 0.1 inch of metal and through much greater thicknesses of soft materials.

As we have already noted, Thomson soon found that beta rays have the same charge-to-mass ratio as cathode rays. The beta rays turned out to be high-speed electrons emitted by the uranium crystal. Rutherford, using similar techniques, showed that alpha rays are *positively* charged particles. By 1906 he had measured their charge-to-mass ratio to be

$$\frac{q}{m} = \frac{1}{2}\frac{e}{m_H}$$

where m_H is the mass of a hydrogen atom. This value could indicate either a singly ionized hydrogen molecule H_2^+ ($q = e, m = 2m_H$) *or* a doubly ionized helium atom He^{++} ($q = 2e, m = 4m_H$).

In an ingenious experiment, Rutherford sealed a sample of radium—an emitter of alpha radiation—into a glass tube. Alpha rays could not penetrate the glass, so the particles were contained within the tube. Several days later, Rutherford used electrodes in the tube to create a discharge and observed the spectrum of the emitted light. He found the characteristic wavelengths of helium, but not those of hydrogen. Alpha rays (or alpha particles, as we now call them) consist of doubly ionized helium atoms (bare helium nuclei) emitted at high speed ($\approx 3 \times 10^7$ m/s) from the sample.

It had been quite a shock to discover that atoms are not indivisible, that they have an inner structure. Now, with the discovery of radioactivity, it appeared that some atoms were not even stable but could spit out various kinds of charged particles! Physics had come a long way from the simple atomic idea of Democritus.

The First Nuclear Physics Experiment

Rutherford soon realized that he could use these high-speed particles to probe inside other atoms. In 1909, Rutherford and his students Hans Geiger and Ernest Marsden set up the experiment shown in Figure 37.11 to shoot alpha particles at very thin metal foils. Some of the alpha particles penetrated the foil, but the beam of alpha particles that did so became somewhat spread out. This was not surprising. The alpha particle is charged, and it experiences forces from the positive and negative charges of the atoms as it passes through the foil. According to Thomson's raisin-cake model of the atom, the forces exerted on the alpha particle by the positive atomic charges were expected to roughly cancel the forces from the negative electrons, causing the alpha particles to be deflected only slightly. Indeed, this was the experimenters' initial observation.

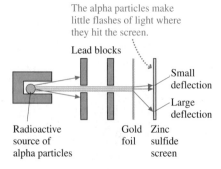

The alpha particles make little flashes of light where they hit the screen.

Lead blocks

Small deflection

Large deflection

Radioactive source of alpha particles

Gold foil   Zinc sulfide screen

FIGURE 37.11 Rutherford's experiment to shoot high-speed alpha particles through a thin gold foil.

At Rutherford's suggestion, Geiger and Marsden set up the apparatus to see if any alpha particles were deflected at *large* angles. It took only a few days to find the answer. Not only were alpha particles deflected at large angles, a very few were reflected almost straight backward toward the source!

How can we understand this result? Figure 37.12a shows that only a small deflection is expected for an alpha particle passing through a Thomson atom. But if an atom has a small, positive core, such as the one in Figure 37.12b, a few of the alpha particles can come very close to the core. Because the electric force varies with the inverse square of the distance, the very large force of this very close approach can cause a large-angle scattering or a backward deflection of the alpha particle. This is what Geiger and Marsden were observing.

I remember two or three days later Geiger coming to me in great excitement and saying, "We have been able to get some of the alpha particles coming backward." It was quite the most incredible event that has ever happened to me in my life. It was almost as if you fired a 15-inch shell at a piece of tissue paper and it came back and hit you. . . . It was then that I had the idea of an atom with a minute massive center, carrying a charge.

Ernest Rutherford

(a)

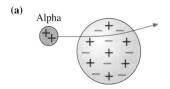

The alpha particle is only slightly deflected by a Thomson atom because forces from the spread-out positive and negative charges nearly cancel.

(b)

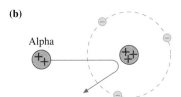

If the atom has a concentrated positive nucleus, some alpha particles will be able to come very close to the nucleus and thus feel a very strong repulsive force.

FIGURE 37.12 Alpha particles interact differently with a concentrated positive nucleus than they would with the spread-out charge in Thomson's model.

Thus the discovery of large-angle scattering of alpha particles led Rutherford to envision an atom in which negative electrons orbit an unbelievably small, massive, positive **nucleus,** rather like a miniature solar system. This is the **nuclear model of the atom.** Notice that nearly all of the atom is merely empty space—the void!

Activ Physics ONLINE 19.1

EXAMPLE 37.3 A nuclear physics experiment

An alpha particle is shot with a speed of 2.0×10^7 m/s directly toward the nucleus of a gold atom. What is the distance of closest approach to the nucleus?

MODEL Energy is conserved in electric interactions. Assume that the gold nucleus, which is much more massive than the alpha particle, does not move. Also recall that the exterior electric field and potential of a sphere of charge can be found by treating the total charge as a point charge at the center.

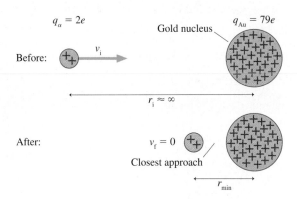

FIGURE 37.13 A before-and-after pictorial representation of an alpha particle colliding with a nucleus.

VISUALIZE Figure 37.13 is a pictorial representation. The motion is in and out along a straight line.

SOLVE We are not interested in how long the collision takes or any of the details of the trajectory, so using conservation of energy rather than Newton's laws is appropriate. Initially, when the alpha particle is very far away, the system has only kinetic energy. At the moment of closest approach, just before the alpha particle is reflected, the charges are at rest and the system has only potential energy. The conservation of energy statement $K_f + U_f = K_i + U_i$ is

$$0 + \frac{1}{4\pi\epsilon_0} \frac{q_\alpha q_{Au}}{r_{min}} = \frac{1}{2}mv_i^2 + 0$$

where q_α is the alpha-particle charge and we've treated the gold nucleus as a point charge q_{Au}. The mass m is that of the alpha particle. The solution for r_{min} is

$$r_{min} = \frac{1}{4\pi\epsilon_0} \frac{2q_\alpha q_{Au}}{mv_i^2}$$

The alpha particle is a helium nucleus, so $m = 4$ u $= 6.64 \times 10^{-27}$ kg and $q_\alpha = 2e = 3.20 \times 10^{-19}$ C. Gold has atomic number 79, so $q_{Au} = 79e = 1.26 \times 10^{-17}$ C. We can then calculate

$$r_{min} = 2.7 \times 10^{-14} \text{ m}$$

This is only about 1/10,000 the size of the atom itself!

ASSESS We ignored the atom's electrons in this example. In fact, they make almost no contribution to the alpha particle's trajectory. The alpha particle is exceedingly massive compared to the electrons, and the electrons are spread out over a distance very large compared to the size of the nucleus. Hence the alpha particle easily pushes them aside without any noticeable change in its velocity.

Rutherford went on to make careful experiments of how the alpha particles scattered at different angles. From these experiments he deduced that the diameter of the atomic nucleus is $\approx 1 \times 10^{-14}$ m $= 10$ fm (1 fm $= 1$ femtometer $= 10^{-15}$ m), increasing a little for elements of higher atomic number and atomic mass.

It may seem surprising to you that the Rutherford model of the atom, with its solar system analogy, was not Thomson's original choice. However, scientists at the time could not imagine matter having the extraordinarily high density implied by a small nucleus. Neither could they understand what holds the nucleus together, why the positive charges do not push each other apart. Thomson's model, in which the positive charge was spread out and balanced by the negative electrons, actually made more sense. It would be several decades before the forces holding the nucleus together began to be understood, but Rutherford's evidence for a very small nucleus was indisputable.

STOP TO THINK 37.2 If the alpha particle has a positive charge, which way will it be deflected in the magnetic field?

a. Up
b. Down
c. Into the page
d. Out of the page

The Electron Volt

The joule is a unit of appropriate size in mechanics and thermodynamics, where we dealt with macroscopic objects, but it is poorly matched to the needs of atomic physics. It will be very useful to have an energy unit appropriate to atomic and nuclear events.

Figure 37.14 shows an electron accelerating from rest across a parallel-plate capacitor with a 1.0 V potential difference. What is the electron's kinetic energy when it reaches the positive plate? We know from energy conservation that $K_f + qV_f = K_i + qV_i$, where $U = qV$ is the electric potential energy. $K_i = 0$ because the electron starts from rest, and the electron's charge is $q = -e$. Thus

$$K_f = -q(V_f - V_i) = -q\Delta V = e\Delta V = (1.60 \times 10^{-19}\,\text{C})(1.0\,\text{V})$$
$$= 1.60 \times 10^{-19}\,\text{J}$$

Let us define a new unit of energy, called the **electron volt,** as

$$1 \text{ electron volt} = 1 \text{ eV} \equiv 1.60 \times 10^{-19}\,\text{J}$$

With this definition, the kinetic energy gained by the electron in our example is

$$K_f = 1 \text{ eV}$$

In other words, **1 electron volt is the kinetic energy gained by an electron (or proton) if it accelerates through a potential difference of 1 volt.**

> **NOTE** ▶ The abbreviation eV uses a lower case e but an upper case V. Units of keV (10^3 eV), MeV (10^6 eV), and GeV (10^9 eV) are common. ◀

The electron volt can be a troublesome unit. One difficulty is its unusual name, which looks less like a unit than, say, "meter" or "second." A more significant

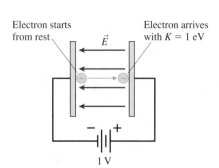

FIGURE 37.14 An electron accelerating across a 1 V potential difference gains 1 eV of kinetic energy.

difficulty is that the name suggests a relationship to volts. But *volts* are units of electric potential, whereas this new unit—with an admittedly confusing name—is a unit of energy! It is crucial to distinguish between the *potential V*, measured in volts, and an *energy* that can be measured either in joules or in electron volts. You can now use electron volts anywhere that you would previously have used joules. Doing so is no different from converting back and forth between pressure units of pascals and atmospheres.

NOTE ▶ To reiterate, the electron volt is a unit of *energy*, convertible to joules, and not a unit of potential. Potential is always measured in volts. However, the joule remains the SI unit of energy. It will be useful to express energies in eV, but you *must* convert this energy to joules before doing most calculations. ◀

EXAMPLE 37.4 **The speed of an alpha particle**
Alpha particles are usually characterized by their kinetic energy in MeV. What is the speed of an 8.30 MeV alpha particle?

SOLVE Alpha particles are helium nuclei, having $m = 4$ u $= 6.64 \times 10^{-27}$ kg. The kinetic energy of this alpha particle is 8.30×10^6 eV. First, convert the energy to joules:

$$K = 8.30 \times 10^6 \text{ eV} \times \frac{1.60 \times 10^{-19} \text{ J}}{1.0 \text{ eV}} = 1.33 \times 10^{-12} \text{ J}$$

Now we can find the speed:

$$K = \frac{1}{2}mv^2 = 1.33 \times 10^{-12} \text{ J}$$

$$v = \sqrt{\frac{2K}{m}} = 2.0 \times 10^7 \text{ m/s}$$

This was the speed of the alpha particle in Example 37.3.

EXAMPLE 37.5 **Energy of an electron**
In a simple model of the hydrogen atom, the electron orbits the proton at 2.19×10^6 m/s in a circle with radius 5.29×10^{-11} m. What is the atom's energy in eV?

MODEL The electron has a kinetic energy of motion, and the electron + proton system (i.e., the atom) has an electric potential energy.

SOLVE The potential energy is that of two point charges, with $q_{proton} = +e$ and $q_{elec} = -e$. Thus

$$E = K + U = \frac{1}{2}m_{elec}v^2 + \frac{1}{4\pi\epsilon_0}\frac{(e)(-e)}{r} = -2.17 \times 10^{-18} \text{ J}$$

Conversion to eV gives

$$E = -2.17 \times 10^{-18} \text{ J} \times \frac{1 \text{ eV}}{1.60 \times 10^{-19} \text{ J}} = -13.6 \text{ eV}$$

ASSESS The negative energy reflects the fact that the electron is *bound* to the proton. You would need to *add* energy to remove the electron.

Using the Nuclear Model

The nuclear model of the atom makes it easy to understand and picture such processes as ionization. Because electrons orbit a positive nucleus, an x-ray photon or a rapidly moving particle, such as another electron, can knock one of the orbiting electrons away, creating a positive ion. Removing one electron makes a singly charged ion, with $q = +e$. Removing two electrons creates a doubly charged ion, with $q = +2e$. This is shown for lithium (atomic number 3) in Figure 37.15.

Nucleus has charge $+3e$

Neutral Li

Singly charged Li$^+$

Doubly charged Li^{++}

FIGURE 37.15 Different ionization stages of the lithium atom ($Z = 3$).

The nuclear model also allows us to understand why, during chemical reactions and when an object is charged by rubbing, electrons are easily transferred but protons are not. The protons are tightly bound in the nucleus, shielded by all the electrons, but outer electrons are easily stripped away. Rutherford's nuclear model has explanatory power that was lacking in Thomson's model.

EXAMPLE 37.6 The ionization energy of hydrogen

What is the minimum energy required to ionize a hydrogen atom? The electron orbits the proton at 2.19×10^6 m/s in a circle with radius 5.29×10^{-11} m.

SOLVE In Example 37.5 we found that the atom's energy is $E_i = -13.6$ eV. Ionizing the atom means removing the electron and taking it very far away. As $r \rightarrow \infty$, the potential energy becomes zero. Further, using the least possible energy to ionize the atom will leave the electron, when it is very far

away, very nearly at rest. Thus the atom's energy after ionization is $E_f = K_f + U_f = 0 + 0 = 0$ eV. This is *larger* than E_i by 13.6 eV, so the minimum energy that is required to ionize a hydrogen atom is 13.6 eV. This is called the atom's *ionization energy*. If the electron receives ≥ 13.6 eV (2.17×10^{-18} J) of energy from a photon, or in a collision with another electron, or by any other means, it will be knocked out of the atom and leave a H^+ ion behind.

STOP TO THINK 37.3 Carbon is the sixth element in the periodic table. How many electrons are in a C^{++} ion?

37.7 Into the Nucleus

Chapter 42 will discuss nuclear physics in more detail, but it will be helpful to give a brief overview of the nucleus. The relative masses of many of the elements were known from chemistry experiments by the mid-19th century. By arranging the elements in order of ascending mass, and noting recurring regularities in their chemical properties, the Russian chemist Dmitri Mendeleev first proposed the periodic table of the elements in 1872. But what did it mean to say that hydrogen was atomic number 1, helium number 2, lithium number 3, and so on?

It soon became known that hydrogen atoms can only be singly ionized, producing H^+. A doubly ionized H^{++} is never observed. Helium, by contrast, can be both singly and doubly ionized, creating He^+ and He^{++}, but He^{+++} is not observed. Once Thomson discovered the electron and Millikan established the fundamental unit of charge, it seemed fairly clear that a hydrogen atom contains only one electron and one unit of positive charge, helium has two electrons and two units of positive charge, and so on. Thus the **atomic number** of an element, which is always an integer, describes the number of electrons (of a neutral atom) and the number of units of positive charge in the nucleus. The atomic number is represented by Z, so hydrogen is $Z = 1$, helium $Z = 2$, and lithium $Z = 3$. Elements are listed in the periodic table by their atomic number.

Rutherford's discovery of the nucleus quickly led to the recognition that the positive charge is associated with a positive subatomic particle called the **proton.** The proton's charge is $+e$, equal in magnitude but opposite in sign to the electron's charge. Further, because nearly all the atomic mass is associated with the nucleus, the proton is much more massive than the electron. According to Rutherford's nuclear model, atoms with atomic number Z consist of Z negative electrons, with net charge $-Ze$, orbiting a massive nucleus that contains protons and has net charge $+Ze$. The Rutherford atom went a large way toward explaining the periodic table.

But there was a problem. Helium, with atomic number 2, has twice as many electrons as hydrogen. Lithium, $Z = 3$, has three electrons. But it was known from chemistry measurements that helium is *four times* as massive as hydrogen and lithium is *seven times* as massive. If a nucleus contains Z protons to balance the Z orbiting electrons, and if nearly all the atomic mass is contained in the nucleus, then helium should be simply twice as massive as hydrogen and lithium three times as massive. Something else must be present in the nucleus to make the atoms more massive than our simple nuclear model predicts.

The Neutron

About 1910, J. J. Thomson and his student Francis Aston developed a device called a **mass spectrometer** for measuring the charge-to-mass ratios of atomic ions. (A mass spectrometer was the subject of homework problem 70 in Chapter 32.) As Aston and others began collecting data, they soon found that many elements consist of atoms of *differing* mass! Neon, for example, had been assigned an atomic mass of 20. But Aston found, as the data of Figure 37.16 show, that while 91% of neon atoms have mass $m = 20$ u, 9% have $m = 22$ u and a very small percentage have $m = 21$ u. Chlorine was found to be a mixture of 75% chlorine atoms with $m = 35$ u and 25% chlorine atoms with $m = 37$ u, both having atomic number $Z = 17$.

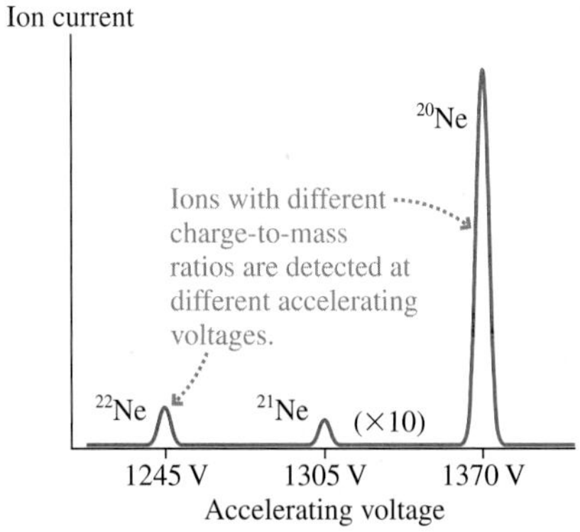

These difficulties were not resolved until the discovery, in 1932, of a third subatomic particle. This particle has essentially the same mass as a proton but *no* electric charge. It is called the **neutron.** Neutrons reside in the nucleus, with the protons, where they contribute to the mass of the atom but not to its charge. As you'll see in Chapter 42, neutrons help provide the "glue" that holds the nucleus together.

The neutron was the missing link needed to explain why atoms of the same element can have different masses. We now know that every atom with atomic number Z has a nucleus containing Z protons with charge $+Ze$. In addition, as shown in Figure 37.17, the nucleus contains N neutrons. There are a *range* of neutron numbers that happily form a nucleus with Z protons, creating a series of nuclei having the same Z-value (i.e., they are all the same chemical element) but different masses. Such a series of nuclei are called **isotopes.**

Chemical behavior is determined by the orbiting electrons. All isotopes of one element have the same number Z of orbiting electrons (if the atoms are electrically neutral) and have the same chemical properties. But different isotopes of the same element can have quite different nuclear properties. In addition, macroscopic behavior that depends on mass, such as the diffusion of a gas, can slightly favor one isotope over another.

An atom's **mass number** A is defined to be $A = Z + N$. It is the total number of protons and neutrons in a nucleus. The mass number, which is dimensionless, is *not* the same thing as the atomic mass m. By definition, A is an integer. But because the proton and neutron masses are both ≈ 1 u, the mass number A is *approximately* the mass in atomic mass units.

The notation used to label isotopes is ^AZ, where the mass number A is given as a *leading* superscript. The proton number Z is not specified by an actual number but, equivalently, by the chemical symbol for that element. The most common isotope of neon has $Z = 10$ protons and $N = 10$ neutrons. Thus it has mass number $A = 20$ and it is labeled ^{20}Ne. The neon isotope ^{22}Ne has $Z = 10$ protons (that's what makes it neon) and $N = 12$ neutrons. Helium has the two isotopes shown in Figure 37.18. The rare ^{3}He is only 0.0001% abundant, but it can be isolated and has important uses in scientific research.

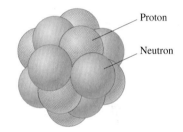

FIGURE 37.16 The mass spectrum of neon.

FIGURE 37.17 The nucleus of an atom contains protons and neutrons.

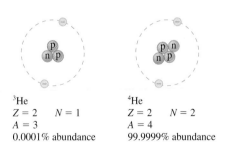

^{3}He
$Z = 2$ $N = 1$
$A = 3$
0.0001% abundance

^{4}He
$Z = 2$ $N = 2$
$A = 4$
99.9999% abundance

FIGURE 37.18 The two isotopes of helium. ^{3}He is only 0.0001% abundant.

STOP TO THINK 37.4 Carbon is the sixth element in the periodic table. How many protons and how many neutrons are there in a nucleus of the isotope ^{14}C?

37.8 The Emission and Absorption of Light

The investigations of cathode rays that led to Thomson's discovery of the electron all followed from Faraday's invention of the gas discharge tube. At the same time, a separate group of scientists was using the gas discharge tube for different purposes. Their discoveries about the emission and absorption of light would also, in the early years of the 20th century, come to bear on the issue of atomic structure.

A gas discharge tube exhibits both a *cathode glow* and a *positive column,* as it is called, that glows with bright color and is different for every gas. The positive column was a hindrance to the study of cathode rays, and those investigators learned that they could eliminate the bright glow by reducing the gas pressure. But other scientists were intrigued by the brightly colored light of the positive column. Why does every gas emit a different color? Can these colors tell us anything about the nature of the atoms and molecules? Fortunately, Faraday's discovery came just at the time that the interference and diffraction of light were first being understood. The production of diffraction gratings was well underway by mid century, and they were the ideal tool to study the light emitted by a discharge tube.

Figure 37.19a shows a typical experimental arrangement for recording the spectrum of light emitted by a gas. The light is focused on the entrance slit of a *spectrometer,* then diffracted by a grating. Different wavelengths in the light are diffracted at different angles, as you learned in Chapter 22, then focused on a film or a photographic plate. A modern spectrometer, widely used today in physics, chemistry, and astronomy, is little changed except that the film has been replaced by an electronic photodetector.

Examples of *emission spectra* are shown in Figure 37.19b. Each line represents one of the wavelengths of light coming from the discharge. These wavelengths can be measured with extremely high accuracy in a well-calibrated instrument. Scientists quickly learned that

1. Gases emit a **discrete spectrum,** consisting of discrete, specific wavelengths of light. This is in contrast to the *continuous* rainbow-like spectrum of the sun or an incandescent light source. Each wavelength in a spectrum is commonly called a **spectral line** because of its appearance in photographs such as Figure 37.19b.
2. Every element in the periodic table emits a unique spectrum.

Substances not only emit light, they can also absorb light. If you look at a lightbulb through a piece of red glass, the bulb looks red because the glass *absorbs* the yellow, green, and blue wavelengths of the white light. Only the red wavelengths make it through to be seen. Similarly, grass and leaves appear green because they absorb both red and blue wavelengths (red and blue wavelengths are the ones that drive photosynthesis), reflecting only the green and yellow wavelengths in the center of the visible spectrum.

NOTE ▶ A peculiarity of the English language is that the process of *absorbing* light is called *absorption,* not "absorbtion." ◀

Do gases absorb light? Indeed they do, although less strongly than solids or liquids because they are less dense. An absorption experiment is shown in Figure 37.20a. Here a white-light source emits a continuous spectrum that, in the absence of a gas, exposes the film completely and uniformly. When a sample of gas is placed in the light's path, any wavelengths absorbed by the gas are missing and the film is dark at that wavelength.

It was discovered that gases not only emit discrete wavelengths, they also absorb discrete wavelengths. But there is an important difference between the emission spectrum and the absorption spectrum of a gas: **Every wavelength that is absorbed by the gas is also emitted, but *not* every emitted wavelength is absorbed.** The wavelengths in the absorption spectrum appear as a subset of the

(a) Measuring an emission spectrum

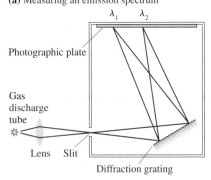

(b)

The spectral lines extend to the series limit at 364.7 nm.

Hydrogen emission spectrum

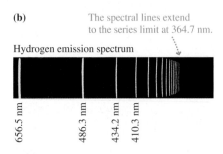

Neon emission spectrum

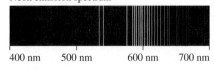

FIGURE 37.19 A grating spectrometer is used to study the emission of light.

wavelengths in the emission spectrum. As an example, Figure 37.20b shows both the emission and the absorption spectra of sodium atoms. All of the absorption wavelengths are prominent in the emission spectrum, but there are many emission lines for which no absorption occurs.

What causes atoms to emit or absorb light? Why a discrete spectrum? Why are some wavelengths emitted but not absorbed? Why is each element different? Nineteenth-century physicists struggled with these questions but could not answer them. Ultimately, their inability to understand the emission and absorption of light forced scientists to the unwelcome realization that classical physics was simply incapable of providing an understanding of atoms.

The only encouraging sign came from an unlikely source. While the spectra of other atoms have dozens or even hundreds of wavelengths, the emission spectrum of hydrogen (Figure 37.19b) is very simple and regular. If any spectrum could be understood, it should be that of the first element in the periodic table. The breakthrough came in 1885, not by an established and recognized scientist but by a Swiss school teacher, Johann Balmer. Balmer showed that the wavelengths in the hydrogen spectrum could be represented by the simple formula

$$\lambda = \frac{91.18 \text{ nm}}{\left(\dfrac{1}{2^2} - \dfrac{1}{n^2}\right)}, \qquad n = 3, 4, 5, \ldots \tag{37.8}$$

Balmer's story was told more completely in Section 24.1, to which you should refer.

Later experimental evidence, as ultraviolet and infrared spectroscopy developed, showed that Balmer's result could be generalized to

$$\lambda = \frac{91.18 \text{ nm}}{\left(\dfrac{1}{m^2} - \dfrac{1}{n^2}\right)}, \qquad m = 1, 2, 3, \ldots \qquad n = m + 1, m + 2, \ldots \tag{37.9}$$

We now refer to Equation 37.9 as the **Balmer formula,** although Balmer himself only suggested the original version of Equation 37.8 in which $m = 2$. Other than at the very highest levels of resolution, where new details appear that need not concern us in this text, the Balmer formula accurately describes *every* wavelength in the emission spectrum of hydrogen.

The Balmer formula is what we call *empirical knowledge*. It is an accurate mathematical representation found empirically—that is, through experimental evidence—but it does not rest on any physical principles or physical laws. Balmer's formula was useful, but no one was able to *derive* Balmer's formula from Newtonian mechanics or the theory of electromagnetism. Yet the formula was so simple that it must, everyone agreed, have a simple explanation. It would take 30 years to find it.

STOP TO THINK 37.5   These spectra are due to the same element. Which one is an emission spectrum and which is an absorption spectrum?

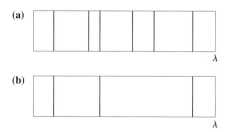

(a) Measuring an absorption spectrum

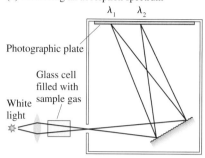

Photographic plate

Glass cell filled with sample gas

White light

(b) Absorption and emission spectra of sodium

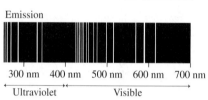

Absorption

Emission

300 nm   400 nm   500 nm   600 nm   700 nm

Ultraviolet   Visible

FIGURE 37.20 Measuring an absorption spectrum. The images in part b are replicas of photographic plates.

37.9 Classical Physics at the Limit

At the start of the 19th century, only a few scientists believed that matter consists of atoms. By century's end, there was substantial evidence not only for atoms but for the existence of charged subatomic particles. The explorations into atomic structure culminated with Rutherford's nuclear model.

Rutherford's nuclear model of the atom matched the experimental evidence about the *structure* of atoms, but it had two serious shortcomings. Electrons orbiting the nucleus in a Rutherford atom are oscillating charged particles. According to Maxwell's theory of electricity and magnetism, these orbiting electrons should act as small antennas and radiate electromagnetic waves. That sounds encouraging, because we know that atoms can emit light, but it was easy to show that a Rutherford atom would radiate a *continuous* rainbow-like spectrum. Thus one failure of Rutherford's model was an inability to predict the discrete nature of emission and absorption spectra.

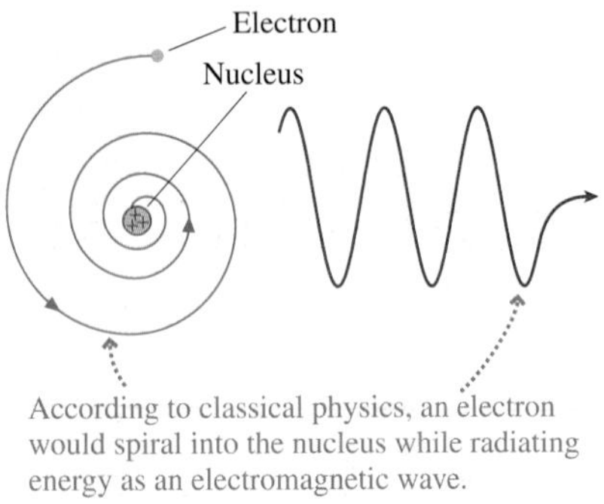

In addition, the atoms would continuously lose energy as they radiated electromagnetic waves. As Figure 37.21 shows, this would cause the electrons to spiral into the nucleus! Calculations showed that a Rutherford atom can last no more than about a microsecond. In other words, classical Newtonian mechanics and electromagnetism predict that an atom in which electrons orbit a nucleus would be highly unstable and would immediately self-destruct. This clearly does not happen.

The experimental efforts of the late 19th century had been impressive, and there could be no doubt about the existence of electrons, about the small positive nucleus, and about the unique discrete spectrum emitted by each atom. But the theoretical framework for understanding such observations had lagged behind. As the new century dawned, physicists could not explain the structure of atoms, could not explain the stability of matter, could not explain discrete spectra or why an element's absorption spectrum differs from its emission spectrum, and could not explain the origin of x rays or radioactivity.

Yet few physicists were willing to abandon the successful and long-cherished theories of classical physics. Despite the attention we have focused on the search for atomic structure, the large majority of scientists were working in other fields—electricity, acoustics, thermodynamics—for which classical physics remained completely satisfactory. Most considered these "problems" with atoms to be minor discrepancies that would soon be resolved. But classical physics had, indeed, reached its limit, and a whole new generation of brilliant young physicists, with new ideas, was about to take the stage. Among the first was an unassuming young man in Berne, Switzerland. His scholastic record had been mediocre, and the best job he could find upon graduation was as a clerk in the patent office, examining patent applications. He needed the job, having recently married a fellow student due, at least in part, to a child conceived out of wedlock. His name was Albert Einstein.

According to classical physics, an electron would spiral into the nucleus while radiating energy as an electromagnetic wave.

FIGURE 37.21 The fate of a Rutherford atom.

SUMMARY

The goal of Chapter 37 has been to understand how scientists discovered the properties of atoms and how these discoveries led to the need for a new theory of light and matter.

IMPORTANT CONCEPTS/EXPERIMENTS

Nineteenth-century scientists were focused on understanding matter, electricity, and light. Faraday's invention of the gas discharge tube launched two important avenues of inquiry.

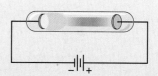

Cathode Rays and Atomic Structure

Thomson found that cathode rays are negative, subatomic particles. These were soon named electrons. Electrons are

- Constituents of atoms.
- The fundamental unit of negative charge.

Rutherford discovered the atomic nucleus. His nuclear model of the atom proposes

- A very small, dense positive nucleus.
- Orbiting negative electrons.

Later, different isotopes were recognized to contain different numbers of **neutrons** in a nucleus with the same number of **protons**.

Atomic Spectra and the Nature of Light

The spectra emitted by the gas in a discharge tube consist of discrete wavelengths.

- Every element has a unique spectrum.
- Every spectral line in an element's absorption spectrum is present in its emission spectrum, but not all emission lines are seen in absorption.

Balmer found that the wavelengths of the hydrogen emission spectrum are

$$\lambda = \frac{91.18 \text{ nm}}{\left(\dfrac{1}{m^2} - \dfrac{1}{n^2}\right)}, \qquad m = 1, 2, 3, \ldots \qquad n = m + 1, m + 2, \ldots$$

The end of classical physics . . .

Atomic spectra had to be related to atomic structure, but no one could understand how. According to all that was known,

- Rutherford's nuclear atom should not be stable.
- Atoms should radiate continuous rather than discrete spectra.

APPLICATIONS

Millikan's oil-drop experiment measured the fundamental unit of charge:

$$e = 1.60 \times 10^{-19} \text{ C}$$

One electron volt (1 eV) is the energy an electron or proton (charge $\pm e$) gains by accelerating through a potential difference of 1 V.

$$1 \text{ eV} = 1.60 \times 10^{-19} \text{ J}$$

TERMS AND NOTATION

electrolysis	subatomic particle	nucleus	neutron
gas discharge tube	electron	nuclear model of the atom	isotope
cathode glow	Millikan oil-drop experiment	electron volt, eV	mass number, A
cathode rays	radioactivity	atomic number, Z	discrete spectrum
Crookes tube	alpha rays	proton	spectral line
crossed-field experiment	beta rays	mass spectrometer	Balmer formula

EXERCISES AND PROBLEMS

Exercises

Section 37.3 Cathode Rays

Section 37.4 J. J. Thomson and the Discovery of the Electron

1. What was the significance of Thomson's experiment in which an off-center electrode was used to collect charge deflected by a magnetic field?

2. What is the evidence by which we know that an electron from an iron atom is identical to an electron from a copper atom?

3. The current in a Crookes tube is 10 nA. How many electrons strike the face of the glass tube each second?

4. An electron in a cathode-ray beam passes between 2.5-cm-long parallel-plate electrodes that are 5.0 mm apart. A 2.0 mT, 2.5-cm-wide magnetic field is perpendicular to the electric field between the plates. The electron passes through the electrodes without being deflected if the potential difference between the plates is 600 V.
   a. What is the electron's speed?
   b. If the potential difference between the plates is set to zero, what is the electron's radius of curvature in the magnetic field?

5. Electrons pass through the parallel electrodes shown in Figure Ex37.5 with a speed of 5.0×10^6 m/s. What magnetic field strength and direction will allow the electrons to pass through without being deflected? Assume that the magnetic field is confined to the region between the electrodes.

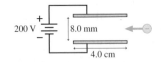

FIGURE EX37.5

Section 37.5 Millikan and the Fundamental Unit of Charge

6. A 0.80-μm-diameter oil droplet is observed between two parallel electrodes spaced 11 mm apart. The droplet hangs motionless if the upper electrode is 20 V more positive than the lower electrode. The density of the oil is 885 kg/m³.
   a. What is the droplet's mass?
   b. What is the droplet's charge?
   c. Does the droplet have a surplus or a deficit of electrons? How many?

7. An oil droplet with 15 excess electrons is observed between two parallel electrodes spaced 12 mm apart. The droplet hangs motionless if the upper electrode is 25 V more positive than the lower electrode. The density of the oil is 860 kg/m³. What is the radius of the droplet?

8. Suppose that in a hypothetical oil-drop experiment you measure the following values for the charges on the drops: 3.99×10^{-19} C, 6.65×10^{-19} C, 2.66×10^{-19} C, 10.64×10^{-19} C, and 9.31×10^{-19} C. What is the largest value of the fundamental unit of charge that is consistent with your measurements?

Section 37.6 Rutherford and the Discovery of the Nucleus

Section 37.7 Into the Nucleus

9. Express in eV (or keV or MeV if more appropriate):
   a. The kinetic energy of an electron moving with a speed of 5.0×10^6 m/s.
   b. The potential energy of an electron and a proton 0.10 nm apart.
   c. The kinetic energy of a proton that has accelerated from rest through a potential difference of 5000 V.

10. Express in eV (or keV or MeV if more appropriate):
   a. The kinetic energy of a Li^{++} ion that has accelerated from rest through a potential difference of 5000 V.
   b. The potential energy of two protons 10 fm apart.
   c. The kinetic energy, just before impact, of a 200 g ball dropped from a height of 1.0 m.

11. Determine:
   a. The speed of a 100 eV electron.
   b. The speed of a 5 MeV neutron.
   c. The specific type of particle that has 2.09 MeV of kinetic energy when moving with a speed of 1.0×10^7 m/s.

12. Determine:
   a. The speed of a 6 MeV proton.
   b. The speed of a 20 MeV helium atom.
   c. The specific type of particle that has 1.14 keV of kinetic energy when moving with a speed of 2.0×10^7 m/s.

13. a. Describe the experimental evidence by which we know that the nucleus is made up not just of protons.
   b. The neutron is not easy to isolate or control because it has no charge that would allow scientists to manipulate it. What evidence allowed scientists to determine that the mass of the neutron is almost the same as the mass of a proton?

14. How many electrons, protons, and neutrons are contained in the following atoms or ions: (a) ^{6}Li, (b) ^{13}C$^+$, and (c) ^{18}O^{++}?

15. How many electrons, protons, and neutrons are contained in the following atoms or ions: (a) ^{9}Be$^+$, (b) ^{12}C, and (c) ^{15}N^{+++}?

16. Write the symbol for an atom or ion with:
    a. three electrons, three protons, and five neutrons.
    b. five electrons, six protons, and eight neutrons.

17. Write the symbol for an atom or ion with:
    a. one electron, one proton, and one neutron.
    b. five electrons, seven protons, and seven neutrons.

18. Consider the gold isotope ^{197}Au.
    a. How many electrons, protons, and neutrons are in a neutral ^{197}Au atom?
    b. The gold nucleus has a diameter of 14.0 fm. What is the density of matter in a gold nucleus?
    c. The density of lead is 11,400 kg/m^3. How many times the density of lead is your answer to part b?

19. Consider the lead isotope ^{207}Pb.
    a. How many electrons, protons, and neutrons are in a neutral ^{207}Pb atom?
    b. The lead nucleus has a diameter of 14.2 fm. What are the electric potential and the electric field strength at the surface of a lead nucleus?

20. Explain how the observation of alpha particles scattered at very large angles led Rutherford to reject Thomson's model of the atom and to propose a nuclear model.

Section 37.8 The Emission and Absorption of Light

21. Figure 37.19b identified the wavelengths of four lines in the spectrum of hydrogen.
    a. Determine the Balmer formula n and m values for these wavelengths.
    b. Predict the wavelength of the fifth line in the spectrum.

22. Figure 37.19b identified the wavelengths of four lines in the spectrum of hydrogen.
    a. Determine the Balmer formula n and m values for these wavelengths.
    b. Figure 37.19b labels a feature called the *series limit*, although no spectral line is present at that point. Verify the wavelength of the series limit.

23. The wavelengths in the hydrogen spectrum with $m = 1$ form a series of spectral lines called the Lyman series. Calculate the wavelengths of the first four members of the Lyman series.

24. Two of the wavelengths emitted by a hydrogen atom are 102.6 nm and 1876 nm.
    a. What are the m and n values for each of these wavelengths?
    b. For each of these wavelengths, is the light infrared, visible, or ultraviolet?

Problems

25. What is the total energy, in MeV, of
    a. A proton traveling at 99% of the speed of light?
    b. An electron traveling at 99% of the speed of light?
    Hint: This problem uses relativity.

26. What is the velocity, as a fraction of c, of
    a. A proton with 500 GeV total energy?
    b. An electron with 2.0 GeV total energy?
    Hint: This problem uses relativity.

27. You learned in Chapter 36 that mass has an equivalent amount of energy. What are the energy equivalents in MeV of the rest masses of an electron and a proton?

28. The factor γ appears in many relativistic expressions. A value $\gamma = 1.01$ implies that relativity changes the Newtonian values by approximately 1% and that relativistic effects can no longer be ignored. At what kinetic energy, in MeV, is $\gamma = 1.01$ for (a) an electron, (b) a proton, and (c) an alpha particle?

29. The fission process n + ^{235}U → ^{236}U → ^{144}Ba + ^{89}Kr + 3n converts 0.185 u of mass into the kinetic energy of the fission products. What is the total kinetic energy in MeV?

30. An electron in a cathode-ray beam passes between 2.5-cm-long parallel-plate electrodes that are 5.0 mm apart. A 1.0 mT, 2.5-cm-wide magnetic field is perpendicular to the electric field between the plates. If the potential difference between the plates is 150 V, the electron passes through the electrodes without being deflected. If the potential difference across the plates is set to zero, through what angle is the electron deflected as it passes through the magnetic field?

31. The two 5.0-cm-long parallel electrodes in Figure P37.31 are spaced 1.0 cm apart. A proton enters the plates from one end, an equal distance from both electrodes. A potential difference $\Delta V = 500$ V across the electrodes deflects the proton so that it strikes the outer end of the lower electrode. What magnetic field strength and direction will allow the proton to pass through undeflected while the 500 V potential difference is applied? Assume that both the electric and magnetic fields are confined to the space between the electrodes.

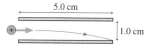

FIGURE P37.31   Trajectory at $\Delta V = 500$ V

32. An unknown charged particle passes without deflection through crossed electric and magnetic fields of strengths 187,500 V/m and 0.125 T, respectively. The particle passes out of the electric field, but the magnetic field continues, and the particle makes a semicircle of diameter 25.05 cm. What is the particle's charge-to-mass ratio? Can you identify the particle?

33. In one of Thomson's experiments he placed a thin metal foil in the electron beam and measured its temperature rise. Consider a cathode-ray tube in which electrons are accelerated through a 2000 V potential difference, then strike a 10 mg copper foil.
    a. How many electrons strike the foil in 10 s if the foil temperature rises 6.0° C?
    b. What is the current of the electron beam?

34. A lithium atom has three electrons. As you will discover in Chapter 41, two of these electrons form an "inner core," but the third—the valence electron—orbits at much larger radius. From the valence electron's perspective, it is orbiting a spherical ball of charge having net charge $+1e$ (i.e., the three protons in the nucleus and the two inner-core electrons). The energy required to ionize a lithium atom is 5.14 eV. According to Rutherford's nuclear model of the atom, what are the orbital radius and speed of the valence electron?
    Hint: Consider the energy needed to remove the electron *and* the force needed to give the electron a circular orbit.

35. The diameter of an atom is 1.2×10^{-10} m and the diameter of its nucleus is 1.0×10^{-14} m. What percent of the atom's volume is occupied by mass and what percent is empty space?

36. Balmer discovered the famous formula that bears his name by inspection and trial-and-error. See if you can discover the formula for each of the following series of wavelengths. Each formula involved an integer n, but, as in the Balmer formula, n may not start with 1.
    a. 125.00, 31.25, 13.90, 7.81, and 5.00 nm.
    b. 375, 900, 1575, 2400, 3375, and 4500 nm.

37. The diameter of an aluminum atom is approximately 1.2×10^{-10} m. The diameter of the nucleus of an aluminum atom is approximately 8×10^{-15} m. The density of solid aluminum is 2700 kg/m^3.
    a. What is the average density of an aluminum atom?
    b. Your answer to part a was similar to but larger than the density of solid aluminum. This suggests that the atoms in solid aluminum have spaces between them rather than being tightly packed together. What is the average volume per atom in solid aluminum? If this volume is a sphere, what is the radius? What can you conclude about the average spacing between atoms compared to the size of the atoms?
    Hint: The volume *per* atom is not the same as the volume *of* an atom.
    c. What is the density of the aluminum nucleus? Compare this to the density of ordinary matter.

38. The charge-to-mass ratio of a nucleus, in units of e/u, is $q/m = Z/A$. For example, a hydrogen nucleus has $q/m = 1/1 = 1$.
    a. Make a graph of charge-to-mass ratio versus proton number Z for nuclei with $Z = 5, 10, 15, 20, \ldots, 90$. For A, use the average atomic mass shown on the periodic table of elements. Show each of these 18 nuclei as a dot, but don't connect the dots together as a curve.
    b. Describe any trend that you notice in your graph.
    c. What's happening in the nuclei that is responsible for this trend?

39. If the nucleus is a few fm in diameter, the distance between the centers of two protons must be ≈ 2 fm.
    a. Calculate the electric force between two protons that are 2.0 fm apart.
    b. Calculate the gravitational force between two protons that are 2.0 fm apart. Could gravity be the force that holds the nucleus together?
    c. Your answers to parts a and b imply that there must be some other force that binds the nucleus together and prevents the protons from pushing each other out. What characteristics of this force can you deduce from the discussion of the atom and the nucleus in this chapter?

40. In a head-on collision, the closest approach of a 6.24 MeV alpha particle to the center of a nucleus is 6.00 fm. The nucleus is in an atom of what element?

41. Through what potential difference would you need to accelerate an alpha particle, starting from rest, so that it will just reach the surface of a 15-fm-diameter ^{238}U nucleus?

42. The oxygen nucleus ^{16}O has a radius of 3.0 fm.
    a. With what speed must a proton be fired toward an oxygen nucleus to have a turning point 1.0 fm from the surface?
    b. What is the proton's kinetic energy in MeV?

43. To initiate a nuclear reaction, an experimental nuclear physicist wants to shoot a proton *into* a ^{12}C nucleus. The proton must impact the nucleus with a kinetic energy of 3.0 MeV. The nuclear radius is 2.75 fm.
    a. With what speed must the proton be fired toward the target?
    b. Through what potential difference must the proton be accelerated from rest to acquire this speed?

44. The cesium isotope ^{137}Cs, with $Z = 55$, is radioactive and decays by beta decay. A beta particle is observed in the laboratory with a kinetic energy of 300 keV. The nucleus of a ^{137}Cs atom has a diameter of 12.4 fm. With what kinetic energy was the beta particle ejected from the ^{137}Cs nucleus?

Challenge Problems

45. An alpha particle approaches a ^{197}Au nucleus with a speed of 1.50×10^7 m/s. As Figure CP37.45 shows, the alpha particle is scattered at a 49° angle at the slower speed of 1.49×10^7 m/s. In what direction does the ^{197}Au nucleus recoil, and with what speed?

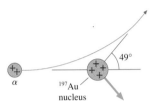

FIGURE CP37.45

46. Physicists first attempted to understand the hydrogen atom by applying the laws of classical physics. Consider an electron of mass m and charge $-e$ in a circular orbit of radius r around a proton of charge $+e$.
    a. Use Newtonian physics to show that the total energy of the atom is $E = -e^2/8\pi\epsilon_0 r$.
    b. Show that the potential energy is -2 times the electron's kinetic energy. This result is called the *virial theorem*.
    c. The minimum energy needed to ionize a hydrogen atom (i.e., to remove the electron) is found experimentally to be 13.6 eV. From this information, what are the electron's speed and the radius of its orbit?

47. Consider an oil droplet of mass m and charge q. We want to determine the charge on the droplet in a Millikan-type experiment. We will do this in several steps. Assume, for simplicity, that the charge is positive and that the electric field between the plates points upward.
    a. An electric field is established by applying a potential difference to the plates. It is found that a field of strength E_0 will cause the droplet to be suspended motionless. Write an expression for the droplet's charge in terms of the suspending field E_0 and the droplet's weight mg.
    b. The field E_0 is easily determined by knowing the plate spacing and measuring the potential difference applied to them. The larger problem is to determine the mass of a microscopic droplet. Consider a mass m falling through viscous medium in which there is a retarding or drag force. For very small particles, the retarding force is given by $F_{drag} = -bv$ where b is a constant and v the droplet's velocity. The sign recognizes that the drag force vector points upward when the droplet is falling (negative v). A falling droplet quickly reaches a *constant* velocity, called the *terminal velocity*. Write an expression for the terminal velocity v_{term} in terms of m, g, and b.

c. A spherical object of radius r moving slowly through the air is known to experience a retarding force $F_{drag} = -6\pi\eta rv$ where η is the *viscosity* of the air. Use this and your answer to part b to show that a spherical droplet of density ρ falling with a terminal velocity v_{term} has a radius

$$r = \sqrt{\frac{9\eta v_{term}}{2\rho g}}$$

d. Oil has a density 860 kg/m^3. An oil droplet is suspended between two plates 1.0 cm apart by adjusting the potential difference between them to 1177 V. When the voltage is removed, the droplet falls and quickly reaches constant speed. It is timed with a stopwatch, and falls 3.00 mm in 7.33 s. The viscosity of air is $1.83 \times 10^{-5} \text{ kg/m s}$. What is the droplet's charge?

e. How many units of the fundamental electric charge does this droplet possess?

STOP TO THINK ANSWERS

Stop to Think 37.1: b. This observation says that all electrons are the same.

Stop to Think 37.2: b. From the right-hand rule with $\vec{v}$ to the right and $\vec{B}$ out of the page.

Stop to Think 37.3: 4. Neutral carbon would have six electrons. C^{++} is missing two.

Stop to Think 37.4: 6 protons and 8 neutrons. The number of protons is the atomic number, which is 6. That leaves $14 - 6 = 8$ neutrons.

Stop to Think 37.5: a is emission, b is absorption. All wavelengths in the absorption spectrum are seen in the emission spectrum, but not all wavelengths in the emission spectrum are seen in the absorption spectrum.

38 Quantization

A scanning tunneling microscope image of a "quantum corral" made from 60 iron atoms.

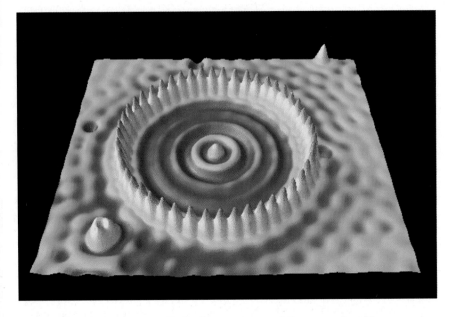

The picture shown here, called a "quantum corral," was made with a scanning tunneling microscope, a device we'll study in Chapter 40. The image shows the electron density in the vicinity of a circle of 60 iron atoms that have been carefully placed on a plane of carbon. But it's not the circle of electrons gathered around the iron atoms that is most interesting. Notice the circular ripple-like rings in the center of the corral. What you're seeing is an *electron standing wave,* rather like the standing wave on the head of a vibrating drum.

Recall from Chapter 24 what Einstein, de Broglie, and others found: that the classical either-or distinction between particles and waves, as useful as it is for macroscopic systems, does not exist in the microscopic world of electrons and atoms. Instead, light and matter exhibit characteristics of *both* particles *and* waves. This new *wave-particle duality,* as it is called, defies our commonsense picture of how things ought to behave. Nonetheless, the experimental evidence for wave-particle duality is now overwhelming and cannot be doubted. Wave-like electrons and particle-like photons of light are no longer just scientific curiosities. Modern engineering devices, such as *quantum-well lasers,* make explicit use of wave-particle duality.

This chapter will explore two critical ideas: Einstein's introduction of a particle-like nature of light and Bohr's development of a quantum atom. We will begin to think about and describe matter and light in terms of a quantum model rather than classical models. In addition, the ideas in this chapter are the final steps we need before introducing quantum mechanics in Chapter 39.

38.1 The Photoelectric Effect

In 1886, Heinrich Hertz was the first to demonstrate that electromagnetic waves can be artificially generated. By verifying the predictions of Maxwell's electromagnetic theory, Hertz cemented the last blocks of classical physics into place.

Yet in one of those ever-present ironies of history, Hertz happened, quite by chance, to discover the very phenomenon that would launch the quantum revolution. He noticed, in the course of his investigations, that a negatively charged electroscope could be discharged by shining ultraviolet light on it.

Hertz's observation caught the attention of J. J. Thomson. Figure 38.1 illustrates Thomson's interpretation of the observation: He inferred that the ultraviolet light was somehow causing the electrode to emit negative charges, thus restoring itself to electric neutrality. In 1899, using techniques similar to those with which he discovered the electron, Thomson showed that the emitted charges had exactly the same charge-to-mass ratio as electrons, and, presumably, were electrons. The emission of electrons from a substance due to light striking its surface came to be called the **photoelectric effect.** The emitted electrons are often called *photoelectrons* to indicate their origin, but they are identical in every respect to all other electrons.

Although this discovery might seem to be a minor footnote in the history of science, it soon became a, or maybe *the,* pivotal event that opened the door to new ideas. We will look at the photoelectric effect in a fair bit of detail. Our goals are to understand how classical physics was unable to explain the details of such a simple experiment and to recognize the startling new concept introduced by Einstein.

Characteristics of the Photoelectric Effect

It was not the discovery itself that dealt the fatal blow to classical physics, but the specific characteristics of the photoelectric effect found around 1900 by one of Hertz's students, Phillip Lenard. Lenard built a glass tube, shown in Figure 38.2, with two facing electrodes and a window. After removing the air from the tube, so that electrons could move freely from one electrode to the other, he allowed light to shine on the cathode.

Lenard found a steady counterclockwise current (clockwise flow of electrons) through the ammeter whenever ultraviolet light was shining on the cathode. There are no junctions in this circuit, so the current must be the same all the way around the loop. The current in the space between the cathode and the anode consists of electrons moving freely through space (i.e., not inside a wire) at the *same rate* (same number of electrons per second) as the current in the wire. There is no current if the electrodes are in the dark, so electrons don't spontaneously leap off the cathode. Instead, the light causes electrons to be ejected from the cathode at a steady rate.

Lenard used a battery to establish an adjustable potential difference ΔV between the two electrodes. He then studied how the current I varied as the potential difference and the light's wavelength and intensity were changed. Lenard found the photoelectric effect to have the following properties.

1. The current I is directly proportional to the light intensity. If the light intensity is doubled, the current also doubles.
2. The current appears without delay when the light is applied. To Lenard, this meant within the ≈ 0.1 s with which his equipment could respond. Later experiments showed that the current begins less than 1 ns after light hits the cathode!
3. Photoelectrons are emitted *only* if the light frequency f exceeds a **threshold frequency** f_0. This is shown in the graph of Figure 38.3.
4. The value of the threshold frequency f_0 depends on the type of metal from which the cathode is made.
5. If the potential difference ΔV is positive (anode positive with respect to the cathode), the current does not change as ΔV is increased. If ΔV is made negative (anode negative with respect to the cathode), by reversing the

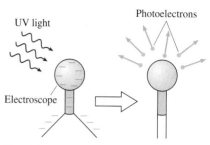

Ultraviolet light discharges a negatively charged electroscope by causing it to emit electrons.

FIGURE 38.1 Ultraviolet light discharges a negatively charged electroscope.

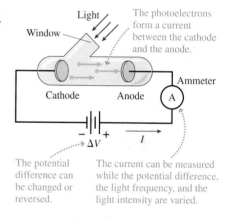

The photoelectrons form a current between the cathode and the anode.

The potential difference can be changed or reversed.

The current can be measured while the potential difference, the light frequency, and the light intensity are varied.

FIGURE 38.2 Lenard's experimental device to study the photoelectric effect.

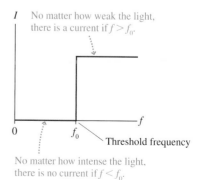

No matter how weak the light, there is a current if $f > f_0$.

Threshold frequency

No matter how intense the light, there is no current if $f < f_0$.

FIGURE 38.3 The photoelectric current as a function of the light frequency f.

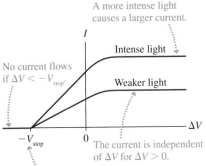

A more intense light causes a larger current.

I

Intense light

No current flows if $\Delta V < -V_{stop}$.

Weaker light

$-V_{stop}$ 0 ΔV

The current is independent of ΔV for $\Delta V > 0$.

The stopping potential is the same for intense light and weak light.

FIGURE 38.4 The photoelectric current as a function of the battery potential difference ΔV.

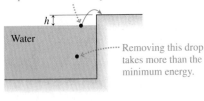

The *minimum* energy to remove a drop of water from the pool is *mgh*.

h

Water

Removing this drop takes more than the minimum energy.

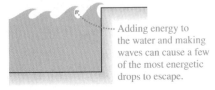

Adding energy to the water and making waves can cause a few of the most energetic drops to escape.

FIGURE 38.5 A swimming pool analogy of electrons in a metal.

TABLE 38.1 The work function for some of the elements

Element	E_0 (eV)
Potassium	2.30
Sodium	2.75
Aluminum	4.28
Tungsten	4.55
Copper	4.65
Iron	4.70
Gold	5.10

battery, the current decreases until, at some voltage $\Delta V = -V_{stop}$ the current reaches zero. The value of V_{stop} is called the **stopping potential.** This behavior is shown in Figure 38.4.

6. The value of V_{stop} is the same for both weak light and intense light. A more intense light causes a larger current, as Figure 38.4 shows, but in both cases the current ceases when $\Delta V = -V_{stop}$.

NOTE ▶ We're defining V_{stop} to be a *positive* number. The potential difference that stops the electrons is $\Delta V = -V_{stop}$, with an explicit minus sign. ◀

Classical Interpretation of the Photoelectric Effect

The mere existence of the photoelectric effect is not, as is sometimes assumed, a difficulty for classical physics. You learned in Chapter 28 that electrons are the charge carriers in a metal and move around freely inside like a sea of negatively charged particles. The electrons are bound inside the metal and do not spontaneously spill out of an electrode at room temperature. But a piece of metal heated to a sufficiently high temperature *does* emit electrons in a process called **thermal emission.** The electron gun in a television or computer display terminal starts with the thermal emission of electrons from a hot tungsten filament.

A useful analogy, shown in Figure 38.5, is the water in a swimming pool. Water molecules do not spontaneously leap out of the pool if the water is calm. To remove a water molecule, you must do *work* on it to lift it upward, against the force of gravity, to the edge of the pool. A minimum energy is needed to extract a water molecule, namely the energy needed to lift a molecule that is right at the surface. Removing a water molecule that is deeper requires more than the minimum energy. People playing in the pool add energy to the water, causing waves. If sufficient energy is added, a small fraction of the water molecules may gain enough energy to splash over the edge and leave the pool.

Similarly, a *minimum* energy is needed to free an electron from a metal. To extract an electron, you would need to exert a force on it and pull it (i.e., do *work* on it) until its speed is large enough to escape. The minimum energy E_0 needed to free an electron is called the **work function** of the metal. Some electrons, like the deeper water molecules, may require more energy than E_0 to escape, but all will require *at least* E_0. The values of the work function E_0 differ among metals according to the densities and crystal structures of the metals. Table 38.1 provides a short list. Notice that work functions are given in electron volts.

Heating a metal, like splashing in the pool, increases the thermal energy of the electrons. At a sufficiently high temperature, the kinetic energy of a small percentage of the electrons may exceed the work function. These electrons can "make it out of the pool" and leave the metal. In practice, the thermal emission of electrons requires $T > 1500°$ C, and there are only a few elements, such as tungsten, for which thermal emission can become significant before the metal melts!

Heating a metal increases the temperature of not only the electrons but also the much more massive crystal lattice of positive ions. But suppose we could raise the temperature of the electrons alone and not the crystal lattice. One possible way to do this is to shine a light wave on the surface. Because electromagnetic waves are absorbed by the conduction electrons, not by the positive ions, the light wave heats only the electrons. Eventually the electrons' energy is transferred to the crystal lattice, via collisions, but if the light is sufficiently intense, the *electron temperature* may be significantly higher than the temperature of the metal. In 1900, it was plausible to think that an intense light source could cause the thermal emission of electrons without melting the metal.

The Stopping Potential

Photoelectrons leave the cathode with kinetic energy. An electron with energy E_{elec} inside the metal loses energy ΔE as it escapes, so it emerges as a photoelectron with kinetic energy $K = E_{elec} - \Delta E$. The work function energy E_0 is the *minimum* energy needed to remove an electron, so the *maximum* possible kinetic energy of a photoelectron is

$$K_{max} = E_{elec} - E_0 \qquad (38.1)$$

The photoelectrons, after leaving the cathode, move out in all directions. Some electrons reach the anode, creating a measurable current, but many do not. However, as Figure 38.6 shows,

- A positive anode attracts *all* of the photoelectrons to the anode. Once all electrons reach the anode, a further increase in ΔV does not cause any further increase in the current I. That is why the graph lines become horizontal on the right side of Figure 38.4.
- A negative anode repels the electrons. However, a photoelectron leaving the cathode with sufficient kinetic energy can still reach the anode, just as a ball hits the ceiling if you toss it upward with sufficient kinetic energy. A slightly negative anode voltage turns back only the slowest electrons. The current steadily decreases as the anode voltage becomes increasingly negative until, at the stopping potential, *all* electrons are turned back and the current ceases. This was the behavior observed on the left side of Figure 38.4.

We can use conservation of energy to analyze the photoelectrons. Let the cathode be the point of zero potential energy, as shown in Figure 38.7. An electron emitted from the cathode with kinetic energy K_i has initial total energy

$$E_i = K_i + U_i = K_i + 0 = K_i$$

When the electron reaches the anode, which is at potential ΔV relative to the cathode, it has potential energy $U = q\Delta V = -e\Delta V$ and final total energy

$$E_f = K_f + U_f = K_f - e\Delta V$$

From conservation of energy, $E_f = E_i$, the electron's final kinetic energy is

$$K_f = K_i + e\Delta V \qquad (38.2)$$

The electron speeds up ($K_f > K_i$) if ΔV is positive. The electron slows down if ΔV is negative, but it still reaches the anode ($K_f > 0$) if K_i is large enough.

An electron with initial kinetic energy K_i will be stopped just as it reaches the anode if the potential difference is $\Delta V = -K_i/e$. The potential difference that will turn back the very fastest electrons, those with $K = K_{max}$, and thus stop the current is

$$\Delta V_{stop\ fastest\ electrons} = -\frac{K_{max}}{e}$$

By definition, the potential difference that causes the electron current to cease is $\Delta V = -V_{stop}$, where V_{stop} is the stopping potential. Thus the stopping potential seen in Figure 38.4 is

$$V_{stop} = \frac{K_{max}}{e} \qquad (38.3)$$

The stopping potential tells us the maximum kinetic energy of the photoelectrons.

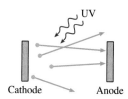
$\Delta V = 0$: The photoelectrons leave the cathode in all directions. Only a few reach the anode.

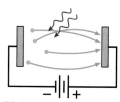
$\Delta V > 0$: Biasing the anode positive creates an electric field that pushes all the photoelectrons to the anode.

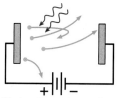
$\Delta V < 0$: Biasing the anode negative repels the electrons. Only the very fastest make it to the anode.

FIGURE 38.6 A positive anode attracts the photoelectrons. A negative anode repels them.

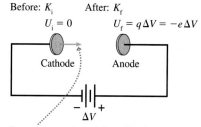
Energy is transformed from kinetic to potential as an electron moves from cathode to anode.

FIGURE 38.7 Energy is conserved.

Activ Physics 17.3

EXAMPLE 38.1 **The classical photoelectric effect**
A photoelectric-effect experiment is performed with an aluminum cathode. An electron inside the cathode has a speed of 1.5×10^6 m/s. If the potential difference between the anode and cathode is -2.00 V, what is the highest possible speed with which this electron could reach the anode?

MODEL Energy is conserved.

SOLVE If this electron succeeds in escaping as a photoelectron, its maximum possible kinetic energy is $K_{max} = E_{elec} - E_0$, where $E_0 = 4.28$ eV is the work function of aluminum. If the electron escapes with the maximum possible kinetic energy, its kinetic energy at the anode will be given by Equation 38.2 with $\Delta V = -2.00$ V.

The electron's initial energy is

$$E_{elec} = \frac{1}{2}mv^2 = \frac{1}{2}(9.11 \times 10^{-31} \text{ kg})(1.5 \times 10^6 \text{ m/s})^2$$

$$= 1.025 \times 10^{-18} \text{ J} = 6.41 \text{ eV}$$

Its maximum possible kinetic energy as it leaves the cathode is

$$K_i = K_{max} = E_{elec} - E_0 = 2.13 \text{ eV}$$

Thus the kinetic energy at the anode is

$$K_f = K_i + e\Delta V = 2.13 \text{ eV} - (e)(2.00 \text{ V}) = 0.13 \text{ eV}$$

Notice that the electron loses 2.00 eV of *energy* as it moves through the *potential* difference of -2.00 V, so we can compute the final kinetic energy in eV without having to convert to joules. However, we must convert K_f to joules to find the final speed:

$$K_f = \frac{1}{2}mv_f^2 = 0.13 \text{ eV} = 2.1 \times 10^{-20} \text{ J}$$

$$v_f = \sqrt{\frac{2K_f}{m}} = 2.1 \times 10^5 \text{ m/s}$$

Limits of the Classical Interpretation

A classical analysis based on the thermal emission of electrons from a metal has provided a possible explanation of observations 1 and 5 above. But nothing in this explanation suggests that there should be a threshold frequency. If a weak intensity at a frequency just slightly above f_0 can generate a current, then certainly a strong intensity at a frequency just slightly below f_0 should be able to do so. There is no reason that a very slight change in frequency should matter. But Lenard found there to be a sharp threshold at f_0.

And what about his observation that the current starts instantly? If the photoelectrons are due to thermal emission, it should take some length of time for the light to raise the electron temperature sufficiently high for some to escape. In fact, fairly straightforward calculations show that, for a light of modest intensity, it should take several minutes before charge starts flowing! The experimental evidence was in sharp disagreement. If $f > f_0$, the current starts instantly for both weak light and intense light.

And last, more intense light would be expected to heat the electrons to a higher temperature. Doing so should increase the maximum kinetic energy of the photoelectrons and thus should increase the stopping potential V_{stop}. But as Lenard found, the stopping potential is the same for strong light as it is for weak light.

Although the mere presence of photoelectrons did not seem surprising, classical physics was unable to explain the observed behavior of the photoelectrons. The threshold frequency and the instant current seemed particularly anomalous.

38.2 Einstein's Explanation

Albert Einstein was a little-known young man of 26 in 1905. A photograph from the time, shown in Figure 38.8, bears little resemblance to the familiar picture of a white-haired older Einstein. He had recently graduated from the Polytechnic Institute in Zurich, Switzerland, with the Swiss equivalent of a Ph.D. in physics. Although his mathematical brilliance was recognized, his overall academic record was mediocre. Rather than pursue an academic career, Einstein took a job with the Swiss Patent Office in Bern. This was a fortuitous choice because it provided him with plenty of spare time to think about physics in his own unique way.

In 1905, within the span of a single year, Einstein published three papers on three different topics, all of which would revolutionize physics. One was his initial paper on the theory of relativity, the subject with which Einstein is most associated in the public mind. Interestingly, this paper received less attention at the time than the other two. A second paper explained a phenomenon called *Brownian motion.* In 1827, the Scottish botanist Robert Brown had used a microscope to examine small pollen grains suspended in water. He observed that the pollen grains jiggled about rather than floating at rest. Einstein, using the techniques of statistical mechanics, provided a convincing explanation that the jiggling resulted from the continual random collisions of water molecules with the pollen grains. His analysis provided one of the most definitive pieces of evidence for the reality of atoms and molecules.

FIGURE 38.8 A young Einstein.

But it is Einstein's third paper of 1905, on the nature of light, in which we are most interested. In it he offered an exceedingly simple but amazingly bold idea to explain Lenard's photoelectric-effect data. A few years earlier, in 1900, the German physicist Max Planck had been trying to understand the details of the rainbow-like spectrum of light emitted by a glowing, incandescent object. This problem didn't seem to yield to a classical physics analysis, but Planck found that he could calculate the spectrum perfectly if he made an unusual assumption. The atoms in a solid vibrate back and forth around their equilibrium positions with frequency f. You learned in Chapter 14 that the energy of a simple harmonic oscillator depends on its amplitude and can have *any* possible value. But to predict the spectrum correctly, Planck had to assume that the oscillating atoms are *not* free to have any possible energy. Instead, the energy of an atom vibrating with frequency f has to be one of the specific energies $E = 0, hf, 2hf, 3hf, \ldots$, where h is a constant. That is, the vibration energies are *quantized.*

Planck was able to determine the value of the constant h by comparing his calculations of the spectrum to experimental measurements. The constant that he introduced into physics is now called **Planck's constant.** Its contemporary value is

$$h = 6.63 \times 10^{-34} \, \text{J s} = 4.14 \times 10^{-15} \, \text{eV s}$$

The first value, with SI units, is the proper one for most calculations, but you will find the second to be useful when energies are expressed in eV.

Planck considered this to be a trick. He was skeptical that the atoms *really* had quantized energy, and he felt sure that further analysis would soon reveal how the trick worked. Einstein was the first to take Planck's idea seriously and to suggest that the quantization is real. Einstein went even further and suggested that **electromagnetic radiation itself is quantized!** That is, light is not really a continuous wave but, instead, arrives in small packets or bundles of energy. Einstein called each packet of energy a **light quantum,** and he postulated that the energy of one light quantum is directly proportional to the frequency of the light. That is, each quantum of light has energy

$$E = hf \tag{38.4}$$

where h is Planck's constant and f is the frequency of the light.

The idea of light quanta is subtle, so let's look at an analogy with raindrops. Although we often think of water as a continuous fluid, such as water in a beaker, rain consists of water that falls in discrete packets called raindrops. Raindrops are analogous to quanta of light. A downpour has a torrent of raindrops, but in a light shower the drops are few. The difference between "intense" rain and "weak" rain is the *rate* at which the drops arrive. An intense rain makes a continuous noise on the roof, so you are not aware of the individual drops, but the individual drops become apparent during a light rain.

For most light sources, the individual quanta are no more discernible than the individual raindrops in a downpour.

Similarly, a great number of light quanta arrive each second when the light is intense, but very weak light consists of only a few quanta per second. And just as raindrops come in different sizes, with larger-mass drops having larger kinetic energy, higher-frequency light quanta have a larger amount of energy. Although this analogy is not perfect, it does provide a useful mental picture of light quanta arriving at a surface.

EXAMPLE 38.2 **The energy of a light quantum**

What is the energy of one quantum of light having a wavelength of 500 nm?

SOLVE Light with a wavelength of 500 nm has frequency

$$f = \frac{v}{\lambda} = \frac{c}{\lambda} = \frac{3.00 \times 10^8 \text{ m/s}}{500 \times 10^{-9} \text{ m}} = 6.00 \times 10^{14} \text{ Hz}$$

One light quantum has energy

$$E = hf = 3.98 \times 10^{-19} \text{ J} = 2.49 \text{ eV}$$

ASSESS Because 500 nm is a typical wavelength for visible light (it would be perceived as green light), you can see that the electron volt is an energy unit of more appropriate size than the joule.

Einstein's Postulates

Einstein framed three postulates about light quanta and their interaction with matter:

1. Light of frequency f consists of discrete quanta, each of energy $E = hf$. Each photon travels at the speed of light c.
2. Light quanta are emitted or absorbed on an all-or-nothing basis. A substance can emit 1 or 2 or 3 quanta, but not 1.5. Similarly, an electron in a metal cannot absorb half a quantum but, instead, only an integer number.
3. A light quantum, when absorbed by a metal, delivers its entire energy to *one* electron.

NOTE ▶ These three postulates—that light comes in chunks, that the chunks cannot be divided, and that the energy of one chunk is delivered to one electron—are crucial for understanding the new ideas that will lead to quantum physics. They are completely at odds with the concepts of classical physics, where energy can be continuously divided and shared, so they deserve careful thought. ◀

Let's look at how Einstein's postulates apply to the photoelectric effect. If Einstein is correct, the light shining on the metal is a torrent of light quanta, each of energy hf. Each quantum is absorbed by *one* electron, giving that electron an energy $E_{elec} = hf$. This leads us to several interesting conclusions:

1. An electron that has just absorbed a quantum of light energy has $E_{elec} = hf$. (The electron's thermal energy at room temperature is so much less than hf that we can neglect it.) Figure 38.9 shows than this electron can escape from the metal, becoming a photoelectron, if

$$E_{elec} = hf \geq E_0 \tag{38.5}$$

In other words, there is a *threshold frequency*

$$f_0 = \frac{E_0}{h} \tag{38.6}$$

for the ejection of photoelectrons. If f is less than f_0, even by just a small amount, none of the electrons will have sufficient energy to escape no matter how intense the light. But even very weak light with $f \geq f_0$ will give a few electrons sufficient energy to escape **because each light quantum delivers all of its energy to one electron.** This threshold behavior is exactly what Lenard observed.

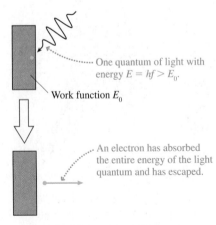

One quantum of light with energy $E = hf > E_0$.

Work function E_0

An electron has absorbed the entire energy of the light quantum and has escaped.

FIGURE 38.9 The creation of a photoelectron.

NOTE ▶ The threshold frequency is directly proportional to the work function. Metals with large work functions, such as iron, copper, and gold, exhibit the photoelectric effect only when illuminated by high-frequency ultraviolet light. Photoemission occurs with lower-frequency visible light for metals with smaller values of E_0, such as sodium and potassium. ◀

2. A more intense light delivers a larger number of light quanta to the surface. These quanta eject a larger number of photoelectrons and cause a larger current, exactly as observed.

3. There is a distribution of kinetic energies, because different photoelectrons require different amounts of energy to escape, but the *maximum* kinetic energy is

$$K_{max} = E_{elec} - E_0 = hf - E_0 \qquad (38.7)$$

As we noted in Equation 38.3, the stopping potential V_{stop} is a measure of K_{max}. Einstein's theory predicts that the stopping potential is related to the light frequency by

$$V_{stop} = \frac{K_{max}}{e} = \frac{hf - E_0}{e} \qquad (38.8)$$

NOTE ▶ The stopping potential does *not* depend on the intensity of the light. Both weak light and intense light will have the same stopping potential, as Lenard had observed but which could not previously be explained. ◀

4. If each light quantum transfers its energy hf to just one electron, that electron *immediately* has enough energy to escape. The current should begin instantly, with no delay, exactly as Lenard had observed.

Using the swimming pool analogy again, Figure 38.10 shows a pebble being thrown into the pool. The pebble increases the energy of the water, but the increase is shared among all the molecules in the pool. The increase in the water's energy is barely enough to make ripples, not nearly enough to splash water out of the pool. But suppose *all* the pebble's energy could go to *one drop* of water that didn't have to share it. That one drop of water would easily have enough energy to leap out of the pool. Einstein's hypothesis that a light quantum transfers all its energy to one electron is equivalent to the pebble transferring all its energy to one drop of water.

A Prediction

Not only do Einstein's hypotheses explain all of Lenard's observations, they make a new prediction. According to Equation 38.8, the stopping potential should be a linearly increasing function of the light's frequency f. We can rewrite Equation 38.8 in terms of the threshold frequency $f_0 = E_0/h$ as

$$V_{stop} = \frac{h}{e}(f - f_0) \qquad (38.9)$$

A graph of the stopping potential V_{stop} versus the light frequency f should start from zero at $f = f_0$, then rise linearly with a slope of h/e. In fact, the slope of the graph provides a way to measure Planck's constant h.

Lenard had not measured the stopping potential for different frequencies, so Einstein offered this as an untested prediction of his postulates. Robert Millikan, who was well known for his oil-drop experiment to measure e, took up the challenge. Some of Millikan's data for a cesium cathode are shown in Figure 38.11. As you can see, Einstein's prediction of a linear relationship between f and V_{stop} was fully confirmed.

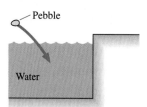
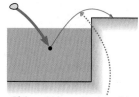

Classically, the energy of the pebble is shaped by all the water molecules. One pebble causes only very small waves.

If the pebble could give *all* its energy to one drop, that drop could easily splash out of the pool.

FIGURE 38.10 A pebble transfers energy to the water.

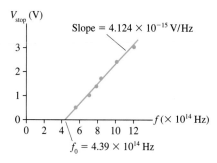

FIGURE 38.11 A graph of Millikan's data for the stopping potential as the light frequency is varied.

Millikan measured the slope of his graph and multiplied it by the value of e (which he had measured a few years earlier in the oil-drop experiment) to find h. His value agreed with the value that Planck had determined in 1900 from an entirely different experiment. Light quanta, whether physicists liked the idea or not, were real.

EXAMPLE 38.3 **The photoelectric threshold frequency**

What are the threshold frequencies and wavelengths for photo emission from sodium and from aluminum?

SOLVE Table 38.1 gives the sodium work function as $E_0 = 2.75$ eV. Aluminum has $E_0 = 4.28$ eV. We can use Equation 38.6, with h in units of eV s, to calculate

$$f_0 = \frac{E_0}{h} = \begin{cases} 6.64 \times 10^{14} \text{ Hz} & \text{sodium} \\ 10.34 \times 10^{14} \text{ Hz} & \text{aluminum} \end{cases}$$

These frequencies are converted to wavelengths with $\lambda = c/f$, giving

$$\lambda = \begin{cases} 452 \text{ nm} & \text{sodium} \\ 290 \text{ nm} & \text{aluminum} \end{cases}$$

ASSESS The photoelectric effect can be observed with sodium for $\lambda < 452$ nm. This includes blue and violet visible light but not red, orange, yellow, or green. Aluminum, with a larger work function, needs ultraviolet wavelengths $\lambda < 290$ nm.

EXAMPLE 38.4 **Maximum photoelectron speed**

What is the maximum photoelectron speed if sodium is illuminated with light of 300 nm?

SOLVE The light frequency is $f = c/\lambda = 1.00 \times 10^{15}$ Hz, so each light quantum has energy $hf = 4.14$ eV. The maximum kinetic energy of a photoelectron is

$$K_{max} = hf - E_0 = 4.14 \text{ eV} - 2.75 \text{ eV} = 1.39 \text{ eV}$$
$$= 2.22 \times 10^{-19} \text{ J}$$

Because $K = \frac{1}{2}mv^2$, where m is the electron's mass, not the mass of the sodium atom, the maximum speed of a photoelectron leaving the cathode is

$$v_{max} = \sqrt{\frac{2K_{max}}{m}} = 6.99 \times 10^5 \text{ m/s}$$

Note that we had to convert K_{max} to SI units of J before calculating a speed in m/s.

STOP TO THINK 38.1 The work function of metal A is 3.0 eV. Metals B and C have work functions of 4.0 eV and 5.0 eV, respectively. Ultraviolet light shines on all three metals, creating photoelectrons. Rank in order, from largest to smallest, the stopping voltages for A, B, and C.

38.3 Photons

Einstein was awarded the Nobel prize in 1921 not for his theory of relativity, as many would suppose, but for his explanation of the photoelectric effect. Although Planck had made the first suggestion, it was Einstein who showed convincingly that energy is quantized and that light, even though it exhibits interference, comes in some kind of particle-like packets of energy. These fundamental units of light energy were later given the name **photons.**

But just what are photons? Although particle-like, they clearly do not mesh with the classical idea of a particle. A classical particle, when faced with Young's double-slit apparatus, would go through one hole or the other. If light consisted of classical particles, we would see two bright spots on the screen. Instead, we see interference fringes behind a double slit. We even observed, in Chapter 24, that the interference pattern can be built up photon by photon if the light intensity is reduced to the point where only one photon at a time is traversing the apparatus. This behavior seems to indicate that a photon must, in some sense, go through *both* slits and interfere with itself! Photons seem to be both wave-like *and* particle-like at the same time.

Photons are sometimes visualized as **wave packets.** The electromagnetic wave shown Figure 38.12 has a wavelength and a frequency, yet it is also discrete and fairly localized. But this cannot be exactly what a photon is because a wave packet would take a finite amount of time to be emitted or absorbed. This is contrary to much evidence that the entire photon is emitted or absorbed in a single instant; there is no point in time at which the photon is "half absorbed." The wave packet idea, although useful, is still too classical to represent a photon.

The bottom line is that there simply is no "true" mental representation of a photon. Analogies such as raindrops or wave packets can be useful, but none are perfectly accurate. We can detect photons, measure the properties of photons, and put photons to practical use, but the ultimate nature of the photon remains a mystery. To paraphrase Gertrude Stein, "A photon is a photon is a photon."

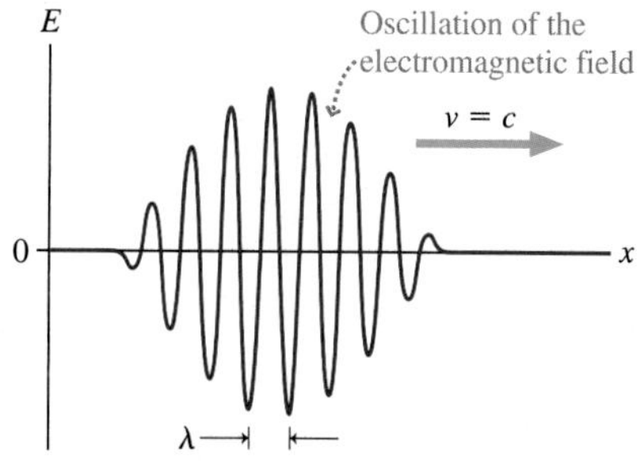

FIGURE 38.12 A wave packet has wave-like and particle-like properties.

The Photon Rate

Light, in the raindrop analogy, consists of a stream of photons. For monochromatic light of frequency f, N photons have a total energy $E_{light} = Nhf$. We are usually more interested in the *power* of the light, or the rate (in joules per second, or watts) at which the light energy is delivered. The power is

$$P = \frac{dE_{light}}{dt} = \frac{dN}{dt}hf = Rhf \qquad (38.10)$$

where $R = dN/dt$ is the *rate* at which photons arrive or, equivalently, the number of photons per second.

EXAMPLE 38.5 The photon rate in a laser beam
The 1.0 mW light beam of a helium-neon laser ($\lambda = 633$ nm) shines on a screen. How many photons strike the screen each second?

SOLVE The light-beam power, or energy delivered per second, is $P = 1.0$ mW $= 0.0010$ J/s. This is a realistic value. The frequency of the light is $f = c/\lambda = 4.74 \times 10^{14}$ Hz. The number of photons striking the screen per second, which is the *rate* of arrival of photons, is

$$R = \frac{P}{hf} = 3.2 \times 10^{15} \text{ photons per second}$$

ASSESS That is a lot of photons per second. No wonder that we are not aware of individual photons!

Photodetectors

Modern photodetectors are descendants of the photoelectric effect. These range from simple "electric eyes" to the detector array in a video camera. Most detectors use what is called a *photodiode* in which the photoelectrons are emitted internally in a semiconductor. Even so, they still have a threshold frequency, a stopping potential, and other attributes of the photoelectric effect.

Very low light levels can be detected photon by photon with a device called a *photomultiplier tube,* or PMT. Figure 38.13a on the next page shows that a PMT consists of a cathode, an anode, and a number of intermediate electrodes sealed inside an evacuated glass tube. The cathode is coated with a low-work-function material, allowing it to respond to most visible wavelengths of light. The cathode is at a fairly high negative voltage and the anode, at the other end, is at essentially zero volts. Steadily descending potentials are applied to the intermediate electrodes.

(a) A photomultiplier tube

Electrode voltages

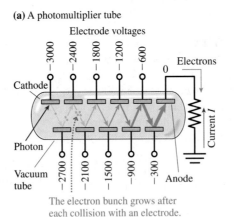

The electron bunch grows after
each collision with an electrode.

(b) The output signal from a single photon

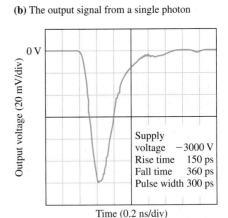

FIGURE 38.13 A photomultiplier tube can detect individual photons.

A photon of light ejects a photoelectron from the cathode. The electric field between the cathode and the first intermediate electrode accelerates that electron through a potential difference of about 300 V, and it then strikes this electrode at high speed. When a fast electron collides with a metal surface, it can kick out two or three other electrons called *secondary electrons*. The secondary electrons of the first electrode are accelerated to the second electrode, where they kick out more electrons. These are accelerated to the third electrode, where they kick out yet more electrons, and so on. There is a chain-reaction *multiplication* of electrons— 1, 2, 4, 8, 16, . . . —as they move from the cathode toward the anode. For a typical PMT, a single photon at the cathode causes an electron bunch with 10^6 or 10^7 electrons to arrive at the anode.

The electrons are collected by the anode and flow through a resistor. Because these are negative charge carriers, we would say that a current pulse I travels upward through the resistor. This creates a *negative* voltage across the resistor, $\Delta V = IR$, for the length of time that the current lasts. Figure 38.12b, an actual measurement, shows a pulse generated by a single photon. The horizontal scale is 0.2 ns/division and the vertical scale is 20 millivolts (mV)/division. You can see that the width of the pulse is ≈0.3 ns and its height (measured downward from the baseline) is ≈120 mV = 0.12 V. This is not a large voltage, even after the multiplication, but it is a voltage easily detected with modern electronics.

NOTE ▶ The 0.3 ns pulse duration is *not* an indication of the duration of a photon. The photon absorption is instantaneous, but as the electron bunch grows in size, the electron-electron repulsion causes the bunch to spread out some. The observed pulse width is an artifact of the PMT, not a characteristic of the photon. ◀

STOP TO THINK 38.2   The intensity of a beam of light is increased but the light's frequency is unchanged. Which one (or perhaps more than one) of the following is true?

a. The photons travel faster.
b. Each photon has more energy.
c. The photons are larger.
d. There are more photons per second.

38.4 Matter Waves and Energy Quantization

Prince Louis-Victor de Broglie was a French graduate student in 1924. It had been 19 years since Einstein had shaken the world of physics by blurring the distinction between a particle and a wave. As de Broglie thought about these issues, it seemed that nature should have some kind of symmetry. If light waves could have a particle-like nature, why shouldn't material particles have some kind of wave-like nature? In other words, could **matter waves** exist?

With no experimental evidence to go on, de Broglie reasoned by analogy with Einstein's equation $E = hf$ for the photon and with some of the ideas of his theory of relativity. The details need not concern us, but they led de Broglie to postulate that *if* a material particle of momentum $p = mv$ has a wave-like nature, its wavelength must be given by

$$\lambda = \frac{h}{p} = \frac{h}{mv} \tag{38.11}$$

where h is Planck's constant. This wavelength is called the **de Broglie wavelength.**

EXAMPLE 38.6 The de Broglie wavelength of an electron
What is the de Broglie wavelength of a 1.0 eV electron?

SOLVE An electron with $1.0\,\text{eV} = 1.6 \times 10^{-19}\,\text{J}$ of kinetic energy has speed

$$v = \sqrt{\frac{2K}{m}} = 5.9 \times 10^5 \text{ m/s}$$

Although fast by macroscopic standards, this is a slow electron because it gains this speed by accelerating through a potential difference of a mere 1 V. Its de Broglie wavelength is

$$\lambda = \frac{h}{mv} = 1.2 \times 10^{-9}\text{ m} = 1.2\text{ nm}$$

ASSESS The electron's wavelength is small, but it is larger than the wavelengths of x rays and larger than the approximately 10^{-10} m spacing of atoms in a crystal.

What would it mean for matter—an electron or a proton or a baseball—to have a wavelength? Would it obey the principle of superposition? Would it exhibit interference and diffraction? These are questions we examined in Chapter 24, where we found that, indeed, matter *does* exhibit interference. For example, Figure 38.14 shows the intensity pattern recorded after 50 keV electrons passed through two slits separated by 1.0 μm. The pattern is clearly a double-slit interference pattern, and the spacing of the fringes is exactly as predicted for a wavelength given by de Broglie's formula. Because the electron beam was weak, with one electron at a time passing through the apparatus, it would appear that each electron somehow went through both slits, then recombined to interfere with itself!

Electrons are fundamental subatomic particles. Perhaps subatomic particles have wave-like aspects, but what about entire atoms, aggregates of many fundamental particles? Amazing as it seems, research during the 1980s demonstrated that whole atoms, and even molecules, can produce interference patterns.

Figure 38.15 on the next page shows an *atom interferometer.* You learned in Chapter 22 that an interferometer, such as the Michelson interferometer, works by dividing a wave front into two waves, sending the two waves along separate paths, then recombining them. For light waves, the wave division is accomplished by sending light through the *periodic* slits in a diffraction grating. In an atom interferometer, the atom's matter wave is divided by sending atoms through the *periodic* intensity of a standing light wave.

FIGURE 38.14 A double-slit interference pattern created with electrons.

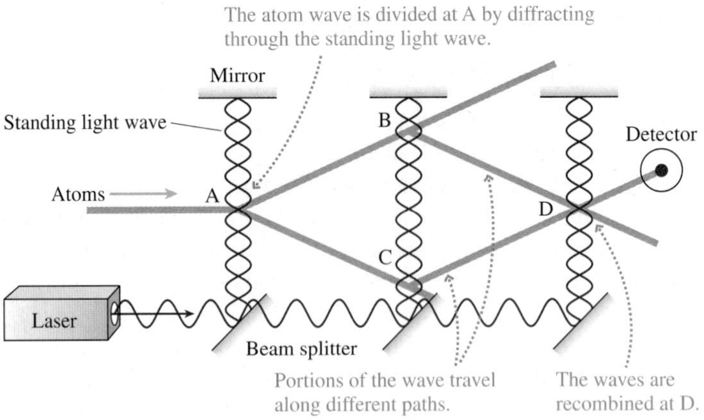

The atom wave is divided at A by diffracting through the standing light wave.

Mirror

Standing light wave

Atoms

Detector

B

A

C

D

Laser

Beam splitter

Portions of the wave travel along different paths.

The waves are recombined at D.

FIGURE 38.15 An atom interferometer.

You can see in the figure that a laser creates three parallel *standing waves* of light, each with nodes spaced a distance $\lambda/2$ apart. The wavelength is chosen so that the light waves exert small forces on an atom in the laser beam. Because the intensity along a standing wave alternates between maximum at the antinodes and zero intensity at the nodes, an atom crossing the laser beam experiences a *periodic* force field. A particle-like atom would be deflected by this periodic force, but a wave is *diffracted*. After being diffracted by the first standing wave at A, an atom is, in some sense, traveling toward both point B *and* point C.

The second standing wave diffracts the atom waves again at points B and C, directing them toward D where, with a third diffraction, they are recombined after having traveled along different paths. Depending on the phases of the waves as they recombine, the detector sometimes records atoms (constructive interference) but at other times does not (destructive interference). Altering one of the paths, such as by applying an electric field in the region around B but not around C, shifts the phases of the atom waves and causes the detector to record interference fringes.

The atom interferometer is fascinating because it completely inverts everything we previously learned about interference and diffraction. The scientists who studied the wave nature of light during the 19th century aimed light (a wave) at a diffraction grating (a periodic structure of matter) and found that it diffracted. Now we aim atoms (matter) at a standing wave (a periodic structure of light) and find that the atoms diffract. The roles of light and matter have been completely reversed!

Quantization of Energy

De Broglie considered a matter wave to be a traveling wave. But suppose that a "particle" of matter is *confined* to a small region of space and cannot travel? How do the wave-like properties manifest themselves?

This is the problem of "a particle in a box" that we looked at in Chapter 24. We will briefly summarize that discussion. Figure 38.16 shows a particle of mass m moving in one dimension as it bounces back and forth with speed v between the ends of a box of length L. A wave, if it reflects back and forth between two fixed points, sets up a standing wave. You learned in Chapter 21 that a standing wave of length L *must* have a wavelength given by

$$\lambda_n = \frac{2L}{n} \qquad n = 1, 2, 3, 4, \ldots \qquad (38.12)$$

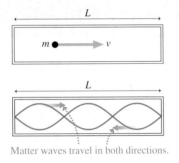

L

m v

L

Matter waves travel in both directions.

FIGURE 38.16 A particle in a box creates a standing de Broglie wave as it reflects back and forth.

If the confined particle has wave-like properties, it should satisfy both Equation 38.12 *and* the de Broglie relationship $\lambda = h/mv$. That is, a particle in a box should obey the relationship

$$\lambda = \frac{h}{mv} = \frac{2L}{n}$$

This can be true only if the particle's speed is

$$v_n = n\left(\frac{h}{2Lm}\right) \qquad n = 1, 2, 3, \ldots \qquad (38.13)$$

In other words, the particle cannot bounce back and forth with just *any* speed. Rather, it can have *only* those specific speeds v_n, given by Equation 38.13, for which the de Broglie wavelength creates a standing wave in the box.

Thus the particle's energy, which is purely kinetic energy, is

$$E_n = \frac{1}{2}mv_n^2 = n^2\frac{h^2}{8mL^2} \qquad n = 1, 2, 3, \ldots \qquad (38.14)$$

De Broglie's hypothesis about the wave-like properties of matter leads us to the remarkable conclusion that **the energy of a confined particle is quantized.** The energy of the particle in the box can be $1(h^2/8mL^2)$, or $4(h^2/8mL^2)$, or $9(h^2/8mL^2)$, but it *cannot* have an energy between these values.

The possible values of the particle's energy are called **energy levels,** and the integer n that characterizes the energy levels is called the **quantum number.** The quantum number can be found by counting the antinodes, just as you learned to do for standing waves on a string. The standing wave shown in Figure 38.16 is $n = 3$, thus its energy is E_3.

We can rewrite Equation 38.14 in the useful form

$$E_n = n^2E_1 \qquad (38.15)$$

where

$$E_1 = \frac{h^2}{8mL^2} \qquad (38.16)$$

is the **fundamental quantum of energy** for a particle in a box. It is analogous to the fundamental frequency f_1 of a standing wave on a string.

EXAMPLE 38.7 **The energy levels of an oil droplet**
What is the fundamental quantum of energy for one of Millikan's 1.0-μm-diameter oil droplets confined in a box of length 10 μm? The density of the oil is 900 kg/m^3.

SOLVE The mass of a droplet is $m = \rho V$, where the volume is $\frac{4}{3}\pi r^3$. A quick calculation shows that a 1.0-μm-diameter droplet has mass $m = 4.7 \times 10^{-16}$ kg. The confinement length is $L = 1.0 \times 10^{-5}$ m. From Equation 38.16, the fundamental quantum of energy is

$$E_1 = \frac{h^2}{8mL^2} = \frac{(6.63 \times 10^{-34} \,\text{J s})^2}{8(4.7 \times 10^{-16} \,\text{kg})(1.0 \times 10^{-5} \,\text{m})^2}$$

$$= 1.2 \times 10^{-42} \,\text{J} = 7.3 \times 10^{-24} \,\text{eV}$$

ASSESS This is such an incredibly small amount of energy that there is no hope of distinguishing between energies of E_1 or $4E_1$ or $9E_1$. For any macroscopic particle, even one this tiny, the allowed energies will *seem* to be perfectly continuous. We will not observe the quantization.

EXAMPLE 38.8 The energy levels of an electron

What are the first three allowed energies for an electron confined in a one-dimensional box of length 0.10 nm, about the size of an atom?

SOLVE We can use Equation 38.16, with $m_{elec} = 9.11 \times 10^{-31}$ kg and $L = 1.0 \times 10^{-10}$ m to find that the fundamental quantum of energy is $E_1 = 6.0 \times 10^{-18}$ J = 3.8 eV. Thus the first three allowed energies of an electron in a 0.10 nm box are

$$E_1 = 3.8 \text{ eV}$$

$$E_2 = 4E_1 = 15.2 \text{ eV}$$

$$E_3 = 9E_1 = 34.2 \text{ eV}$$

ASSESS These energies are similar to the energy we calculated in Example 38.2 for a photon with a wavelength of 500 nm.

We see that confining a wave-like particle creates a standing de Broglie wave, and we know that a standing wave has only certain discrete wavelengths. Thus we find that a confined particle can have only certain discrete energies. In other words, **the confinement of a particle leads directly to the quantization of its energy.** The particle in a box, although not a realistic model of an atom, is a simple example to illustrate these ideas. An electron confined in a real atom will need to be a much more complex three-dimensional standing wave. But, just like the simple particle in a box, it will have quantized energies. Furthermore, we expect a typical energy difference between adjacent energy levels will be a few electron volts.

Now, this is an intriguing result. We found that visible and ultraviolet photons of light have energies of a few electron volts. We also know that atoms emit *discrete* wavelengths of visible and ultraviolet light, with photon energies of a few electron volts. Now we see that an electron confined into an atomic-size box has energy levels spaced a few electron volts apart. Might there be a connection between these phenomena? We will explore this topic in the next section.

STOP TO THINK 38.3   What is the quantum number of this particle confined in a box?

38.5 Bohr's Model of Atomic Quantization

18.1   Activ
Physics
ONLINE

Thomson's electron and Rutherford's nucleus made it clear that the atom has a *structure* of some sort. The challenge at the beginning of the 20th century was to deduce, from experimental evidence, the correct structure. The difficulty of this task cannot be exaggerated. The evidence about atoms, such as observations of atomic spectra, was very indirect, and experiments were carried out with only the simplest of measuring devices. Most observations were made by eye, and all calculations were carried out by hand. Using observations as a guide, physicists were attempting to construct a *model* of the atom that could successfully explain the various experiments.

Rutherford's nuclear model was the most successful of various proposals, but Rutherford's model failed to explain why atoms are stable or why their spectra are discrete. A missing piece of the puzzle, although not recognized as such for a few years, was Einstein's 1905 introduction of light quanta. If light comes in discrete packets of energy, which we now call photons, and if atoms emit and absorb light, what does that imply about the structure of the atoms?

This was the question posed by Niels Bohr. Bohr, shown as young man in Figure 38.17, was born, educated, and spent most of his life in Denmark. He later established an institute in Copenhagen that, for many decades, was the leading center for the development of quantum physics. Although few discoveries bear Bohr's name, he was the intellectual driving force behind the development of quantum mechanics and the mentor of many of the young physicists who reshaped physics in the 1920s and 1930s.

After receiving his doctoral degree in physics in 1911, Bohr went to England to work in Rutherford's laboratory. Rutherford had just, within the previous year, completed his development of the nuclear model of the atom. Rutherford's model certainly contained a kernel of truth, but Bohr wanted to understand how a solar-system-like atom could be stable and not radiate away all its energy. He soon recognized that Einstein's light quanta had profound implications about the structure of atoms. In 1913, Bohr proposed a radically new model of the atom in which he added quantization to Rutherford's nuclear atom.

The basic assumptions of the **Bohr model of the atom** are as follows:

1. An atom consists of negative electrons orbiting a very small positive nucleus, as in the Rutherford model.
2. Atoms can exist only in certain **stationary states.** Each stationary state corresponds to a particular set of electron orbits around the nucleus. These states are distinct and can be numbered $n = 1, 2, 3, 4, \ldots$ where n is the *quantum number.*
3. Each stationary state has a discrete, well-defined energy E_n. That is, atomic energies are *quantized.* The stationary states of an atom are numbered in order of increasing energy: $E_1 < E_2 < E_3 < E_4 < \cdots$
4. The lowest energy state of the atom, with energy E_1, is *stable* and can persist indefinitely. It is called the **ground state** of the atom. Other stationary states with energies $E_2, E_3, E_4, \ldots$ are called **excited states** of the atom.
5. An atom can "jump" from one stationary state to another by emitting or absorbing a photon of frequency

$$f_{\text{photon}} = \frac{\Delta E_{\text{atom}}}{h} \qquad (38.17)$$

where h is Planck's constant and $\Delta E_{\text{atom}} = |E_f - E_i|$. E_i and E_f are the energies of the initial and final states. Such a jump is called a **transition** or, sometimes, a **quantum jump.** Figure 38.18a is a schematic view of the emission and absorption of photons in an atom with stationary states.
6. An atom can move from a lower energy state to a higher energy state by absorbing energy $\Delta E_{\text{atom}} = E_f - E_i$ in an inelastic collision with an electron or another atom. This process, called **collisional excitation,** is shown in Figure 38.18b.
7. Atoms will seek the lowest energy state. An atom in an excited state, if left alone, will jump to lower and lower energy states until it reaches the ground state.

Bohr's model builds upon Rutherford's model, but it adds two new ideas that are derived from Einstein's ideas of quanta. The first, expressed in assumption 2, is that only certain electron orbits are "allowed" or can exist. The second, expressed in assumption 5, is that **the atom can jump from one state to another by emitting or absorbing a photon of just the right frequency to conserve energy.**

According to Einstein, a photon of frequency f has energy $E_{\text{photon}} = hf$. If an atom jumps from an initial state with energy E_i to a final state with *lower* energy E_f, energy will be conserved if the atom emits a photon with $E_{\text{photon}} = \Delta E_{\text{atom}}$. This photon must have exactly the frequency given by Equation 38.17 if it is to carry away exactly the right amount of energy. Similarly, an atom can jump to a higher energy state, for which additional energy is needed, by absorbing a photon

FIGURE 38.17 Niels Bohr.

(a) Emission and absorption of light

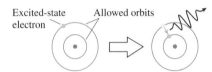

The electron jumps to a lower-energy stationary state and emits a photon.

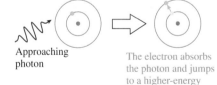

The electron absorbs the photon and jumps to a higher-energy stationary state.

(b) Collisional excitation

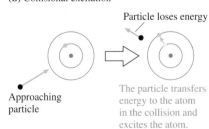

The particle transfers energy to the atom in the collision and excites the atom.

FIGURE 38.18 An atom can change stationary states by emitting or absorbing a photon or by undergoing a collision.

of frequency $f_{\text{photon}} = \Delta E_{\text{atom}}/h$. The total energy of the atom-plus-light system is conserved.

NOTE ▶ When an atom is excited to a higher energy level by absorbing a photon, the photon vanishes. Thus energy conservation requires $E_{\text{photon}} = \Delta E_{\text{atom}}$. When an atom is excited to a higher energy level in a collision with a particle, such as an electron or another atom, the particle still exists after the collision and still has energy. Thus energy conservation requires the less stringent condition $E_{\text{particle}} \geq \Delta E_{\text{atom}}$. ◀

The implications of Bohr's model are profound. In particular:

1. **Matter is stable.** Once an atom is in its ground state, there are no states of any lower energy to which it can jump. It can remain in the ground state forever.
2. **Atoms emit and absorb a *discrete spectrum*.** Only those photons whose frequencies match the energy *intervals* between the stationary states can be emitted or absorbed. Photons of other frequencies cannot be emitted or absorbed without violating energy conservation.
3. **Emission spectra can be produced by collisions.** In a gas discharge tube, the current-carrying electrons moving through the tube occasionally collide with the atoms. A collision transfers energy to an atom and can kick it to an excited state. Once the atom is in an excited state, it can emit photons of light—a discrete emission spectrum—as it jumps back down to lower-energy states.
4. **Absorption wavelengths are a subset of the wavelengths in the emission spectrum.** Recall that all the lines seen in an absorption spectrum are also seen in emission, but many emission lines are *not* seen in absorption. According to Bohr's model, most atoms, most of the time, are in their lowest energy state, the $n = 1$ ground state. Thus the absorption spectrum consists of *only* those transitions such as $1 \rightarrow 2$, $1 \rightarrow 3$, ... in which the atom jumps from $n = 1$ to a higher value of n by absorbing a photon. Transitions such as $2 \rightarrow 3$ are *not* observed because there are essentially no atoms in $n = 2$ at any instant of time. On the other hand, atoms that have been excited to the $n = 3$ state by collisions can emit photons corresponding to transitions $3 \rightarrow 1$ *and* $3 \rightarrow 2$. Thus the wavelength corresponding to $\Delta E_{\text{atom}} = E_3 - E_1$ is seen in both emission and absorption, but transitions with $\Delta E_{\text{atom}} = E_3 - E_2$ occur in emission only.
5. **Each element in the periodic table has a unique spectrum.** The energies of the stationary states are just the energies of the orbiting electrons. The atom has no other form of energy. Different elements, with different numbers of electrons, will have different stable orbits and thus different stationary states. States with different energies will emit and absorb photons of different wavelengths.

EXAMPLE 38.9 **The wavelength of an emitted photon**

An atom has stationary states with energies $E_j = 4.0$ eV and $E_k = 6.0$ eV. What is the wavelength of a photon emitted in a quantum jump from state k to state j?

MODEL To conserve energy, the emitted photon must have exactly the energy lost by the atom in the quantum jump.

SOLVE The atom can jump from the higher energy state k to the lower energy state j by emitting a photon. The atom's change in energy is $\Delta E_{\text{atom}} = -2.0$ eV, so the photon energy must be $E_{\text{photon}} = 2.0$ eV.

The photon frequency is

$$f = \frac{E_{\text{photon}}}{h} = \frac{2.0 \text{ eV}}{4.14 \times 10^{-15} \text{ eV s}} = 4.83 \times 10^{14} \text{ Hz}$$

The wavelength of this photon is

$$\lambda = \frac{c}{f} = 621 \text{ nm}$$

ASSESS 621 nm is a visible-light wavelength.

Energy-Level Diagrams

An **energy-level diagram,** such as the one shown in Figure 38.19, is a useful pictorial representation of the stationary-state energies. An energy-level diagram is less a graph than it is a picture. The vertical axis represents energy, but the horizontal axis is not a scale. Think of this as a picture of a ladder in which the energies are the rungs of the ladder. The lowest rung, with energy E_1 is the ground state. Higher rungs are labeled by their quantum numbers, $n = 2, 3, 4, \ldots$

Energy-level diagrams are especially useful for showing transitions, or quantum jumps, in which a photon of light is emitted or absorbed. As examples, Figure 38.19 shows upward transitions in which a photon is absorbed by a ground-state atom ($n = 1$) and downward transitions in which a photon is emitted from an $n = 4$ excited state.

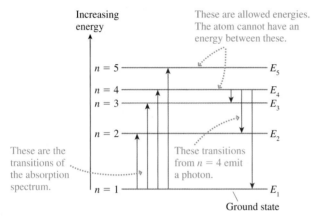

FIGURE 38.19 An energy-level diagram.

EXAMPLE 38.10 Emission and absorption

An atom has stationary states $E_1 = 0.0$ eV, $E_2 = 3.0$ eV, and $E_3 = 5.0$ eV. What wavelengths are observed in the absorption spectrum and in the emission spectrum of this atom?

MODEL Photons are emitted when an atom undergoes a quantum jump from a higher energy level to a lower energy level. Photons are absorbed in a quantum jump from a lower energy level to a higher energy level. But most of the atoms are in the $n = 1$ ground state, so the only quantum jumps seen in the absorption spectrum start from the $n = 1$ state.

VISUALIZE Figure 38.20 shows an energy-level diagram for the atom.

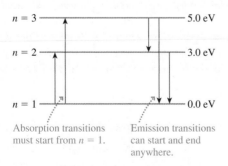

FIGURE 38.20 The atom's energy-level diagram.

SOLVE This atom will absorb photons on the $1 \rightarrow 2$ and $1 \rightarrow 3$ transitions, with $\Delta E_{1 \rightarrow 2} = 3.0$ eV and $\Delta E_{1 \rightarrow 3} = 5.0$ eV. From $f = \Delta E_{atom}/h$ and $\lambda = c/f$, we find that the wavelengths in the absorption spectrum are

$$1 \rightarrow 2 \quad f = 3.0 \text{ eV}/h = 7.25 \times 10^{14} \text{ Hz}$$

$$\lambda = 414 \text{ nm (blue)}$$

$$1 \rightarrow 3 \quad f = 5.0 \text{ eV}/h = 1.21 \times 10^{15} \text{ Hz}$$

$$\lambda = 248 \text{ nm (ultraviolet)}$$

The emission spectrum will also have the 414 nm and 248 nm wavelengths due to the $2 \rightarrow 1$ and $3 \rightarrow 1$ quantum jumps from excited states 2 and 3 to the ground state. In addition, the emission spectrum will contain the $3 \rightarrow 2$ quantum jump with $\Delta E_{3 \rightarrow 2} = -2.0$ eV that is *not* seen in absorption because there are too few atoms in the $n = 2$ state to absorb. We found in Example 38.9 that a 2.0 eV transition corresponds to a wavelength of 621 nm. Thus the emission wavelengths are

$$2 \rightarrow 1 \quad \lambda = 414 \text{ nm (blue)}$$

$$3 \rightarrow 1 \quad \lambda = 248 \text{ nm (ultraviolet)}$$

$$3 \rightarrow 2 \quad \lambda = 621 \text{ nm (orange)}$$

A photon with a wavelength of 414 nm has energy $E_{\text{photon}} =$ 3.0 eV. Do you expect to see a spectral line with $\lambda = 414$ nm in the emission spectrum of the atom represented by this energy-level diagram? If so, what transition or transitions will emit it? Do you expect to see a spectral line with $\lambda = 414$ nm in the absorption spectrum? If so, what transition or transitions will absorb it?

$n = 3$ ————————————— 6.0 eV
$n = 2$ ————————————— 5.0 eV

$n = 1$ ————————————— 2.0 eV

$n = 0$ ————————————— 0.0 eV

38.6 The Bohr Hydrogen Atom

Bohr's hypothesis was a bold new idea, yet there was still one enormous stumbling block: What *are* the stationary states of an atom? Everything in Bohr's model hinges on the existence of these stationary states, of there being only certain electron orbits that are allowed. But nothing in classical physics provides any basis for such orbits. And Bohr's model describes only the *consequences* of having stationary states, not how to find them. If such states really exist, we will have to go beyond classical physics to find them.

To address this problem, Bohr did an explicit analysis of the hydrogen atom. The hydrogen atom, with only a single electron, was known to be the simplest atom. Furthermore, as we discussed in Chapters 24 and 37, Balmer had discovered a fairly simple formula that characterized the wavelengths in the hydrogen emission spectrum. Anyone with a successful model of an atom was going to have to *derive* Balmer's formula for the hydrogen atom.

Bohr's paper followed a rather circuitous line of reasoning. That is not surprising, because he had little to go on at the time. But our goal is a clear explanation of the ideas, not a historical study of Bohr's methods, so we are going to follow a different analysis using de Broglie's matter waves. De Broglie did not propose matter waves until 1924, 11 years after Bohr's paper, but with the clarity of hindsight we can see that treating the electron as a wave provides a more straightforward analysis of the hydrogen atom. Although our route will be different from Bohr's, we will arrive at the same point, and, in addition, we will be in a much better position to understand the work that came after Bohr.

NOTE ▶ Bohr's analysis of the hydrogen atom is sometimes called the *Bohr atom* or the *Bohr model*. It's important not to confuse this analysis, which applies only to hydrogen, with the more general postulates of the *Bohr model of the atom*. Those postulates, which we looked at in the previous section, apply to any atom. To make the distinction clear, we'll call Bohr's analysis of hydrogen the *Bohr hydrogen atom.* ◀

The Stationary States of the Hydrogen Atom

Figure 38.21 shows a Rutherford hydrogen atom, with a single electron orbiting a nucleus that consists of a single proton. We will assume a circular orbit of radius r and speed v. We will also assume, to keep the analysis manageable, that the pro-

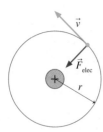

FIGURE 38.21 A Rutherford hydrogen atom. The size of the nucleus is greatly exaggerated.

ton remains stationary while the electron revolves around it. This is a reasonable assumption because the proton is roughly 1800 times as massive as the electron. With these assumptions, the atom's energy is the kinetic energy of the electron plus the potential energy of the electron-proton interaction. This is

$$E = K + U = \frac{1}{2}mv^2 + \frac{1}{4\pi\epsilon_0}\frac{q_{elec}q_{proton}}{r} = \frac{1}{2}mv^2 - \frac{e^2}{4\pi\epsilon_0 r} \quad (38.18)$$

where we used $q_{elec} = -e$ and $q_{proton} = +e$.

NOTE ▶ m is the mass of the electron, *not* the mass of the entire atom. ◀

Now, the electron, as we are coming to understand it, has both particle-like and wave-like properties. First, let us treat the electron as a charged particle. The proton exerts a Coulomb electric force on the electron,

$$\vec{F}_{elec} = \left(\frac{1}{4\pi\epsilon_0}\frac{e^2}{r^2}, \text{ toward center}\right) \quad (38.19)$$

This force gives the electron an acceleration $\vec{a}_{elec} = \vec{F}_{elec}/m$ that also points to the center. This is a centripetal acceleration, causing the particle to move in its circular orbit. The centripetal acceleration of a particle moving in a circle of radius r at speed v *must* be v^2/r, thus

$$a_{elec} = \frac{F_{elec}}{m} = \frac{e^2}{4\pi\epsilon_0 mr^2} = \frac{v^2}{r} \quad (38.20)$$

Rearranging, we find

$$v^2 = \frac{e^2}{4\pi\epsilon_0 mr} \quad (38.21)$$

Equation 38.21 is a *constraint* on the motion. The speed v and radius r must obey Equation 38.21 if the electron is to move in a circular orbit. This constraint is not unique to atoms. We earlier found a similar relationship between v and r for orbiting satellites.

Now let's treat the electron as a de Broglie wave. In Section 38.4 we found that a particle confined to a one-dimensional box sets up a standing wave as it reflects back and forth. A standing wave, you will recall, consists of two traveling waves moving in opposite directions. When the round-trip distance in the box is equal to an integer number of wavelengths ($2L = n\lambda$), the two oppositely traveling waves interfere constructively to set up the standing wave.

Suppose that, instead of traveling back and forth along a line, our wave-like particle travels around the circumference of a circle. The particle will set up a standing wave, just like the particle in the box, if there are waves traveling in both directions and if the round-trip distance is an integer number of wavelengths. This is the idea we want to carry over from the particle in a box. As an example, Figure 38.22 shows a standing wave around a circle with $n = 10$ wavelengths.

The mathematical condition for a circular standing wave is found by replacing the round-trip distance $2L$ in a box with the round-trip distance $2\pi r$ on a circle. Thus a circular standing wave will occur when

$$2\pi r = n\lambda \qquad n = 1, 2, 3, \ldots \quad (38.22)$$

But the de Broglie wavelength for a particle *has* to be $\lambda = h/p = h/mv$. Thus the standing-wave condition for a de Broglie wave is

$$2\pi r = n\frac{h}{mv}$$

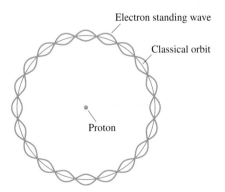

FIGURE 38.22 An $n = 10$ electron standing wave around the orbit's circumference.

This condition is true only if the electron's speed is

$$v_n = \frac{nh}{2\pi mr} \qquad n = 1, 2, 3, \ldots \qquad (38.23)$$

The quantity $h/2\pi$ occurs so often in quantum physics that it is customary to give it a special name. We define the quantity $\hbar$, pronounced "h bar," as

$$\hbar \equiv \frac{h}{2\pi} = 1.055 \times 10^{-34}\,\mathrm{J\,s} = 6.58 \times 10^{-16}\,\mathrm{eV\,s}$$

With this definition, we can write Equation 38.23 as

$$v_n = \frac{n\hbar}{mr} \qquad n = 1, 2, 3, \ldots \qquad (38.24)$$

This, like Equation 38.21, is another relationship between v and r. This is the constraint that arises from treating the electron as a wave.

Now if the electron can act as both a particle *and* a wave, then both the Equation 38.21 *and* Equation 38.24 constraints have to be obeyed. That is, v^2 as given by the Equation 38.21 particle constraint has to equal v^2 of the Equation 38.24 wave constraint. Equating these gives

$$v^2 = \frac{e^2}{4\pi\epsilon_0 mr} = \frac{n^2\hbar^2}{m^2 r^2}$$

We can solve this equation to find that the radius r is

$$r_n = n^2 \frac{4\pi\epsilon_0\hbar^2}{me^2} \qquad n = 1, 2, 3, \ldots \qquad (38.25)$$

where we have added a subscript n to the radius r to indicate that it depends on the integer n.

The right-hand side of Equation 38.25, except for the n^2, is just a collection of constants. Let's group them all together and define the **Bohr radius** a_B as

$$a_B = \text{Bohr radius} \equiv \frac{4\pi\epsilon_0\hbar^2}{me^2} = 5.29 \times 10^{-11}\,\mathrm{m} = 0.0529\,\mathrm{nm}$$

With this definition, Equation 38.25 for the radius of the electron's orbit becomes

$$r_n = n^2 a_B \qquad n = 1, 2, 3, \ldots \qquad (38.26)$$

The first few allowed values of r_n are

$$r_n = \begin{cases} 0.053\,\mathrm{nm} & n = 1 \\ 0.212\,\mathrm{nm} & n = 2 \\ 0.476\,\mathrm{nm} & n = 3 \\ \vdots & \vdots \end{cases}$$

We have discovered stationary states! That is, **a hydrogen atom can exist *only* if the radius of the electron's orbit is one of the values given by Equation 38.26.** Intermediate values of the radius, such as $r = 0.100$ nm, cannot exist because the electron cannot set up a standing wave around the circumference. The possible orbits are *quantized,* with only certain orbits allowed.

The key step leading to Equation 38.26 was the requirement that the electron have wave-like properties in addition to particle-like properties. This requirement leads to quantized orbits, or what Bohr called stationary states. The integer n is thus the *quantum number* that numbers the various stationary states.

Hydrogen Atom Energy Levels

Now we can make progress quickly. Knowing the possible radii, we can return to Equation 38.23 and find the possible electron speeds to be

$$v_n = \frac{n\hbar}{mr_n} = \frac{1}{n}\frac{\hbar}{ma_B} = \frac{v_1}{n} \qquad n = 1, 2, 3, \ldots \qquad (38.27)$$

where $v_1 = \hbar/ma_B = 2.19 \times 10^6$ m/s is the electron's speed in the $n = 1$ orbit. The speed decreases as n increases.

Finally, we can determine the energies of the stationary states. From Equation 38.18 for the energy, with Equations 38.26 and 38.27 for r and v, we have

$$E_n = \frac{1}{2}mv_n^2 - \frac{e^2}{4\pi\epsilon_0 r_n} = \frac{1}{2}m\left(\frac{\hbar^2}{m^2 a_B^2 n^2}\right) - \frac{e^2}{4\pi\epsilon_0 n^2 a_B} \qquad (38.28)$$

As a homework problem, you can show that this rather messy expression simplifies to

$$E_n = -\frac{1}{n^2}\left(\frac{1}{4\pi\epsilon_0}\frac{e^2}{2a_B}\right) \qquad (38.29)$$

Let's define

$$E_1 \equiv \frac{1}{4\pi\epsilon_0}\frac{e^2}{2a_B} = 13.60 \text{ eV}$$

We can then write the energy levels of the stationary states of the hydrogen atom as

$$E_n = -\frac{E_1}{n^2} = -\frac{13.60 \text{ eV}}{n^2} \qquad n = 1, 2, 3, \ldots \qquad (38.30)$$

This has been a lot of math, so we need to see where we are and what we have learned. Table 38.2 shows values of r_n, v_n, and E_n evaluated for quantum number values $n = 1$ to 5. We do indeed seem to have discovered stationary states of the hydrogen atom. Each state, characterized by its quantum number n, has a unique radius, speed, and energy. These are displayed graphically in Figure 38.23, in which the orbits are drawn to scale. Notice how the atom's diameter increases very rapidly as n increases. At the same time, the electron's speed decreases.

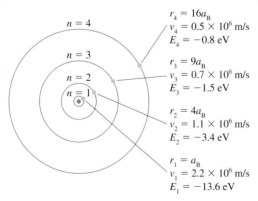

FIGURE 38.23 The first four stationary states, or allowed orbits, of the Bohr hydrogen atom drawn to scale.

TABLE 38.2 Radii, speeds, and energies for the first five states of the Bohr hydrogen atom

n	r_n (nm)	v_n (m/s)	E_n (eV)
1	0.053	2.19×10^6	-13.60
2	0.212	1.09×10^6	-3.40
3	0.476	0.73×10^6	-1.51
4	0.846	0.55×10^6	-0.85
5	1.322	0.44×10^6	-0.54

EXAMPLE 38.11 Stationary states of the hydrogen atom

Can an electron in a hydrogen atom have a speed of 3.60×10^5 m/s? If so, what are its energy and the radius of its orbit? What about a speed of 3.65×10^5 m/s?

SOLVE To be in a stationary state, the electron must have speed

$$v_n = \frac{v_1}{n} = \frac{2.19 \times 10^6 \text{ m/s}}{n}$$

where n is an integer. A speed of 3.60×10^5 m/s would require quantum number

$$n = \frac{2.19 \times 10^6 \text{ m/s}}{3.60 \times 10^5 \text{ m/s}} = 6.08$$

This is not an integer, so the electron can *not* have this speed. But if $v = 3.65 \times 10^5$ m/s, then

$$n = \frac{2.19 \times 10^6 \text{ m/s}}{3.65 \times 10^5 \text{ m/s}} = 6$$

This is the speed of an electron in the $n = 6$ excited state. An electron in this state has energy

$$E_6 = -\frac{13.60 \text{ eV}}{6^2} = -0.38 \text{ eV}$$

and the radius of its orbit is

$$r_6 = 6^2(5.29 \times 10^{-11} \text{ nm}) = 1.90 \times 10^{-9} \text{ m} = 1.90 \text{ nm}$$

Binding Energy and Ionization Energy

It is important to understand why the energies of the stationary states are negative. Because the potential energy of two charged particles is $U = q_1 q_2 / 4\pi\epsilon_0 r$, the zero of potential energy occurs at $r = \infty$ where the particles are infinitely far apart. The state of zero total energy corresponds to having the electron at rest ($K = 0$) and infinitely far from the proton ($U = 0$). This situation, which is the case of two "free particles," occurs in the limit $n \rightarrow \infty$, for which $r_n \rightarrow \infty$ and $v_n \rightarrow 0$.

An electron and a proton bound into an atom have *less* energy than two free particles. We know this because we would have to do work (i.e., add energy) to pull the electron and proton apart. If the bound atom has less energy than two free particles, and if the total energy of two free particles is zero, then it must be the case that the atom has a *negative* amount of energy.

Thus $|E_n|$ is the **binding energy** of the electron in stationary state n. In the ground state, where $E_1 = -13.60$ eV, we would have to add 13.60 eV to the electron to free it from the proton and reach the zero energy state of two free particles. We can say that the electron in the ground state is "bound by 13.60 eV." An electron in an $n = 3$ orbit, where it is farther from the proton and moving more slowly, is bound by only 1.51 eV. That is the amount of energy you would have to supply to remove the electron from an $n = 3$ orbit.

Removing the electron entirely leaves behind a positive ion, H^+ in the case of a hydrogen atom. (The fact that H^+ happens to be a proton does not alter the fact that it is also an atomic ion.) Because nearly all atoms are in their ground state, the binding energy $|E_1|$ of the ground state is called the **ionization energy** of an atom. Bohr's analysis predicts that the ionization energy of hydrogen is 13.60 eV. Figure 38.24 illustrates the ideas of binding energy and ionization energy.

We can test this prediction by shooting a beam of electrons at hydrogen atoms. A projectile electron can knock out an atomic electron if its kinetic energy K is greater than the atom's ionization energy, leaving an ion behind. But a projectile electron will be unable to cause ionization if its kinetic energy is less than the atom's ionization energy. This is a fairly straightforward experiment to carry out, and the evidence shows that the ionization energy of hydrogen is, indeed, 13.60 eV.

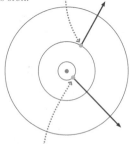

The *binding energy* is the energy needed to remove an electron from its orbit.

The *ionization energy* is the energy needed to create an ion by removing a ground-state electron.

FIGURE 38.24 Binding energy and ionization energy.

Quantization of Angular Momentum

The angular momentum of a particle in circular motion, whether it is a planet or an electron, is

$$L = mvr$$

You will recall that angular momentum is conserved in orbital motion because a force directed toward a central point exerts no torque on the particle. Bohr used conservation of energy explicitly in his analysis of the hydrogen atom, but what role does conservation of angular momentum play?

The condition that a de Broglie wave for the electron set up a standing wave around the circumference was given, in Equation 38.22, as

$$2\pi r = n\lambda = n\frac{h}{mv}$$

We can rewrite this equation as

$$mvr = n\frac{h}{2\pi} = n\hbar \tag{38.31}$$

But mvr is the angular momentum L for a particle in a circular orbit. It appears that the angular momentum of an orbiting electron cannot have just any value. Instead, it must satisfy

$$L = n\hbar \qquad n = 1, 2, 3, \ldots \tag{38.32}$$

Thus angular momentum is quantized! The atom's angular momentum must be an integer multiple of Planck's constant $\hbar$.

The quantization of angular momentum is a direct consequence of this wave-like nature of the electron. We will find that the quantization of angular momentum plays a major role in the behavior of more complex atoms, leading to the idea of electron shells that you likely have studied in chemistry.

STOP TO THINK 38.5 What is the quantum number of this hydrogen atom?

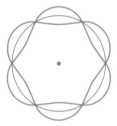

38.7 The Hydrogen Spectrum

Our analysis of the hydrogen atom has revealed stationary states, but how do we know whether the results make any sense? The most important experimental evidence that we have about the hydrogen atom is its spectrum, so the primary test of the Bohr hydrogen atom is whether it correctly predicts the spectrum.

The Hydrogen Energy-Level Diagram

Figure 38.25 is an energy-level diagram for the hydrogen atom. As we noted earlier, the energies are like the rungs of a ladder. The lowest rung is the ground state, with $E_1 = -13.6$ eV. The top rung, with $E = 0$ eV, corresponds to a hydrogen ion in the limit $n \to \infty$. This top rung is called the **ionization limit.** In principle there are an infinite number of rungs, but only the lowest few are shown. The higher values of n are all crowded together just below the ionization limit at $n = \infty$.

The figure shows a $1 \to 4$ transition in which a photon is absorbed and a $4 \to 2$ transition in which a photon is emitted. For two quantum states m and n, where $n > m$ and E_n is the higher-energy state, an atom can *emit* a photon in an $n \to m$ transition or *absorb* a photon in an $m \to n$ transition.

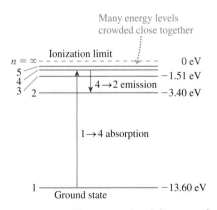

FIGURE 38.25 The energy-level diagram of the hydrogen atom.

The Emission Spectrum

According to the fifth assumption of Bohr's model of atomic quantization, the frequency of the photon emitted in an $n \rightarrow m$ transition is

$$f = \frac{\Delta E_{\text{atom}}}{h} = \frac{E_n - E_m}{h} \tag{38.33}$$

We can use Equation 38.29 for the energies E_n and E_m, to predict that the emitted photon has frequency

$$f = \frac{1}{h} \left\{ \left[-\frac{1}{n^2} \left(\frac{1}{4\pi\epsilon_0} \frac{e^2}{2a_{\text{B}}} \right) \right] - \left[-\frac{1}{m^2} \left(\frac{1}{4\pi\epsilon_0} \frac{e^2}{2a_{\text{B}}} \right) \right] \right\}$$

$$= \frac{1}{4\pi\epsilon_0} \frac{e^2}{2ha_{\text{B}}} \left(\frac{1}{m^2} - \frac{1}{n^2} \right) \tag{38.34}$$

The frequency is a positive number because $m < n$ and thus $1/m^2 > 1/n^2$.

We are more interested in wavelength than frequency, because wavelengths are the quantity measured by experiment. The wavelength of the photon emitted in an $n \rightarrow m$ quantum jump is

$$\lambda_{n \rightarrow m} = \frac{c}{f} = \frac{8\pi\epsilon_0 hc a_{\text{B}}/e^2}{\left(\dfrac{1}{m^2} - \dfrac{1}{n^2} \right)} \tag{38.35}$$

This looks rather gruesome, but notice that the numerator is simply a collection of various constants. The value of the numerator, which we can call λ_0, is

$$\lambda_0 = \frac{8\pi\epsilon_0 hc a_{\text{B}}}{e^2} = 9.112 \times 10^{-8} \text{ m} = 91.12 \text{ nm}$$

With this definition, our prediction for the wavelengths in the hydrogen emission spectrum is

$$\lambda_{n \rightarrow m} = \frac{\lambda_0}{\left(\dfrac{1}{m^2} - \dfrac{1}{n^2} \right)} \quad m = 1, 2, 3, \ldots \quad n = m+1, m+2, \ldots \tag{38.36}$$

This should look familiar. It is the Balmer formula from Chapter 37! However, there is one *slight* difference: Bohr's analysis of the hydrogen atom has predicted $\lambda_0 = 91.12$ nm whereas Balmer found, from experiment, that $\lambda_0 = 91.18$ nm. Could Bohr have come this close but then fail to predict the Balmer formula correctly?

The problem, it turns out, is in our assumption that the proton remains at rest while the electron orbits it. In fact, *both* particles rotate about their common center of mass, rather like a dumbbell with a big end and a small end. The center of mass is very close to the proton, which is far more massive than the electron, but the proton is not entirely motionless. The good news is that a more advanced analysis can account for the proton's motion. It changes the energies of the stationary states ever so slightly—about 1 part in 2000—but that is precisely what is needed to give a revised value:

$$\lambda_0 = 91.18 \text{ nm when corrected for the nuclear motion}$$

It works! Unlike all previous atomic models, **the Bohr hydrogen atom correctly predicts the discrete spectrum of the hydrogen atom.** Figure 38.26 shows the *Balmer series* and the *Lyman series* transitions on an energy-level

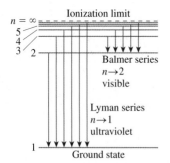

FIGURE 38.26 Transitions producing the Lyman series and the Balmer series of lines in the hydrogen spectrum.

diagram. Only the Balmer series, consisting of transitions ending on the $m = 2$ state, gives visible wavelengths, and this is the series that Balmer initially analyzed. The Lyman series, ending on the $m = 1$ ground state, is in the ultraviolet region of the spectrum and was not measured until later. These series, as well as others in the infrared, are observed in a discharge tube where collisions with electrons excite the atoms upward from the ground state to state n. They then decay downward by emitting photons. Only the Lyman series is observed in the absorption spectrum because, as noted previously, essentially all the atoms in a quiescent gas are in the ground state.

EXAMPLE 38.12 **Hydrogen absorption**
Whenever astronomers look at distant galaxies, they find that the light has been strongly absorbed at the wavelength of the $1 \rightarrow 2$ transition in the Lyman series of hydrogen. This absorption tells us that interstellar space is filled with vast clouds of hydrogen left over from the Big Bang. What is the wavelength of the $1 \rightarrow 2$ absorption in hydrogen?

SOLVE Equation 38.36 predicts the *absorption* spectrum of hydrogen if we let $m = 1$. The absorption seen by astronomers is from the ground state of hydrogen ($m = 1$) to its first excited state ($n = 2$). The wavelength is

$$\lambda_{1 \rightarrow 2} = \frac{91.18 \text{ nm}}{\left(\frac{1}{1^2} - \frac{1}{2^2}\right)} = 121.6 \text{ nm}$$

This wavelength is far into the ultraviolet. Ground-based astronomy cannot observe this region of the spectrum because the wavelengths are strongly absorbed by the atmosphere, but with space-based telescopes, first widely used in the 1970s, astronomers see 121.6 nm absorption in nearly every direction they look.

Hydrogen-Like Ions

An ion with a *single* electron orbiting Z protons in the nucleus is called a **hydrogen-like ion.** Z is the atomic number and describes the number of protons in the nucleus. He^+, with one electron circling a $Z = 2$ nucleus, and Li^{++}, with one electron and a $Z = 3$ nucleus, are hydrogen-like ions. So is U^{+91}, with one lonely electron orbiting a $Z = 92$ uranium nucleus.

Any hydrogen-like ion is simply a variation on the Bohr hydrogen atom. The only difference between a hydrogen-like ion and neutral hydrogen is that the potential energy $-e^2/4\pi\epsilon_0 r$ becomes, instead, $-Ze^2/4\pi\epsilon_0 r$. Hydrogen itself is the $Z = 1$ case. If we repeat the analysis of the previous sections with this one change, we find:

$$r_n = \frac{n^2 a_B}{Z}$$

$$v_n = Z\frac{v_1}{n}$$

$$E_n = -\frac{13.60 Z^2 \text{ eV}}{n^2}$$

$$\lambda_0 = \frac{91.18 \text{ nm}}{Z^2}$$

(38.37)

As the nuclear charge increases, the electron moves in to a smaller-diameter, higher-speed orbit. Its ionization energy $|E_1|$ increases significantly, and its spectrum shifts to shorter wavelengths. Table 38.3 compares the ground-state atomic diameter $2r_1$, the ionization energy $|E_1|$, and the first wavelength $3 \rightarrow 2$ in the Balmer series for hydrogen and the first two hydrogen-like ions.

TABLE 38.3 Comparison of hydrogen-like ions with $Z = 1, 2,$ and 3

Ion	Diameter $2r_1$	Ionization energy $|E_1|$	Wavelength of $3 \rightarrow 2$
H ($Z = 1$)	0.106 nm	13.6 eV	656 nm
He$^+$ ($Z = 2$)	0.053 nm	54.4 eV	164 nm
Li^{++} ($Z = 3$)	0.035 nm	125.1 eV	73 nm

Success and Failure

Bohr's analysis of the hydrogen atom seemed to be a resounding success. By introducing stationary states, together with Einstein's ideas about light quanta, Bohr was able to provide the first solid understanding of discrete spectra and, in particular, to predict the Balmer formula for the wavelengths in the hydrogen spectrum. And the Bohr hydrogen atom, unlike Rutherford's model, was stable. There was clearly some validity to the idea of stationary states.

But Bohr was completely unsuccessful at explaining the spectra of any other atom. His method did not work even for helium, the second element in the periodic table with a mere two electrons. Something inherent in Bohr's assumptions seemed to work correctly for a single electron but not in situations with two or more electrons.

It is important to make a distinction between the Bohr model of atomic quantization, described in Section 38.5, and the Bohr hydrogen atom. The Bohr model assumes that stationary states exist, but it does not say how to find them. We found the stationary states of a hydrogen atom by requiring that an integer number of de Broglie waves fit around the circumference of the orbit, setting up standing waves. The difficulty with more complex atoms is not the Bohr model but the method of finding the stationary states. Bohr's model of the atomic quantization remains valid, and we will continue to use it, but the procedure of fitting standing waves to a circle is just too simple to find the stationary states of complex atoms. We need to find a better procedure.

Einstein, de Broglie, and Bohr carried physics into uncharted waters. Their successes made it clear that the microscopic realm of light and atoms is governed by quantization, discreteness, and a blurring of the distinction between particles and waves. Although Bohr was clearly on the right track, his inability to extend the Bohr hydrogen atom to more complex atoms made it equally clear that the complete and correct theory remained to be discovered. Bohr's theory was what we now call "semiclassical," a hybrid of classical Newtonian mechanics with the new ideas of quanta. Still missing was a complete theory of motion and dynamics in a quantized universe—a *quantum* mechanics.

SUMMARY

The goal of Chapter 38 has been to understand the quantization of energy for light and matter.

GENERAL PRINCIPLES

Light has particle-like properties

- The energy of a light wave comes in discrete packets called light quanta or photons.
- For light of frequency f, the energy of each photon is $E = hf$, where h is **Planck's constant.**
- For a light wave that delivers power P, photons arrive at rate R such that $P = Rhf$.
- Photons are "particle-like" but are not classical particles.

Matter has wave-like properties

- The **de Broglie wavelength** of a "particle" of mass m is $\lambda = h/mv$.
- The wave-like nature of matter is seen in the interference patterns of electrons, neutrons, and entire atoms.
- When a particle is confined, it sets up a de Broglie standing wave. The fact that standing waves have only certain allowed wavelengths leads to the conclusion that a confined particle has only certain allowed energies. That is, energy is quantized.

IMPORTANT CONCEPTS

Einstein's Model of Light

- Light consists of quanta of energy $E = hf$.
- Quanta are emitted and absorbed on an all-or-nothing basis.
- When a light quantum is absorbed, it delivers all its energy to *one* electron.

Bohr's Model of the Atom

- An atom can exist in only certain stationary states. The allowed energies are quantized. State n has energy E_n.
- An atom can jump from one stationary state to another by emitting or absorbing a photon with $E_{photon} = hf = \Delta E_{atom}$.
- Atoms can be excited in inelastic collisions.
- Atoms seek the $n = 1$ **ground state.** Most atoms, most of the time, are in the ground state.

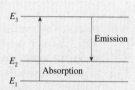

APPLICATIONS

Photoelectric effect

Light can eject electrons from a metal only if $f \geq f_0 = E_0/h$, where E_0 is the metal's **work function.**

The **stopping potential** that stops even the fastest electrons is

$$V_{stop} = \frac{h}{e}(f - f_0)$$

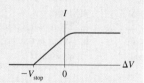

Particle in a box

A particle confined to a one-dimensional box of length L sets up de Broglie standing waves. The allowed energies are

$$E_n = \tfrac{1}{2}mv_n^2 = n^2\frac{h^2}{8mL^2} \qquad n = 1, 2, 3, \ldots$$

The Bohr hydrogen atom

The stationary states are found by requiring an integer number of de Broglie wavelengths to fit around the circumference of the electron's orbit: $2\pi r = n\lambda$.

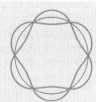

This leads to energy quantization with

$$r_n = n^2 a_B \qquad v_n = \frac{v_1}{n} \qquad E_n = -\frac{13.60\,\text{eV}}{n^2}$$

where $a_B = 0.0529$ nm is the **Bohr radius.** These energies successfully predict the Balmer formula for the hydrogen spectrum. Angular momentum is also quantized, with $L = n\hbar$.

TERMS AND NOTATION

photoelectric effect	de Broglie wavelength	quantum jump
threshold frequency, f_0	quantized	collisional excitation
stopping potential, V_{stop}	energy level	energy-level diagram
thermal emission	quantum number, n	Bohr radius, a_B
work function, E_0	fundamental quantum of energy, E_1	binding energy
Planck's constant, h or $\hbar$	Bohr model of the atom	ionization energy, $|E_1|$
light quantum	stationary state	ionization limit
photon	ground state	hydrogen-like ion
wave packet	excited state	
matter wave	transition	

EXERCISES AND PROBLEMS

Exercises

Section 38.1 The Photoelectric Effect

1. a. Explain why the graphs of Figure 38.4 are horizontal for $\Delta V > 0$.
   b. Explain why photoelectrons are ejected from the cathode with a range of kinetic energies, rather than all electrons having the same kinetic energy.
   c. Explain the reasoning by which we claim that the stopping potential V_{stop} measures the maximum kinetic energy of the electrons.

2. How would the graph of Figure 38.3 look *if* classical physics provided the correct description of the photoelectric effect? Draw the graph and explain your reasoning. Assume that the light intensity remains constant as its frequency and wavelength are varied.

3. How would the graphs of Figure 38.4 look *if* classical physics provided the correct description of the photoelectric effect? Draw the graph and explain your reasoning. Include curves for both weak light and intense light.

4. Figure Ex38.4 is the current-versus-potential-difference graph for a photoelectric-effect experiment with an unknown metal. *If* classical physics provided the correct description of the photoelectric effect, how would the graph look if:
   a. The light was replaced by an equally intense light with a shorter wavelength? Draw it.
   b. The metal was replaced by a different metal with a smaller work function? Draw it.

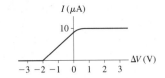

FIGURE EX38.4

5. How many photoelectrons are ejected per second in the experiment represented by the graph of Figure Ex38.4?

Section 38.2 Einstein's Explanation

6. Which metals in Table 38.1 exhibit the photoelectric effect for (a) light with $\lambda = 400$ nm and (b) light with $\lambda = 250$ nm?

7. Photoelectrons are observed when a metal is illuminated by light with a wavelength less than 388 nm. What is the metal's work function?

8. Electrons in a photoelectric-effect experiment emerge from a copper surface with a maximum kinetic energy of 1.10 eV. What is the wavelength of the light?

9. You need to design a photodetector that can respond to the entire range of visible light. What is the maximum possible work function of the cathode?

10. Use Millikan's photoelectric-effect data in Figure 38.11 to determine:
    a. The work function, in eV, of cesium.
    b. An experimental value of Planck's constant.

11. A photoelectric-effect experiment finds a stopping potential of 2.0 V when light of 200 nm is used to illuminate the cathode.
    a. From what metal is the cathode made?
    b. What is the stopping potential if the intensity of the light is doubled?

Section 38.3 Photons

12. a. Determine the energy, in eV, of a photon with a 700 nm wavelength.
    b. Determine the wavelength of a 5.0 keV x-ray photon.

13. What is the wavelength, in nm, of a photon with energy (a) 0.30 eV, (b) 3.0 eV, and (c) 30 eV? For each, is this wavelength visible, ultraviolet, or infrared light?

14. What is the energy, in eV, of (a) a 100 MHz radio-frequency photon, (b) a visible-light photon with a wavelength of 500 nm, and (c) an x-ray photon with a wavelength of 0.10 nm?

15. For what wavelength of light does a 100 mW laser deliver 2.5×10^{17} photons per s?

Section 38.4 Matter Waves and Energy Quantization

16. At what speed is an electron's de Broglie wavelength (a) 1.0 pm, (b) 1.0 nm, (c) 1.0 μm, and (d) 1.0 mm?

17. Through what potential difference must an electron be accelerated from rest to have a wavelength of 500 nm?

18. The diameter of the nucleus is about 10 fm. What is the kinetic energy, in MeV, of a proton with a de Broglie wavelength of 10 fm?

19. What is the length of a one-dimensional box in which an electron in the $n = 1$ state has the same energy as a photon with a wavelength of 600 nm?

20. The diameter of the nucleus is about 10 fm. A simple model of the nucleus is that protons and neutrons are confined within a one-dimensional box of length 10 fm. What are the first three energy levels, in MeV, for a proton in such a box?

21. An electron confined in a one-dimensional box is observed, at different times, to have energies of 12 eV, 27 eV, and 48 eV. What is the length of the box?

Section 38.5 Bohr's Model of Atomic Quantization

22. Figure Ex38.22 is an energy-level diagram for a simple atom. What wavelengths appear in the atom's (a) emission spectrum and (b) absorption spectrum?

$$n = 3 \rule{2cm}{0.4pt} E_3 = 4.0 \text{ eV}$$

$$n = 2 \rule{2cm}{0.4pt} E_2 = 1.5 \text{ eV}$$

FIGURE EX38.22 $n = 1 \rule{2cm}{0.4pt} E_1 = 0.0 \text{ eV}$

23. An electron with 2.0 eV of kinetic energy collides with the atom shown in Figure Ex38.22.
    a. Is the electron able to kick the atom to an excited state? Why or why not?
    b. If your answer to part a was yes, what is the electron's kinetic energy after the collision?

24. The allowed energies of a simple atom are 0.0 eV, 4.0 eV, and 6.0 eV.
    a. Draw the atom's energy-level diagram. Label each level with the energy and the quantum number.
    b. What wavelengths appear in the atom's emission spectrum?
    c. What wavelengths appear in the atom's absorption spectrum?

25. The allowed energies of a simple atom are 0.0 eV, 4.0 eV, and 6.0 eV. An electron traveling with a speed of 1.3×10^6 m/s collides with the atom. Can the electron excite the atom to the $n = 2$ stationary state? The $n = 3$ stationary state? Explain.

Section 38.6 The Bohr Hydrogen Atom

26. Show, by actual calculation, that the Bohr radius is 0.0529 nm and that the ground-state energy of hydrogen is -13.60 eV.

27. a. What quantum number of the hydrogen atom comes closest to giving a 500-nm-diameter electron orbit?
    b. What are the electron's speed and energy in this state?

28. a. Calculate the de Broglie wavelength of the electron in the $n = 1, 2,$ and 3 states of the hydrogen atom. Use the information in Table 38.2.

    b. Show numerically that the circumference of the orbit for each of these stationary states is exactly equal to n de Broglie wavelengths.
    c. Sketch the de Broglie standing wave for the $n = 3$ orbit.

29. How much energy does it take to ionize a hydrogen atom that is in its first excited state?

30. Show, by calculation, that the first three states of the hydrogen atom have angular momenta $\hbar$, $2\hbar$, and $3\hbar$, respectively.

31. Show that Planck's constant $\hbar$ has units of angular momentum.

Section 38.7 The Hydrogen Spectrum

32. Determine the wavelengths of all the possible photons that can be emitted from the $n = 4$ state of a hydrogen atom.

33. What is the wavelength of the series limit (i.e., the shortest possible wavelength) of the Lyman series in hydrogen?

34. Is a spectral line with wavelength 656.5 nm seen in the absorption spectrum of hydrogen atoms? Why or why not?

35. a. Find the radius of the electron's orbit, the electron's speed, and the energy of the atom for the first three stationary states of He$^+$.
    b. Show, by calculation, that the angular momentum in each state is equal to $n\hbar$.

Problems

36. An AM radio station broadcasts with a power of 10 kW at a frequency of 1000 kHz.
    a. How many photons does the antenna emit each second?
    b. Should the broadcast be treated as an electromagnetic wave or discrete photons? Explain.

37. A red laser with a wavelength of 650 nm and a blue laser with a wavelength of 450 nm emit laser beams with the same light power. How do their rates of photon emission compare? Answer this by computing R_{red}/R_{blue}.

38. A 100 W lightbulb emits about 5 W of visible light. (The other 95 W are emitted as infrared radiation or lost as heat to the surroundings.) The average wavelength of the visible light is about 600 nm, so make the simplifying assumption that all the light has this wavelength.
    a. What is the frequency of the emitted light?
    b. How many visible-light photons does the bulb emit per second?
    c. Should your answers to parts a and b be the same? Explain.
    d. How much *mass* does the lightbulb lose per second?

39. A ruby laser emits an intense pulse of light that lasts a mere 10 ns. The light has a wavelength of 690 nm, and each pulse has an energy of 500 mJ.
    a. How many photons are emitted in each pulse?
    b. What is the *rate* of photon emission, in photons per second, during the 10 ns that the laser is "on"?

40. In practice it is easier to measure the wavelength of light than to measure its frequency. Draw a graph of the stopping potential of a metal as a function of the wavelength of the incident light. Consider the full range of wavelengths from zero to infinity. Be sure to identify the threshold wavelength λ_0.

41. Potassium and gold cathodes are used in a photoelectric-effect experiment. For each cathode, find:
    a. The threshold frequency.
    b. The threshold wavelength.
    c. The maximum photoelectron ejection speed if the light has a wavelength of 220 nm.
    d. The stopping potential if the wavelength is 220 nm.

42. The maximum kinetic energy of photoelectrons is 2.8 eV. When the wavelength of the light is increased by 50%, the maximum energy decreases to 1.1 eV. What are (a) the work function of the cathode and (b) the initial wavelength?

43. In a photoelectric-effect experiment, the stopping potential at a wavelength of 400 nm is 25.7% of the stopping potential at a wavelength of 300 nm. Of what metal is the cathode made?

44. The graph in Figure P38.44 was measured in a photoelectric-effect experiment.
    a. What is the work function (in eV) of the cathode?
    b. What experimental value of Planck's constant is obtained from these data?

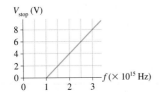

FIGURE P38.44

45. Figure P38.45 shows the stopping potential versus the light frequency for a metal cathode used in a photoelectric-effect experiment. Suppose this cathode is now illuminated with 10 μW of 300 nm light and that the efficiency of converting photons to photoelectrons is 10%. Draw a graph showing current I versus potential difference ΔV for potential difference values from -3 V to $+3$ V. Include a numerical scale on both axes.

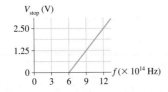

FIGURE P38.45

46. In a photoelectric-effect experiment, the stopping potential was measured for several different wavelengths of incident light. The data are shown below. Analyze these data to determine:
    a. The metal used for the cathode.
    b. An experimental value for Planck's constant. Your value should be found using *all* the data.

λ (nm)	V_{stop} (volts)
500	0.19
450	0.48
400	0.83
350	1.28
300	1.89
250	2.74

47. The relationship between momentum and energy from Einstein's theory of relativity is $E^2 - (pc)^2 = E_0^2$, where, in this context, $E_0 = mc^2$ is the rest energy rather than the work function.
    a. A photon is a massless particle. What is a photon's momentum p in terms of its energy E?
    b. Einstein also claimed that the energy of a photon is related to its frequency by $E = hf$. Use this and your result from part a to write an expression for the wavelength of a photon in terms of its momentum p.
    c. Your result for part b is for a "particle-like wave." Suppose you thought this expression should also apply to a "wave-like particle." What is your expression for λ if you replace p with the classical-mechanics expression for the momentum of a particle of mass m? Is this a familiar-looking expression?

48. The electron interference pattern of Figure 38.14 was made by shooting electrons with 50 keV of kinetic energy through two slits spaced 1.0 μm apart. The fringes were recorded on a detector 1.0 m behind the slits.
    a. What was the speed of the electrons? (The speed is large enough to justify using relativity, but for simplicity do this as a nonrelativistic calculation.)
    b. Figure 38.14 is greatly magnified. What was the actual spacing on the detector between adjacent bright fringes?

49. The neutron interference pattern of Figure P38.49 was made by shooting neutrons with a speed of 200 m/s through two slits spaced 0.10 mm apart.
    a. What was the energy, in eV, of the neutrons?
    b. What was the de Broglie wavelength of the neutrons?
    c. The pattern was recorded by using a neutron detector to measure the neutron intensity at different positions. Notice the 100 μm scale on the figure. By making appropriate measurements directly *on the figure*, determine how far the detector was behind the slits.

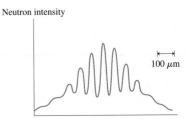

FIGURE P38.49

50. The electron beam in a cathode-ray tube is accelerated through a potential difference of 250 V. The electrons then pass through a small circular hole and are viewed on the screen. You observe that the central bright spot on the screen is the base of a cone with its apex at the hole. The outer edge of the cone makes a 0.50° angle with the original direction of the electron beam. What is the diameter of the hole?

51. An electron confined in a one-dimensional box emits a 200 nm photon in a quantum jump from $n = 2$ to $n = 1$. What is the length of the box?

52. A proton confined in a one-dimensional box emits a 2.0 MeV gamma-ray photon in a quantum jump from $n = 2$ to $n = 1$. What is the length of the box?

53. Imagine that the horizontal box of Figure 38.16 is instead oriented vertically. Also imagine the box to be on a neutron star where the gravitational field is so strong that the particle in the

box slows significantly, nearly stopping, before it hits the top of the box. Make a *qualitative* sketch of the $n = 3$ de Broglie standing wave of a particle in this box.

Hint: The nodes are *not* uniformly spaced.

54. The absorption spectrum of an atom consists of the wavelengths 200 nm, 300 nm, and 500 nm.
    a. Draw the atom's energy-level diagram.
    b. What wavelengths are seen in the atom's emission spectrum?

55. The first three energy levels of the fictitious element X are shown in Figure P38.55.
    a. What is the ionization energy of element X?
    b. What wavelengths are observed in the absorption spectrum of element X? Express your answers in nm.
    c. State whether each of your wavelengths in part b corresponds to ultraviolet, visible, or infrared light.
    d. An electron with a speed of 1.4×10^6 m/s collides with an atom of element X. Shortly afterward, the atom emits a 1240 nm photon. What was the electron's speed after the collision? Assume that, because the atom is so much more massive than the electron, the recoil of the atom is negligible.

    Hint: The energy of the photon is *not* the energy transferred to the atom in the collision.

FIGURE P38.55

56. Starting from Equation 38.28, derive Equation 38.29.
57. What is the energy of a hydrogen atom with a 5.18 nm diameter?
58. Calculate *all* the wavelengths of *visible* light in the emission spectrum of the hydrogen atom.

    Hint: There are infinitely many wavelengths in the spectrum, so you'll need to develop a strategy for this problem rather than using trial and error.

59. A hydrogen atom in the ground state absorbs a 12.75 eV photon. Immediately after the absorption, the atom undergoes a quantum jump to the next-lowest energy level. What is the wavelength of the photon emitted in this quantum jump?

60. a. What wavelength photon does a hydrogen atom emit in a $200 \rightarrow 199$ transition?
    b. What is the *difference* in the wavelengths absorbed in a $2 \rightarrow 199$ transition and a $2 \rightarrow 200$ transition?

61. a. Calculate the orbital radius and the speed of an electron in both the $n = 99$ and the $n = 100$ state of hydrogen.
    b. Determine the orbital frequency of the electron in each of these states.
    c. Calculate the frequency of a photon emitted in a $100 \rightarrow 99$ transition.
    d. Compare the photon frequency of part c to the *average* of your two orbital frequencies from part b. By what percent do they differ?

62. Draw an energy-level diagram, similar to Figure 38.25, for the He^+ ion. On your diagram:
    a. Show the first five energy levels. Label each with the values of n and E_n.
    b. Show the ionization limit.
    c. Show all possible emission transitions from the $n = 4$ energy level.
    d. Calculate the wavelengths (in nm) for each of the transitions in part c and show them alongside the appropriate arrow.

63. What are the wavelengths of the transitions $3 \rightarrow 2$, $4 \rightarrow 2$, and $5 \rightarrow 2$ in the hydrogen-like ion O^{+7}? In what spectral range do these lie?

64. Two hydrogen atoms collide head on. The collision brings both atoms to a halt. Immediately after the collision, both atoms emit a 121.6 nm photon. What was the speed of each atom just before the collision?

65. Ultraviolet light with a wavelength of 70 nm shines on a gas of hydrogen atoms in their ground states. Some of the atoms are ionized by the light. What is the kinetic energy of the electrons that are freed in this process?

66. A beam of electrons is incident upon a gas of hydrogen atoms.
    a. What minimum speed must the electrons have to cause the emission of 656 nm light from the $3 \rightarrow 2$ transition of hydrogen?
    b. Through what potential difference must the electrons be accelerated to have this speed?

Challenge Problems

67. The photomultiplier tube (PMT) of Figure 38.13 consists of a cathode, which the photon strikes; an anode, where the electrons are collected; and a number of intermediate electrodes called *dynodes*. The tube shown in the figure has nine dynodes, but consider a PMT with N dynodes. The cathode, when struck by a photon, ejects a single photoelectron. That electron is accelerated to the first dynode, where it causes (on average) the ejection of ϵ secondary electrons. The quantity ϵ is called the *secondary emission coefficient*. Each of these electrons ejects, on average, ϵ electrons from the second dynode, each of which in turn ejects ϵ electrons from the third dynode, and so on until a large pulse of electrons is collected by the anode.
    a. Write an expression, in terms of ϵ and N, for the average number of electrons arriving at the anode due to a single photon striking the cathode. This is called the *gain* of the PMT.
    b. The graph in Figure 38.13b shows the voltage pulse generated when the electron current flowed through a 50 Ω resistor. The baseline of the pulse is zero volts, and the voltage scale is 20 mV per division. What is the maximum *current* of this pulse?
    c. Because $I = dQ/dt$, the amount of charge delivered by a pulse of current is $Q = \int I\,dt$. This can be interpreted geometrically as the area under the I-versus-t curve. The area of a pulse is reasonably well approximated as its height multiplied by its width measured at half of its maximum height. Estimate the number of electrons in the current pulse shown in Figure 38.13b.
    d. The PMT that produced this pulse had 14 dynodes. By comparing your answers to parts a and c, determine the secondary emission coefficient for this PMT.

68. In the atom interferometer experiment of Figure 38.15, laser-cooling techniques were used to cool a dilute vapor of sodium atoms to a temperature of 0.001 K = 1 mK. The ultracold atoms passed through a series of collimating apertures to form the *atomic beam* you see entering the figure from the left. The standing light waves were created from a laser beam with a wavelength of 590 nm.
    a. What is the rms speed v_{rms} of a sodium atom ($A = 23$) in a gas at temperature 1 mK?
    b. By treating the laser beam as if it were a diffraction grating, calculate the first-order diffraction angle of a sodium atom traveling with the rms speed of part a.
    c. How far apart are points B and C if the second standing wave is 10 cm from the first?
    d. Because interference is observed between the two paths, each individual atom is apparently present at both point B *and* point C. Describe, in your own words, what this experiment tells you about the nature of matter.

69. Consider a hydrogen atom in stationary state n.
    a. Show that the orbital period of the electron is $T = n^3 T_1$, and find a numerical value for T_1.
    b. On average, an atom stays in the $n = 2$ state for 1.6 ns before undergoing a quantum jump to the $n = 1$ state. On average, how many revolutions does the electron make before the quantum jump?

70. Consider an electron undergoing cyclotron motion in a magnetic field. According to Bohr, the electron's angular momentum must be quantized in units of $\hbar$.
    a. Show that allowed radii for the electron's orbit are given by $r_n = (n\hbar/eB)^{1/2}$, where $n = 1, 2, 3, \ldots$
    b. Compute the first four allowed radii in a 1.0 T magnetic field.
    c. Find an expression for the allowed energy levels E_n in terms of $\hbar$ and the cyclotron frequency f_{cyc}.

71. The *muon* is a subatomic particle with the same charge as an electron but with a mass that is 207 times greater: $m_\mu = 207 m_e$. Physicists think of muons as "heavy electrons." However, the muon is not a stable particle; it decays with a half-life of 1.5 μs into an electron plus two neutrinos. Muons from cosmic rays are sometimes "captured" by the nuclei of the atoms in a solid. A captured muon orbits this nucleus, like an electron, until it decays. Because the muon is often captured into an excited orbit ($n > 1$), its presence can be detected by observing the photons emitted in transitions such as $2 \rightarrow 1$ and $3 \rightarrow 1$.

    Consider a muon captured by a carbon nucleus ($Z = 6$). Because of its large mass, the muon orbits well *inside* the electron cloud and is not affected by the electrons. Thus the muon "sees" the full nuclear charge Ze and acts like the electron in a hydrogen-like ion.
    a. What are the orbital radius and speed of a muon in the $n = 1$ ground state? Note that the mass of a muon differs from the mass of an electron.
    b. What is the wavelength of the $2 \rightarrow 1$ muon transition?
    c. Is the photon emitted in the $2 \rightarrow 1$ transition infrared, visible, ultraviolet, or x ray?
    d. How many orbits will the muon complete during 1.5 μs? Is this a sufficiently large number that the Bohr model "makes sense," even though the muon is not stable?

STOP TO THINK ANSWERS

Stop to Think 38.1: $V_A > V_B > V_C$. For a given wavelength of light, electrons are ejected faster from metals with smaller work functions because it takes less energy to remove an electron. Faster electrons need a larger negative voltage to stop them.

Stop to Think 38.2: d. Photons always travel at c, and a photon's energy depends only on the light's frequency, not its intensity.

Stop to Think 38.3: $n = 4$. There are four antinodes.

Stop to Think 38.4: Not in absorption. In emission from the $n = 2$ to $n = 1$ transition. The photon energy has to match the energy *difference* between two energy levels. Absorption is from the ground state, at $E_1 = 0$ eV. There's no energy level at 3 eV to which the atom could jump.

Stop to Think 38.5: $n = 3$. Each antinode is half a wavelength, so this standing wave has three full wavelengths in one circumference.

39 Wave Functions and Uncertainty

The surface of graphite, as imaged with atomic resolution by a scanning tunneling microscope. The hexagonal ridges show the most probable locations of the electrons.

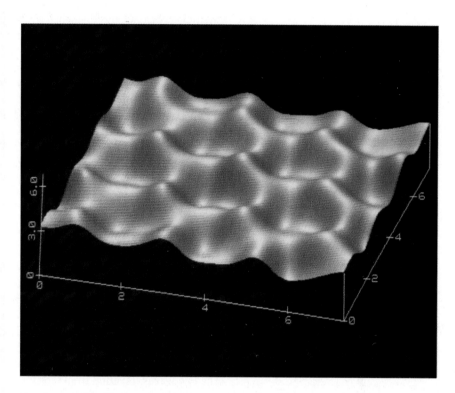

▶ **Looking Ahead**

The goal of Chapter 39 is to introduce the wave-function description of matter and learn how it is interpreted. In this chapter you will learn to:

- Connect the particle and wave descriptions of matter.
- Use basic ideas about probability.
- Use the wave function to calculate the probabilities of detecting particles.
- Recognize the limitations on knowledge imposed by the Heisenberg uncertainty principle.

◀ **Looking Back**

The ideas developed in this chapter are highly dependent on understanding the double-slit interference experiment for both light and matter. Please review:

- Sections 21.8 and 22.2 Interference, beats, and the double-slit experiment.
- Sections 24.3–24.4 Photons and matter waves.
- Section 38.4 The de Broglie wavelength and wave-particle duality.

You learned in the last two chapters that classical mechanics and electromagnetism were unable to explain the new phenomena associated with light, electrons, and atoms. Scientific theories that had triumphed during the 18th and 19th centuries stumbled over the smallest specks of matter. Many scientists refused to accept these limitations, thinking that it was only a question of time until someone discovered how to apply classical ideas to atoms. Their hopes were to go unfulfilled.

At the same time that classical physics was reaching its limits, the new ideas put forward by Einstein, Bohr, and de Broglie began pointing the way toward a new theory of light and matter. **Quantum mechanics,** as the theory came to be called, did not reach its completed form until the mid-1920s, but it has since proven to be the most successful physical theory ever devised.

This chapter and the next will introduce the essential ideas of quantum mechanics in one dimension. Although the full theory is beyond the scope of this textbook, we can delve far enough into quantum mechanics to learn how it solves the problems of atomic and nuclear structure. Our goal in this chapter is to introduce the concept of the *wave function*. The wave function, which reconciles the wave-like and particle-like aspects of matter, characterizes microscopic particles in terms of the *probability* of finding them at various points in space. This scanning tunneling microscope image of graphite shows that the electrons are most likely found along the ring-like structure created by the carbon-carbon bonds.

Interference fringes in an optical double-slit interference experiment.

39.1 Waves, Particles, and the Double-Slit Experiment

You may feel surprise at how slowly we have been building up to quantum mechanics. Why not just write it down and start using it? There are two reasons. First, quantum mechanics explains microscopic phenomena that we cannot directly sense or experience. It was important to begin by learning how light and atoms behave. Otherwise, how would you know if quantum mechanics explains anything? Second, the concepts we'll need in quantum mechanics are rather abstract. Before launching into the mathematics, we need to establish a connection between theory and experiment.

We will make the connection by returning to the double-slit interference experiment, an experiment that goes right to the heart of wave-particle duality. The significance of the double-slit experiment arises from the fact that both light and matter exhibit the same interference pattern. Regardless of whether photons, electrons, or neutrons pass through the slits, their arrival at a detector is a particle-like event. That is, they make a collection of discrete dots on a detector. Yet our understanding of how interference "works" is based on the properties of waves. Our goal is to find the connection between the wave description and the particle description of interference.

A Wave Analysis of Interference

The interference of light can be analyzed from either a wave perspective or a photon perspective. Let's start with a wave analysis. Figure 39.1 shows light waves passing through a double slit with slit separation d. You should recall that the lines in a wave-front diagram represent wave crests, spaced one wavelength apart. The bright fringes of constructive interference occur where two crests or two troughs overlap. The graphs and pictures below the detection screen (notice that they're aligned vertically) show the outcome of the experiment.

You studied interference and the double-slit experiment in Chapters 21 and 22. The two waves traveling from the slits to the viewing screen are traveling waves with displacements

$$D_1 = a\sin(kr_1 - \omega t)$$
$$D_2 = a\sin(kr_2 - \omega t)$$

where a is the amplitude of each wave, $k = 2\pi/\lambda$ is the wave number, and r_1 and r_2 are the distances from the two slits. The "displacement" of a light wave is not a physical displacement, as in a water wave, but a change in the electromagnetic field.

According to the principle of superposition, these two waves add together where they meet at a point on the screen to give a wave with net displacement $D = D_1 + D_2$. Previously (see Equations 21.24 and 22.12) we found that the amplitude of their superposition is

$$A(x) = 2a\cos\left(\frac{\pi d x}{\lambda L}\right) \tag{39.1}$$

where x is the horizontal coordinate on the screen, measured from $x = 0$ in the center.

The function $A(x)$, the top graph in Figure 39.1, is called the *amplitude function*. It describes the amplitude A of the light wave as a function of the position x on the viewing screen. The amplitude function has maxima where two crests from individual waves overlap and add constructively to make a larger wave with amplitude $2a$. $A(x)$ is zero at points where the two individual waves are out of phase and interfere destructively.

Approaching wave fronts

Double slit

Screen

$A(x)$

Wave amplitude along the screen

I

Interference fringes

Photon arrival positions

FIGURE 39.1 The double-slit experiment with light.

If you carry out a double-slit experiment in the lab, what you observe on the screen is the light's *intensity,* not its amplitude. A wave's intensity I is proportional to the *square* of the amplitude. That is, $I \propto A^2$, where the symbol $\propto$ means "is proportional to." Using Equation 39.1 for the amplitude at each point, the intensity $I(x)$ as a function of position x on the screen is

$$I(x) = C\cos^2\!\left(\frac{\pi d x}{\lambda L}\right) \tag{39.2}$$

where C is a proportionality constant.

The lower graph in Figure 39.1 shows the intensity as a function of position along the screen. This graph shows the alternating bright and dark interference fringes that you see in the laboratory. In other words, the intensity of the wave is the *experimental reality* that you observe and measure.

Probability

Before discussing photons, we need to introduce some ideas about probability. Imagine throwing darts at a dart board while blindfolded. Figure 39.2 shows how the board might look after your first 100 throws. From this information, can you predict where your 101st throw is going to land?

No. The position of any individual dart is *unpredictable.* No matter how hard you try to reproduce the previous throw, a second dart will not land at the same place as the first. Yet there is clearly an overall *pattern* to the where the darts strike the board. Even blindfolded, you had a general sense of where the center of the board was, so each dart was *more likely* to land near the center than at the edge.

Although we can't predict where any individual dart will land, we can use the information in Figure 39.2 to determine the *probability* that your next throw will land in region A or region B or region C. Because 45 out of 100 throws landed in region A, we could say that the *odds* of hitting region A are 45 out of 100, or 45%.

Now, 100 throws isn't all that many. If you throw another 100 darts, perhaps only 43 will land in region A. Then maybe 48 of the next 100 throws. Imagine that the total number of throws N_{tot} becomes extremely large. Then the **probability** that any particular throw lands in region A is defined to be

$$P_A = \lim_{N_{tot} \to \infty} \frac{N_A}{N_{tot}} \tag{39.3}$$

In other words, the probability that the outcome will be A is the fraction of outcomes that are A in an enormously large number of trials. Similarly, $P_B = N_B/N_{tot}$ and $P_C = N_C/N_{tot}$ as $N_{tot} \to \infty$. We can give probabilities either as a decimal fraction or a percentage. In this example, $P_A \approx 45\%$, $P_B \approx 35\%$, and $P_C \approx 20\%$. We've used $\approx$ rather than $=$ because 100 throws isn't enough to determine the probabilities with great precision.

What is the probability that a dart lands in either region A *or* region B? The number of darts landing in either A *or* B is $N_{A \text{ or } B} = N_A + N_B$, so we can use the definition of probability to learn that

$$
\begin{aligned}
P_{A \text{ or } B} &= \lim_{N_{tot} \to \infty} \frac{N_{A \text{ or } B}}{N_{tot}} = \lim_{N_{tot} \to \infty} \frac{N_A + N_B}{N_{tot}} \\
&= \lim_{N_{tot} \to \infty} \frac{N_A}{N_{tot}} + \lim_{N_{tot} \to \infty} \frac{N_B}{N_{tot}} = P_A + P_B
\end{aligned}
\tag{39.4}
$$

That is, **the probability that the outcome will be A *or* B is the sum of P_A and P_B.** This important conclusion is a general property of probabilities.

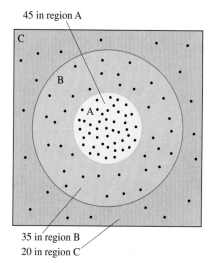

45 in region A

35 in region B
20 in region C

FIGURE 39.2 One hundred throws at a dart board.

Each dart lands *somewhere* on the board. Consequently, the probability that a dart lands in A *or* B *or* C must be 100%. And, in fact,

$$P_{\text{somewhere}} = P_{\text{A or B or C}} = P_{\text{A}} + P_{\text{B}} + P_{\text{C}} = 0.45 + 0.35 + 0.20 = 1.00$$

Thus another important property of probabilities is that **the sum of the probabilities of all possible outcomes must equal 1.**

Suppose exhaustive trials have established that the probability of a dart landing in region A is P_{A}. If you throw N darts, how many do you *expect* to land in A? This value, called the **expected value**, is

$$N_{\text{A expected}} = NP_{\text{A}} \qquad (39.5)$$

The expected value is your best possible prediction of the outcome of an experiment.

If $P_{\text{A}} = 0.45$, your *best prediction* is that 27 of 60 throws (45% of 60) will land in A. Of course, predicting 27 and actually getting 27 aren't the same thing. You would predict 30 heads in 60 flips of a coin, but you wouldn't be surprised if the actual number were 28 or 31. Similarly, the number of darts landing in region A might be 24 or 29 instead of 27. In general, the agreement between actual values and expected values improves as you throw more darts.

STOP TO THINK 39.1 Suppose you roll a die 30 times. What is the expected numbers of 1's *and* 6's?

A Photon Analysis of Interference

Now let's look at the double-slit results from a photon perspective. We know, from experimental evidence, that the interference pattern is built up photon by photon. The bottom portion of Figure 39.1 shows the pattern made on a detector after the arrival of the first few dozen photons. It is clearly a double-slit interference pattern, but it's made, rather like a newspaper photograph, by piling up dots in some places but not others.

The arrival position of any particular photon is *unpredictable*. That is, nothing about how the experiment is set up or conducted allows us to predict exactly where the dot of an individual photon will appear on the detector. Yet there is clearly an overall pattern. There are some positions at which a photon is *more likely* to be detected, other positions at which it is *less likely* to be found.

If we record the arrival positions of many thousands of photons, we will be able to determine the *probability* that a photon will be detected at any given location. If 50 out of 50,000 photons land in one small area of the screen, then each photon has a probability of 50/50,000 = 0.001 = 0.1% of being detected there. The probability will be zero at the interference minima because no photons at all arrive at those points. Similarly, the probability will be a maximum at the interference maxima. The probability will have some in-between value on the sides of the interference fringes.

Figure 39.3a shows a narrow strip, with width δx and height H. (We will assume that δx is very small in comparison with the fringe spacing, so the light's intensity over δx is very nearly constant.) Think of this strip as a very narrow detector that can detect and count the photons landing on it. Suppose we place the narrow strip at position x. We'll use the notation $N(\text{in } \delta x \text{ at } x)$ to indicate the number of photons that hit the detector at this position. The value of $N(\text{in } \delta x \text{ at } x)$ varies from point to point. $N(\text{in } \delta x \text{ at } x)$ is large if x happens to be near the center of a bright fringe; $N(\text{in } \delta x \text{ at } x)$ is small if x is in a dark fringe.

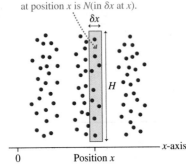

(a) The number of photons in this narrow strip when it is at position x is $N(\text{in } \delta x \text{ at } x)$.

δx

H

x-axis

0   Position x

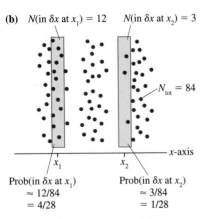

(b) $N(\text{in } \delta x \text{ at } x_1) = 12$ $N(\text{in } \delta x \text{ at } x_2) = 3$

$N_{\text{tot}} = 84$

x-axis

x_1 x_2

Prob(in δx at x_1)
$\approx 12/84$
$= 4/28$

Prob(in δx at x_2)
$\approx 3/84$
$= 1/28$

FIGURE 39.3 A strip of width δx at position x.

Suppose N_{tot} photons are fired at the slits. The *probability* that any one photon ends up in the strip at position x is

$$\text{Prob(in } \delta x \text{ at } x) = \lim_{N_{tot} \to \infty} \frac{N(\text{in } \delta x \text{ at } x)}{N_{tot}} \tag{39.6}$$

As Figure 39.3b shows, Equation 39.6 is an empirical method for determining the probabilities of the photons hitting a particular spot on the detector.

Alternatively, suppose we can calculate the probabilities from a theory. In that case, the *expected value* for the number of photons landing in the narrow strip when it is at position x is

$$N(\text{in } \delta x \text{ at } x) = N \times \text{Prob(in } \delta x \text{ at } x) \tag{39.7}$$

We cannot predict what any individual photon will do, but we can predict the fraction of the photons that should land in this little region of space. Prob(in δx at x) is the probability that it will happen.

39.2 Connecting the Wave and Photon Views

The wave model of light describes the interference pattern in terms of the wave's intensity $I(x)$, a continuous-valued function. The photon model describes the interference pattern in terms of the probability Prob(in δx at x) of detecting a photon. These two models are very different, yet Figure 39.1 shows a clear correlation between the *intensity of the wave* and the *probability of detecting photons*. That is, photons are more likely to be detected at those points where the wave intensity is high, less likely to be detected at those points where the wave intensity is low.

The intensity of a wave is $I = P/A$, the ratio of light power P (joules per second) to the area A on which the light falls. The narrow strip in Figure 39.3a has area $A = H\delta x$. If the light intensity at position x is $I(x)$, the amount of light energy falling onto this narrow strip during each second is

$$E(\text{in } \delta x \text{ at } x) = I(x)A = I(x)H\delta x \tag{39.8}$$

The notation $E(\text{in } \delta x \text{ at } x)$ refers to the energy landing on this narrow strip if you place it at position x.

From the photon perspective, energy E is due to the arrival of N photons, each of which has energy hf. The number of photons that arrive in the strip each second is

$$N(\text{in } \delta x \text{ at } x) = \frac{E(\text{in } \delta x \text{ at } x)}{hf} = \frac{H}{hf}I(x)\,\delta x \tag{39.9}$$

We can then use the Equation 39.6 definition of probability to write the *probability* that a photon lands in the narrow strip δx at position x as

$$\text{Prob(in } \delta x \text{ at } x) = \frac{N(\text{in } \delta x \text{ at } x)}{N_{tot}} = \frac{H}{hfN_{tot}}I(x)\,\delta x \tag{39.10}$$

Equation 39.10 is the link between the wave model and the photon model.

As a final step, recall that the light intensity $I(x)$ is proportional to $|A(x)|^2$, the square of the amplitude function. Consequently,

$$\text{Prob(in } \delta x \text{ at } x) \propto |A(x)|^2 \delta x \tag{39.11}$$

where the various constants in Equation 39.10 have all been incorporated into the unspecified proportionality constant of Equation 39.11.

In other words, **the probability of detecting a photon at a particular point is directly proportional to the square of the light-wave amplitude function at that point.** If the wave amplitude at point A is twice that at point B, then a photon is four times as likely to land in a narrow strip at A as it is to land in an equal-width strip at B.

NOTE ▶ Equation 39.11 is the connection between the particle perspective and the wave perspective. It relates the probability of observing a particle-like event—the arrival of a photon—to the amplitude of a continuous, classical wave. This connection will become the basis of how we interpret the results of quantum-physics calculations. ◀

Probability Density

We need one last definition. Recall that the mass of a wire or string of a length L can be expressed in terms of the linear mass density μ as $m = \mu L$. Similarly, the charge along a length L of wire can be expressed in terms of the linear charge density λ as $Q = \lambda L$. If the length had been very short—in which case we might have denoted it as δx, and if the density varied from point to point—we could have written

$$\text{mass(in length } \delta x \text{ at } x) = \mu(x)\,\delta x$$

$$\text{charge(in length } \delta x \text{ at } x) = \lambda(x)\,\delta x$$

where $\mu(x)$ and $\lambda(x)$ are the linear densities at position x. Writing the mass and charge this way separates the role of the density from the role of the small length δx.

Equation 39.11 looks similar. Using the mass and charge densities as analogies, as shown in Figure 39.4, let us define the **probability density** $P(x)$ such that

$$\text{Prob(in } \delta x \text{ at } x) = P(x)\,\delta x \qquad (39.12)$$

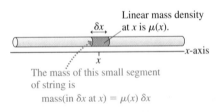

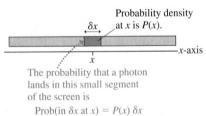

Linear mass density at x is $\mu(x)$.

The mass of this small segment of string is
mass(in δx at x) = $\mu(x)\,\delta x$

Probability density at x is $P(x)$.

The probability that a photon lands in this small segment of the screen is
Prob(in δx at x) = $P(x)\,\delta x$

FIGURE 39.4 The probability density is analogous to the linear mass density.

Probability density has SI units of m^{-1}. Thus the probability density multiplied by a length, as in Equation 39.12, yields a dimensionless probability.

NOTE ▶ $P(x)$ itself is *not* a probability, just as the linear mass density λ is not, by itself, a mass. You must multiply the probability density by a length, as shown in Equation 39.12, to find an actual probability. ◀

By comparing Equation 39.12 to Equation 39.11, you can see that the photon probability density is directly proportional to the square of the light-wave amplitude:

$$P(x) \propto |A(x)|^2 \qquad (39.13)$$

The probability density, unlike the probability itself, is independent of the width δx and depends only on the position x.

Although we were inspired by the double-slit experiment, nothing in our analysis actually depends on the double-slit geometry. Consequently, Equation 39.13 is quite general. It says that for *any* experiment in which we detect photons, **the probability density for detecting a photon is directly proportional to the square of the amplitude function of the corresponding electromagnetic wave.** We now have an explicit connection between the wave-like and the particle-like properties of the light.

EXAMPLE 39.1 **Calculating the probability density**

In an experiment, 6000 out of 600,000 photons are detected in a 1.0-mm-wide strip located at position $x = 50$ cm. What is the probability density at $x = 50$ cm?

SOLVE The probability that a photon arrives at this particular strip is

$$\text{Prob(in 1.0 mm at } x = 50 \text{ cm)} = \frac{6000}{600{,}000} = 0.010$$

Thus the probability density $P(x) = \text{Prob(in } \delta x \text{ at } x)/\delta x$ at this position is

$$P(50 \text{ cm}) = \frac{\text{Prob(in 1.0 mm at } x = 50 \text{ cm)}}{0.0010 \text{ m}} = \frac{0.010}{0.0010 \text{ m}}$$

$$= 1.0 \times 10^{-5} \text{ m}^{-1}$$

STOP TO THINK 39.2 The figure shows the detection of photons in an optical experiment. Rank in order, from largest to smallest, the square of the amplitude function of the electromagnetic wave at positions A, B, C, and D.

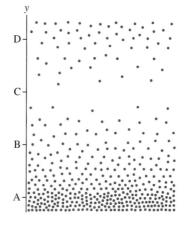

39.3 The Wave Function

Now let's look at the interference of matter. Electrons passing through a double-slit apparatus create the same interference patterns as photons. The pattern is built up electron by electron, but there is no way to predict where any particular electron will be detected. Even so, we can establish the *probability* of an electron landing in a narrow strip of width δx by measuring the positions of many individual electrons. The probability, as you might guess, turns out to be exactly the same as the arrival probability of a photon with the same wavelength.

For light, we were able to relate the photon probability density $P(x)$ to the amplitude of an electromagnetic wave. But there is no wave for electrons that is analogous to electromagnetic waves for light. So how do we find the probability density for electrons? We have reached the point where we must make an inspired leap beyond classical physics. Let us *assume* that there is some kind of continuous, wave-like function for matter that plays a role analogous to the electromagnetic amplitude function $A(x)$ for light. We will call this function the **wave function** $\psi(x)$, where ψ is a lowercase Greek psi. The wave function is a function of position, which is why we write it as $\psi(x)$.

To connect the wave function to the real world of experimental measurements, we will interpret $\psi(x)$ in terms of the *probability* of detecting a particle at position x. If a matter particle, such as an electron, is described by the wave function $\psi(x)$, then the probability $\text{Prob(in } \delta x \text{ at } x)$ of finding the particle within a narrow region of width δx at position x is

$$\text{Prob(in } \delta x \text{ at } x) = |\psi(x)|^2 \delta x = P(x)\, \delta x \qquad (39.14)$$

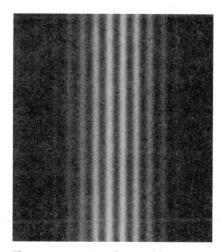

Electrons create interference fringes after passing through a double slit.

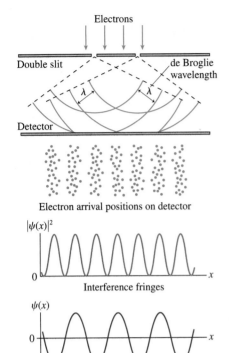

Electrons

Double slit

de Broglie
wavelength

λ λ

Detector

Electron arrival positions on detector

$|\psi(x)|^2$

0

Interference fringes

$\psi(x)$

0

Electron wave function

FIGURE 39.5 The double-slit experiment with electrons.

(a) Wave function

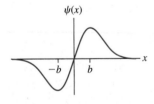

$\psi(x)$

$-b$ b

x

(b) Probability density

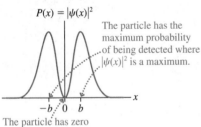

$P(x) = |\psi(x)|^2$

The particle has the maximum probability of being detected where $|\psi(x)|^2$ is a maximum.

$-b$ 0 b x

The particle has zero probability of being detected where $|\psi(x)|^2 = 0$.

FIGURE 39.6 The square of the wave function is the probability density for detecting the electron at various values of the position x.

That is, the probability density $P(x)$ for finding the particle is

$$P(x) = |\psi(x)|^2 \qquad (39.15)$$

With Equations 39.14 and 39.15, we are *defining* the wave function $\psi(x)$ to play the same role for material particles that the amplitude function $A(x)$ does for photons. The only difference is that $P(x) = |\psi(x)|^2$ is for particles, whereas Equation 39.13 for photons is $P(x) \propto |A(x)|^2$. The difference is due to the fact that the electromagnetic field amplitude $A(x)$ had previously been defined through the laws of electricity and magnetism. $|A(x)|^2$ is *proportional* to the probability density for finding a photon, but it is not directly *the* probability density. In contrast, we do not have any preexisting definition for the wave function $\psi(x)$. Thus we are free to define $\psi(x)$ so that $|\psi(x)|^2$ is *exactly* the probability density. That is why we used $=$ rather than $\propto$ in Equation 39.15.

Figure 39.5 shows the double-slit experiment with electrons. This time we will work backward. From the observed distribution of electrons, which represents the probabilities of their landing in any particular location, we can deduce that $|\psi(x)|^2$ has alternating maxima and zeros. The oscillatory wave function $\psi(x)$ is the square root *at each point* of $|\psi(x)|^2$. Notice the very close analogy with the amplitude function $A(x)$ in Figure 39.1.

> **NOTE** ▶ $|\psi(x)|^2$ is uniquely determined by the data, but the wave function $\psi(x)$ is *not* unique. The alternative wave function $\psi'(x) = -\psi(x)$—an upside-down version of the graph in Figure 39.5—would be equally acceptable. ◀

Figure 39.6a is a different example of a wave function. After squaring it *at each point,* as shown in Figure 39.6b, we see that this wave function represents a particle most likely to be detected very near $x = -b$ or $x = +b$. These are the points where $|\psi(x)|^2$ is a maximum. There is zero likelihood of finding the particle right in the center. The particle is more likely to be detected at some positions than at others, but we cannot predict its exact location. The wave function, from which we can calculate probabilities, is all we know about the particle.

> **NOTE** ▶ One of the difficulties in learning to use the concept of a wave function is coming to grips with the fact that there is no "thing" that is waving. There is no disturbance associated with a physical medium. The wave function $\psi(x)$ is simply a *wave-like function* (i.e., it oscillates between positive and negative values) that can be used to make probabilistic predictions about atomic particles. ◀

A Little Science Methodology

Equation 39.14 defines the wave function $\psi(x)$ for a particle in terms of the probability of finding the particle at different positions x. But our interests go beyond merely characterizing experimental data. We would like to develop a new *theory* of matter. But just what is a theory? Although this is not a book on scientific methodology, we can loosely say that a physical theory needs two basic ingredients:

1. A *descriptor.* This is a mathematical quantity used to describe our knowledge of a physical object.
2. One or more *laws* that govern the behavior of the descriptor.

For example, Newtonian mechanics is a theory of motion. The primary descriptor in Newtonian mechanics is a particle's *position* $x(t)$ as a function of time. This describes our knowledge of the particle at all times. The position is governed by *Newton's laws.* These laws, especially the second law, are mathematical statements of how the descriptor changes in response to forces. If we predict $x(t)$ for a known set of forces, we feel confident that an experiment carried out at time t will find the particle right where predicted.

Newton's theory of motion *assumes* that a particle's position is well-defined at every instant of time. The difficulty facing physicists early in the 20th century was the astounding discovery that **the position of an atomic-size particle is *not* well-defined.** An electron in a double-slit experiment must, in some sense, go through *both* slits to produce an electron interference pattern. It simply does not have a well-defined position as it interacts with the slits. But if the position function $x(t)$ is not a valid descriptor for matter at the atomic level, what is?

We will assert that the wave function $\psi(x)$ is the *descriptor* of a particle in quantum mechanics. In other words, the wave function tells us everything we can know about the particle. The wave function $\psi(x)$ plays the same leading role in quantum mechanics that the position function $x(t)$ plays in classical mechanics.

Whether this hypothesis has any merit will not be known until we see if it leads to predictions that can be verified. And before we can do that, we need to learn the "law of psi." What new law of physics determines the wave function $\psi(x)$ in a given situation? We will answer this question in the next chapter.

It may seem to you, as we go along, that we are simply "making up" ideas. Indeed, that is at least partially true. The inventors of entirely new theories use their existing knowledge as a guide, but ultimately they have to make an inspired guess as to what a new theory should look like. Newton and Einstein both made such leaps, and the inventors of quantum mechanics had to make such a leap. We can attempt to make the new ideas *plausible,* but ultimately a new theory is simply a bold new assertion that must be tested against experimental reality. The wave-function theory of quantum mechanics passed the only test that really matters in science—it works!

| STOP TO THIS 39.3 | This is the wave function of a neutron. At what value of x is the neutron most likely to be found? |

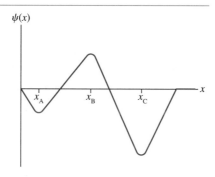

39.4 Normalization

In our discussion of probability we noted that the dart has to hit the wall *somewhere.* The mathematical statement of this idea is the requirement that $P_A + P_B + P_C = 1$. That is, the probabilities of all the mutually exclusive outcomes *must* add up to 1.

Similarly, a photon or electron has to land *somewhere* on the detector after passing through an experimental apparatus. Consequently, the probability that it will be detected at *some* position is 100%. To make use of this requirement, consider an experiment in which an electron is detected on the x-axis. As Figure 39.7 shows, we can divide the region between position x_L and x_R into N adjacent narrow strips of width δx.

The probability that any particular electron lands in the narrow strip i at position x_i is

$$\text{Prob(in } \delta x \text{ at } x_i) = P(x_i)\,\delta x$$

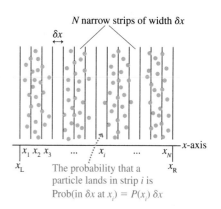

FIGURE 39.7 Dividing the entire detector into many small strips of width δx.

where $P(x_i) = |\psi(x_i)|^2$ is the probability density at x_i. The probability that the electron lands in the strip at x_1 or x_2 or x_3 or . . . is the sum

$$
\begin{aligned}
\text{Prob(between } x_L \text{ and } x_R) &= \text{Prob(in } \delta x \text{ at } x_1) \\
&\quad + \text{Prob(in } \delta x \text{ at } x_2) + \cdots \\
&= \sum_{i=1}^{N} P(x_i)\,\delta x = \sum_{i=1}^{N} |\psi(x_i)|^2 \delta x
\end{aligned}
\tag{39.16}
$$

That is, **the probability that the electron lands *somewhere* between x_L and x_R is the sum of the probabilities of landing in each narrow strip.**

If we let the strips become narrower and narrower, then $\delta x \to dx$ and the sum becomes an integral. Thus the probability of finding the particles in the range $x_L \leq x \leq x_R$ is

$$
\text{Prob(in range } x_L \leq x \leq x_R) = \int_{x_L}^{x_R} P(x)\,dx = \int_{x_L}^{x_R} |\psi(x)|^2 dx
\tag{39.17}
$$

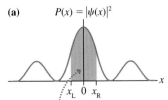

(a) $P(x) = |\psi(x)|^2$

The area under the curve between x_L and x_R is the probability of finding the particle between x_L and x_R.

As Figure 39.8a shows, we can interpret Prob(in range $x_L \leq x \leq x_R$) as the area under the probability density curve between x_L and x_R.

NOTE ▶ The integral of Equation 39.17 is needed when the probability density changes over the range x_L to x_R. For sufficiently narrow intervals, over which $P(x)$ remains essentially constant, the expression Prob(in δx at x) = $P(x)\delta x$ is still valid and is easier to use. ◀

Now let the detector become infinitely wide, so that the probability that the electron will arrive *somewhere* on the detector becomes 100%. The statement that the electron has to land *somewhere* on the x-axis is expressed mathematically as

$$
\int_{-\infty}^{\infty} P(x)\,dx = \int_{-\infty}^{\infty} |\psi(x)|^2 dx = 1
\tag{39.18}
$$

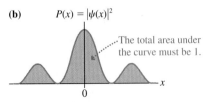

(b) $P(x) = |\psi(x)|^2$

The total area under the curve must be 1.

FIGURE 39.8 The area under the probability density curve is a probability.

Equation 39.18 is called the **normalization condition.** Any wave function $\psi(x)$ must satisfy this condition; otherwise we would not be able to interpret $|\psi(x)|^2$ as a probability density. As Figure 39.8b shows, Equation 39.18 tells us that the total area under the probability density curve must be 1.

NOTE ▶ The normalization condition integrates the *square* of the wave function. We don't have any information about what the integral of $\psi(x)$ might be. ◀

EXAMPLE 39.2 Normalizing and interpreting a wave function

Figure 39.9 shows the wave function of a particle confined within the region between $x = 0$ nm and $x = L = 1.0$ nm. The wave function is zero outside this region.

a. Determine the value of the constant c.
b. Draw a graph of the probability density $P(x)$.
c. Draw a dot picture showing where the first 40 or 50 particles might be found.
d. Calculate the probability of finding the particle in a region of width $\delta x = 0.01$ nm at positions $x_1 = 0.05$ nm, $x_2 = 0.50$ nm, and $x_3 = 0.95$ nm.

MODEL The probability of finding the particle is determined by the probability density $P(x)$.

VISUALIZE The wave function is shown in Figure 39.9.

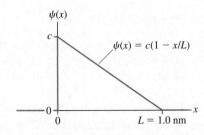

FIGURE 39.9 The wave function of Example 39.2.

SOLVE

a. The wave function is $\psi(x) = c(1 - x/L)$. This is a function that decreases linearly from $\psi = c$ at $x = 0$ to $\psi = 0$ at $x = L$. The constant c is the height of this wave function.

The particle *has* to be in the region $0 \le x \le L$ with probability 1, and only one value of c will make it so. We can determine c by using Equation 39.18, the normalization condition. Because the wave function is zero outside the interval from 0 to L, the integration limits are 0 to L. Thus

$$1 = \int_0^L |\psi(x)|^2 dx = c^2 \int_0^L \left(1 - \frac{x}{L}\right)^2 dx$$

$$= c^2 \int_0^L \left(1 - \frac{2x}{L} + \frac{x^2}{L^2}\right) dx$$

$$= c^2 \left[x - \frac{x^2}{L} + \frac{x^3}{3L^2}\right]_0^L = \frac{1}{3}c^2 L$$

The solution for c is

$$c = \sqrt{\frac{3}{L}} = \sqrt{\frac{3}{1 \text{ nm}}} = 1.732 \text{ nm}^{-1/2}$$

Note the unusual units for c. Although these are not SI units, we can correctly compute probabilities as long as δx has units of nm. A multiplicative constant such as c is often called a *normalization constant*.

b. The wave function is

$$\psi(x) = (1.732 \text{ nm}^{-1/2})\left(1 - \frac{x}{1.0 \text{ nm}}\right)$$

Thus the probability density is

$$P(x) = |\psi(x)|^2 = (3.0 \text{ nm}^{-1})\left(1 - \frac{x}{1.0 \text{ nm}}\right)^2$$

This probability density is graphed in Figure 39.10a.

c. Particles are most likely to be detected at the left edge of the interval, where the probability density $P(x)$ is maximum. The probability steadily decreases across the interval, becoming zero at $x = 1.0$ nm. Figure 39.10b shows how a group of particles described by this wave function might appear on a detection screen.

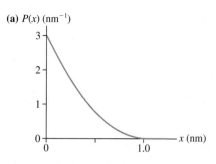

(a) $P(x)$ (nm^{-1})

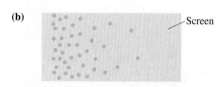

(b) Screen

FIGURE 39.10 The probability density $P(x)$ and the detected positions of particles described by this probability density.

d. $P(x)$ is essentially constant over the small interval $\delta x = 0.01$ nm, so we can use

$$\text{Prob(in } \delta x \text{ at } x) = P(x)\delta x = |\psi(x)|^2 \delta x$$

for the probability of finding the particle in a region of width δx at the position x. We need to evaluate $|\psi(x)|^2$ at the three positions $x_1 = 0.05$ nm, $x_2 = 0.50$ nm, and $x_3 = 0.95$ nm. Doing so gives

$$\text{Prob(in 0.01 nm at } x_1 = 0.05 \text{ nm)} = c^2(1 - x_1/L)^2 \delta x$$
$$= 0.0270 = 2.70\%$$

$$\text{Prob(in 0.01 nm at } x_2 = 0.50 \text{ nm)} = c^2(1 - x_2/L)^2 \delta x$$
$$= 0.0075 = 0.75\%$$

$$\text{Prob(in 0.01 nm at } x_3 = 0.95 \text{ nm)} = c^2(1 - x_3/L)^2 \delta x$$
$$= 0.00008 = 0.008\%$$

EXAMPLE 39.3 The probability of finding a particle

A particle is described by the wave function

$$\psi(x) = \begin{cases} 0 & x < 0 \\ ce^{-x/L} & x \ge 0 \end{cases}$$

where $L = 1$ nm.

a. Determine the value of the constant c.
b. Draw graphs of $\psi(x)$ and the probability density $P(x)$.
c. Calculate the probability of finding the particle in the region $x \ge 1$ nm.

MODEL The probability of finding the particle is determined by the probability density $P(x)$.

SOLVE

a. The wave function is an exponential $\psi(x) = ce^{-x/L}$ that extends from $x = 0$ to $x = +\infty$. Equation 39.18, the normalization condition, is

$$1 = \int_{-\infty}^{\infty} |\psi(x)|^2 dx = c^2 \int_0^\infty e^{-2x/L} dx = -\frac{c^2 L}{2}e^{-2x/L}\Big|_0^\infty = \frac{c^2}{2L}$$

We can solve this for the normalization constant c:

$$c = \sqrt{\frac{2}{L}} = \sqrt{\frac{2}{1 \text{ nm}}} = 1.414 \text{ nm}^{-1/2}$$

b. The probability density is

$$P(x) = |\psi(x)|^2 = (2.0 \text{ nm}^{-1})e^{-2x/(1.0 \text{ nm})}$$

The wave function and the probability density are graphed in Figure 39.11.

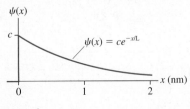

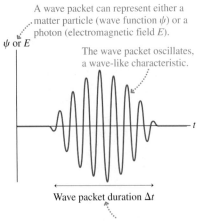

FIGURE 39.11 The wave function and probability density of Example 39.3.

c. The probability of finding the particle in the region $x \geq 1$ nm is the shaded area under the probability density curve in Figure 39.11. We must use Equation 39.17 and integrate to find a numerical value. The probability is

$$\text{Prob}(x \geq 1 \text{ nm}) = \int_{1 \text{ nm}}^{\infty} |\psi(x)|^2 dx$$

$$= (2.0 \text{ nm}^{-1}) \int_{1 \text{ nm}}^{\infty} e^{-2x/(1.0 \text{ nm})} dx$$

$$= (2.0 \text{ nm}^{-1}) \left(-\frac{1.0 \text{ nm}}{2} \right) e^{-2x/(1.0 \text{ nm})} \Big|_{1 \text{ nm}}^{\infty}$$

$$= e^{-2} = 0.135 = 13.5\%$$

ASSESS There is a 13.5% chance of finding the particle beyond 1 nm and thus an 86.5% chance of finding it within the interval $0 \leq x \leq 1$ nm. Unlike classical physics, we cannot make an exact prediction of the particle's position.

STOP TO THINK 39.4 The value of the constant a is

a. $a = 2.0 \text{ mm}^{-1}$.
b. $a = 1.0 \text{ mm}^{-1}$.
c. $a = 0.5 \text{ mm}^{-1}$.
d. $a = 2.0 \text{ mm}^{-1/2}$.
e. $a = 1.0 \text{ mm}^{-1/2}$.
f. $a = 0.5 \text{ mm}^{-1/2}$.

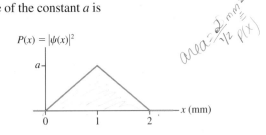

39.5 Wave Packets

The experimental evidence is overwhelming that light has particle-like characteristics and that matter has wave-like characteristics. These observations are completely at odds with the classical models of particles and waves that we used to study the physics of baseballs and sound waves. In particular, the classical ideas of particles and waves are mutually exclusive. An object can be one or the other, but not both. The classical models fail to describe the wave-particle duality seen at the atomic level. An alternative model with both particle and wave characteristics is a *wave packet*.

Consider the wave shown in Figure 39.12. Unlike the sinusoidal waves we have considered previously, which stretch through time and space, this wave is bunched up, or localized. The localization is a particle-like characteristic. The oscillations are wave-like. Such a localized wave is called a **wave packet.**

A wave packet travels through space with constant speed v, just like a photon in a light wave or an electron in a force-free region. A wave packet has a wavelength, hence it will undergo interference and diffraction. But because it is also localized, a wave packet has the possibility of making a "dot" when it strikes a detector. We can visualize a light wave as consisting of a very large number of these wave packets moving along together. Similarly, we can think of a beam of electrons as a series of wave packets spread out along a line.

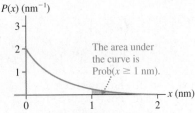

FIGURE 39.12 History graph of a wave packet with duration Δt.

Wave packets are not a perfect model of photons or electrons (we need the full treatment of quantum physics to get a more accurate description), but they do provide a useful way of thinking about photons and electrons in many circumstances.

You might have noticed that the wave packet in Figure 39.12 looks very much like one cycle of a beat pattern. You will recall that beats occur if we superimpose two waves of frequencies f_1 and f_2 where the two frequencies are very similar: $f_1 \approx f_2$. Figure 39.13, which is copied from Chapter 21 where we studied beats, shows that the loud, soft, loud, soft, . . . pattern of beats is a series of wave packets.

In Chapter 21, the beat frequency (number of pulses per second) was found to be

$$f_{\text{beat}} = f_1 - f_2 = \Delta f \tag{39.19}$$

where Δf is the *range* of frequencies that are superimposed to form the wave packet. Figure 39.13 defines Δt as the duration of each beat or each wave packet. This interval of time is equivalent to the *period* T_{beat} of the beat. Because period and frequency are inverses of each other, the duration Δt is

$$\Delta t = T_{\text{beat}} = \frac{1}{f_{\text{beat}}} = \frac{1}{\Delta f}$$

We can rewrite this as

$$\Delta f \Delta t = 1 \tag{39.20}$$

Equation 39.20 is nothing new; we are simply writing what we already knew in a different form. Equation 39.20 is a combination of three things: the relationship $f = 1/T$ between period and frequency, writing T_{beat} as Δt, and the specific knowledge that the beat frequency f_{beat} is the difference Δf of the two frequencies contributing to the wave packet. As the frequency separation gets smaller, the duration of each beat gets longer.

When we superimpose two frequencies to create beats, the wave packet repeats over and over. A more advanced treatment of waves, called Fourier analysis, reveals that a single, *nonrepeating* wave packet can be created through superposition of *many* waves of very similar frequency. Figure 39.14 illustrates this idea. At one instant of time, all the waves interfere constructively to produce the maximum amplitude of the wave packet. At other times, the individual waves get out of phase and their superposition tends toward zero.

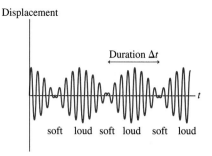

Displacement

FIGURE 39.13 Beats are a series of wave packets.

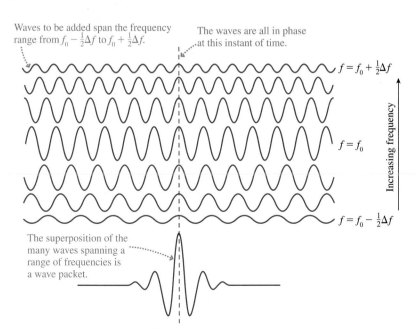

FIGURE 39.14 A single wave packet is the superposition of many component waves of similar wavelength and frequency.

Suppose a single nonrepeating wave packet of duration Δt is created by the superposition of *many* waves that span a range of frequencies Δf. We'll not prove it, but Fourier analysis shows that for *any* wave packet

$$\Delta f \Delta t \approx 1 \tag{39.21}$$

The relationship between Δf and Δt for a general wave packet is not as precise as Equation 39.20 for beats. There are two reasons for this:

1. Wave packets come in a variety of shapes. The exact relationship between Δf and Δt depends somewhat on the shape of the wave packet.
2. We have not given a precise definition of Δt and Δf for a general wave packet. The quantity Δt is "about how long the wave packet lasts," while Δf is "about the range of frequencies needing to be superimposed to produce this wave packet." For our purposes, we will not need to be any more precise than this.

Equation 39.21 is a purely classical result that applies to waves of any kind. It tells you the range of frequencies you need to superimpose to construct a wave packet of duration Δt. Alternatively, Equation 39.21 tells you that a wave packet created as a superposition of various frequencies cannot be arbitrarily short but *must* last for a time interval $\Delta t \approx 1/\Delta f$.

EXAMPLE 39.4 Creating radio-frequency pulses

A short-wave radio station broadcasts at a frequency of 10.000 MHz. What is the range of frequencies of the waves that must be superimposed to broadcast a radio-wave pulse lasting 0.800 μs?

MODEL A pulse of radio waves is an electromagnetic wave packet, hence it must satisfy the relationship $\Delta f \Delta t \approx 1$.

VISUALIZE Figure 39.15 shows the pulse.

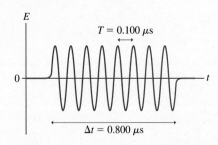

FIGURE 39.15 A pulse of radio waves.

SOLVE The period of a 10.000 MHz oscillation is 0.100 μs. A pulse 0.800 μs in duration is 8 oscillations of the wave. Although the station broadcasts at a nominal frequency of 10.000 MHz, this pulse is not a pure 10.000 MHz oscillation. Instead, the pulse has been created by the superposition of many waves whose frequencies span

$$\Delta f \approx \frac{1}{\Delta t} = \frac{1}{0.800 \times 10^{-6} \text{ s}} = 1.250 \times 10^6 \text{ Hz} = 1.250 \text{ MHz}$$

This range of frequencies will be centered at the 10.000 MHz broadcast frequency, so the waves that must be superimposed to create this pulse span the frequency range

$$9.375 \text{ MHz} \le f \le 10.625 \text{ MHz}$$

Bandwidth

Short-duration pulses, like the one in Example 39.4, are used to transmit digital information. Digital signals are sent over a phone line by brief tone pulses, over satellite links by brief radio pulses like the one in the example, and through optical fibers by brief laser-light pulses. Regardless of the type of wave and the medium through which it travels, any wave pulse must obey the fundamental relationship $\Delta f \Delta t \approx 1$.

Sending data at a higher rate (i.e., more pulses per second) requires that the pulse duration Δt be shorter. But a shorter-duration pulse must be created by the superposition of a *larger* range of frequencies. Thus the medium through which a shorter-duration pulse travels must be physically able to transmit the full range of frequencies.

The range of frequencies that can be transmitted through a medium is called the **bandwidth** Δf_B of the medium. The shortest possible pulse that can be transmitted through a medium is

$$\Delta t_{\min} \approx \frac{1}{\Delta f_B} \tag{39.22}$$

A pulse shorter than this would require a larger range of frequencies than the medium can support.

The concept of bandwidth is extremely important in digital communications. A higher bandwidth transmits shorter pulses and allows a higher data rate. A standard telephone line does not have a very high bandwidth, and that is why a modem is limited to sending data at the rate of roughly 50,000 pulses per second. A $0.80~\mu s$ pulse can't be sent over a phone line simply because the phone line won't transmit the range of frequencies that would be needed.

An optical fiber is a high-bandwidth medium. A fiber has a bandwidth $\Delta f_B > 1$ GHz and thus can transmit laser-light pulses with duration $\Delta t < 1$ ns. More than 10^9 pulses per second can be sent along an optical fiber, which is why optical-fiber networks form the backbone of the Internet.

Uncertainty

There is another way of thinking about the time-frequency relationship $\Delta f \Delta t \approx 1$. Suppose you want to determine *when* a wave packet arrives at a specific point in space, such as at a detector of some sort. At what instant of time can you say that the wave packet is detected? When the front edge arrives? When the maximum amplitude arrives? When the back edge arrives? Because a wave packet is spread out in time, there is not a unique and well-defined time t at which the packet arrives. All we can say is that it arrives within some interval of time Δt. We are *uncertain* about the exact arrival time.

Similarly, suppose you would like to know the oscillation frequency of a wave packet. There is no precise value for f because the wave packet is constructed from many waves within a range of frequencies Δf. All we can say is that the frequency is within this range. We are *uncertain* about the exact frequency.

The time-frequency relationship $\Delta f \Delta t \approx 1$ tells us that the uncertainty in our knowledge about the arrival time of the wave packet is related to our uncertainty about the packet's frequency. The more precisely and accurately we know one quantity, the less precisely we will be able to know the other.

Figure 39.16 shows two different wave packets. The wave packet of Figure 39.16a is very narrow and thus very localized in time. As it travels, our knowledge of when it will arrive at a specified point is fairly precise. But a very wide range of frequencies Δf is required to create a wave packet with a very small Δt. The price we pay for being fairly certain about the time is a very large uncertainty Δf about the frequency of this wave packet.

Figure 39.16b shows the opposite situation: The wave packet oscillates many times and the frequency of these oscillations is pretty clear. Our knowledge of the frequency is good, with minimal uncertainty Δf. But such a wave packet is so spread out that there is a very large uncertainty Δt as to its time of arrival.

In practice, $\Delta f \Delta t \approx 1$ is really a lower limit. Technical limitations may cause the uncertainties in our knowledge of f and t to be even larger than this relationship implies. Consequently, a better statement about our knowledge of a wave packet is

$$\Delta f \Delta t \geq 1 \tag{39.23}$$

The fact that waves are spread out makes it meaningless to specify an exact frequency and an exact arrival time simultaneously. This is an inherent feature of waviness that applies to all waves.

(a)

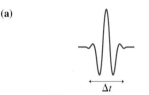

This wave packet has a large frequency uncertainty Δf.

(b)

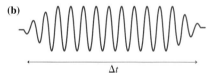

This wave packet has a small frequency uncertainty Δf.

FIGURE 39.16 Two wave packets with different Δt.

$100 \times 10^{-9} = \frac{1}{\Delta f B}$

What minimum bandwidth must a medium have to transmit a 100-ns-long pulse?

a. 1 MHz   b. 10 MHz   c. 100 MHz   d. 1000 MHz

39.6 The Heisenberg Uncertainty Principle

17.6, 17.7   Activ ONLINE Physics

If matter has wave-like aspects and a de Broglie wavelength, then the expression $\Delta f \Delta t \geq 1$ must somehow apply to matter. How? And what are the implications?

Consider a particle with velocity v_x as it travels along the x-axis with de Broglie wavelength $\lambda = h/p_x$. Figure 39.12 showed a *history graph* (ψ versus t) of a wave packet that might represent the particle as it passes a point on the x-axis. It will be more useful to have a *snapshot graph* (ψ versus x) of the wave packet traveling along the x-axis.

The time interval Δt is the duration of the wave packet as the particle passes a point in space. During this interval, the packet moves forward

$$\Delta x = v_x \Delta t = \frac{p_x}{m} \Delta t \qquad (39.24)$$

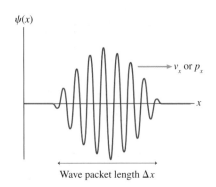

$\psi(x)$

v_x or p_x

x

Wave packet length Δx

FIGURE 39.17 A snapshot graph of a wave packet.

where $p_x = mv_x$ is the x-component of the particle's momentum. The quantity Δx, shown in Figure 39.17, is the length or spatial extent of the wave packet. Conversely, we can write the wave packet's duration in terms of its length as

$$\Delta t = \frac{m}{p_x} \Delta x \qquad (39.25)$$

You will recall that any wave with sinusoidal oscillations must satisfy the wave condition $\lambda f = v$. For a material particle, where λ is the de Broglie wavelength, the frequency f is

$$f = \frac{v}{\lambda} = \frac{(p_x/m)}{(h/p_x)} = \frac{p_x^2}{hm}$$

A small range of frequencies Δf is related to a small range of momenta Δp_x by

$$\Delta f = \frac{2 p_x \Delta p_x}{hm} \qquad (39.26)$$

where we have assumed that $\Delta f \ll f$ and $\Delta p_x \ll p_x$ (a reasonable assumption) and thus treated the small ranges Δf and Δp_x as if they were differentials df and dp_x.

Multiplying together these expressions for Δt and Δf, we find that

$$\Delta f \Delta t = \frac{2 p_x \Delta p_x}{hm} \frac{m \Delta x}{p_x} = \frac{2}{h} \Delta x \Delta p_x \qquad (39.27)$$

Because $\Delta f \Delta t \geq 1$ for any wave, one last rearrangement of Equation 39.27 shows that a matter wave must obey the condition

$$\Delta x \Delta p_x \geq \frac{h}{2} \qquad \text{(Heisenberg uncertainty principle)} \qquad (39.28)$$

This statement about the relationship between the position and momentum of a particle was proposed by Werner Heisenberg, creator of one of the first successful quantum theories. Physicists often just call it the **uncertainty principle.**

NOTE ▶ In statements of the uncertainty principle, the right side is sometimes $h/2$, as we have it, but other times it is just h or contains various factors of π. The specific number is not especially important because it depends on exactly how Δx and Δp are defined. The important idea is that the product of Δx and Δp_x for a particle cannot be significantly less than Planck's constant h. A similar relationship for $\Delta y \Delta p_y$ applies along the y-axis. ◀

What Does It Mean?

Heisenberg's uncertainty principle is a statement about our *knowledge* of the properties of a particle. If we want to know *where* a particle is located, we measure its position x. That measurement is not absolutely perfect, but has some uncertainty Δx. Likewise, if we want to know *how fast* the particle is going we need to measure its velocity v_x or, equivalently, its momentum p_x. This measurement also has some uncertainty Δp_x.

Uncertainties are associated with all experimental measurements, but better procedures and techniques can reduce those uncertainties. Newtonian physics places no limits on how small the uncertainties can be. A Newtonian particle at any instant of time has an exact position x and an exact momentum p_x, and with sufficient care we can measure both x and p_x with such precision that the product $\Delta x \Delta p_x \rightarrow 0$. There are no inherent limits to our knowledge about a classical, or Newtonian, particle.

Heisenberg, however, made the bold and original statement that our knowledge has real limitations. No matter how clever you are, and no matter how good your experiment, you *cannot* measure both x and p_x simultaneously with arbitrarily good precision. Any measurements you make are limited by the condition that $\Delta x \Delta p_x \geq h/2$. **Our knowledge about a particle is *inherently* uncertain.**

Why? Because of the wave-like nature of matter. The "particle" is spread out in space, so there simply is not a precise value of its position x. Similarly, the de Broglie relationship between momentum and wavelength implies that we cannot know the momentum of a wave packet any more exactly than we can know its wavelength or frequency. Our belief that position and momentum have precise values is tied to our classical concept of a particle. As we revise our ideas of what atomic particles are like, we will also have to revise our old ideas about position and momentum.

EXAMPLE 39.5 The uncertainty of a dust particle

A 1.0-μm-diameter dust particle ($m \approx 10^{-15}$ kg) is confined within a 10-μm-long box. Can we know with certainty if the particle is at rest? If not, within what range is its velocity likely to be found?

MODEL All matter is subject to the Heisenberg uncertainty principle.

SOLVE If we know *for sure* that the particle is at rest, then $p_x = 0$ with no uncertainty. That is, $\Delta p_x = 0$. But then, according to the uncertainty principle, the uncertainty in our knowledge of the particle's position would have to be $\Delta x \rightarrow \infty$. In other words, we would have no knowledge at all about the particle's position—it could be anywhere! But that is not the case. We know the particle is *somewhere* in the box, so the uncertainty in our knowledge of its position is at most $\Delta x = L = 10~\mu$m. With a finite Δx, the uncertainty Δp_x *cannot* be zero. We cannot know with certainty if the particle is at rest inside the box. No matter how hard we try to bring the particle to rest, the

uncertainty in our knowledge of the particle's momentum will be $\Delta p_x \approx h/(2\Delta x) = h/2L$. We've assumed the most accurate measurements possible so that the $\geq$ in Heisenberg's uncertainty principle becomes $\approx$. Consequently, the range of possible velocities is

$$\Delta v_x = \frac{\Delta p_x}{m} \approx \frac{h}{2mL} \approx 3.0 \times 10^{-14}~\text{m/s}$$

This range of possible velocities will be centered on $v_x = 0$ m/s if we have done our best to have the particle be at rest. Thus all we can know with certainty is that the particle's velocity is somewhere within the interval -1.5×10^{-14} m/s $\leq v \leq 1.5 \times 10^{-14}$ m/s.

ASSESS For practical purposes you might consider this to be a satisfactory definition of "at rest." After all, a particle moving with a speed of 1.5×10^{-14} m/s would need 6×10^{10} s to move a mere 1 mm. That is about 2000 years! Nonetheless, we can't know if the particle is "really" at rest.

EXAMPLE 39.6 **The uncertainty of an electron**

What range of velocities might an electron have if confined to a 0.10-nm-wide region, about the size of an atom?

MODEL Electrons are subject to the Heisenberg uncertainty principle.

SOLVE The analysis is the same as in Example 39.5. If we know that the electron's position is located within an interval $\Delta x \approx 0.1$ nm, then the best we can know is that its velocity is within the range

$$\Delta v_x = \frac{\Delta p_x}{m} \approx \frac{h}{2mL} \approx 4 \times 10^6 \text{ m/s}$$

Because the *average* velocity is zero, the best we can say is that the electron's velocity is somewhere in the interval -2×10^6 m/s $\leq v \leq 2 \times 10^6$ m/s. It is simply not possible to know the electron's velocity any more precisely than this.

ASSESS Unlike the situation in Example 39.5, where Δv was so small as to be of no practical consequence, our uncertainty about the electron's velocity is enormous—about 1% of the speed of light!

Once again, we see that even the smallest of macroscopic objects behaves very much like a classical Newtonian particle. Perhaps a 1-μm-diameter particle is slightly fuzzy and has a slightly uncertain velocity, but it is far beyond the measuring capabilities of even the very best instruments to detect this wave-like behavior. In contrast, the effects of the uncertainty principle at the atomic scale are stupendous. We are unable to determine the velocity of an electron in an atom-size container to any better accuracy than about 1% of the speed of light.

STOP TO THINK 39.6 Which of these particles, A or B, can you locate more precisely?

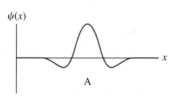

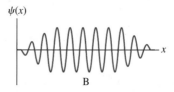

SUMMARY

The goal of Chapter 39 has been to introduce the wave-function description of matter and learn how it is interpreted.

GENERAL PRINCIPLES

Wave Functions and the Probability Density

We cannot predict the exact trajectory of an atomic-level particle such as an electron. The best we can do is to predict the **probability** that a particle will be found in some region of space. The probability is determined by the particle's wave function $\psi(x)$.

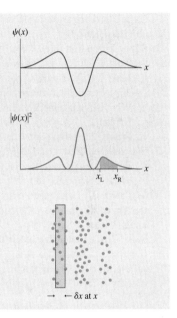

- $\psi(x)$ is a continuous, wave-like (i.e., oscillatory) function.

- The probability that a particle will be found in the narrow interval δx at position x is
  Prob(in δx at x) $= |\psi(x)|^2\, \delta x$.

- $|\psi(x)|^2$ is the probability density $P(x)$.

- For the probability interpretation of $\psi(x)$ to make sense, the wave function must satisfy the normalization condition

$$\int_{-\infty}^{\infty} P(x)\,dx = \int_{-\infty}^{\infty} |\psi(x)|^2\, dx = 1$$

That is, it is certain that the particle is *somewhere* on the x-axis.

- For an extended interval

$$\text{Prob}(x_L \leq x \leq x_R) = \int_{x_L}^{x_R} |\psi(x)|^2\, dx = \text{area under the curve}$$

$\rightarrow$ $\leftarrow \delta x$ at x

Heisenberg Uncertainty Principle

A particle with wave-like characteristics does not have a precise value of position x or a precise value of momentum p_x. Both are uncertain. The position uncertainty Δx and momentum uncertainty Δp_x are related by $\Delta x\, \Delta p_x \geq h/2$. The more you try to pin down the value of one, the less precisely the other can be known.

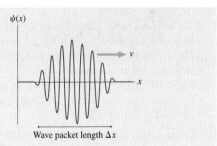

Wave packet length Δx

IMPORTANT CONCEPTS

The probability that a particle is found in region A is

$$P_A = \lim_{N_{tot}\to\infty} \frac{N_A}{N_{tot}}$$

If the probability is known, the expected number of A outcomes in N trials is $N_A = NP_A$.

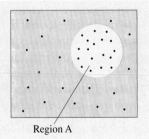

Region A

A wave packet of duration Δt can be created by the superposition of many waves spanning the frequency range Δf. These are related by

$$\Delta f \Delta t \approx 1$$

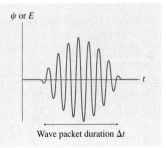

Wave packet duration Δt

TERMS AND NOTATION

quantum mechanics	wave function, $\psi(x)$	bandwidth, Δf_B
probability	normalization condition	uncertainty principle
probability density, $P(x)$	wave packet	

EXERCISES AND PROBLEMS

Exercises

Section 39.1 Waves, Particles, and the Double-Slit Experiment

1. An experiment has four possible outcomes, labeled A to D. The probability of A is $P_A = 40\%$ and of B is $P_B = 30\%$. Outcome C is twice as probable as outcome D. What are the probabilities P_C and P_D?

2. Suppose you toss three coins into the air and let them fall on the floor. Each coin shows either a head or a tail.
   a. Make a table in which you list all the possible outcomes of this experiment. Call the coins A, B, and C.
   b. What is the probability of getting two heads and one tail? Explain.
   c. What is the probability of getting *at least* two heads?

3. Suppose you draw a card from a regular deck of 52 cards.
   a. What is the probability that you draw an ace?
   b. What is the probability that you draw a spade?

4. You are dealt 1 card each from 1000 decks of cards. What is the expected number of picture cards (jacks, queens, and kings)?

5. Make a table in which you list all possible outcomes of rolling two dice. Call the dice A and B. What is the probability of rolling (a) a 7, (b) any double, and (c) a 6 or an 8? You can give the probabilities as fractions, such as 3/36.

Section 39.2 Connecting the Wave and Photon Views

6. In one experiment, 2000 photons are detected in a 0.10-mm-wide strip where the amplitude of the electromagnetic wave is 10 V/m. How many photons are detected in a nearby 0.10-mm-wide strip where the amplitude is 30 V/m?

7. 1.0×10^{10} photons pass through an experimental apparatus. How many of them land in a 0.10-mm-wide strip where the probability density is 20 m^{-1}?

Section 39.3 The Wave Function

8. What are the units of ψ? Explain.

9. What is the difference between the probability and the probability density?

10. For the electron wave function shown in Figure Ex39.10, at what position or positions is the electron most likely to be found? Least likely to be found? Explain.

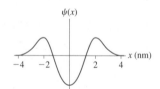

FIGURE EX39.10

11. Figure Ex39.11 shows the probability density for an electron that has passed through an experimental apparatus. If 1.0×10^6 electrons are used, what is the expected number that will land in a 0.010-mm-wide strip at (a) $x = 0.000$ mm and (b) $x = 2.000$ mm?

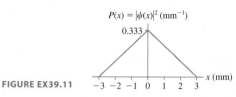

FIGURE EX39.11

12. In an interference experiment with electrons, you find the most intense fringe is at $x = 7.0$ cm. There are slightly weaker fringes at $x = 6.0$ and 8.0 cm, still weaker fringes at $x = 4.0$ and 10.0 cm, and two very weak fringes at $x = 1.0$ and 13.0 cm. No electrons are detected at $x < 0$ cm or $x > 14$ cm.
   a. Sketch a graph of $|\psi(x)|^2$ for these electrons.
   b. Sketch a possible graph of $\psi(x)$.
   c. Are there other possible graphs for $\psi(x)$? If so, draw one.

13. Figure Ex39.13 shows the probability density for an electron that has passed through an experimental apparatus. What is the probability that the electron will land in a 0.010-mm-wide strip at (a) $x = 0.000$ mm, (b) $x = 0.500$ mm, (c) $x = 1.000$ mm, and (d) $x = 2.000$ mm?

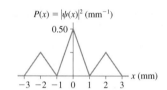

FIGURE EX39.13

Section 39.4 Normalization

14. Figure Ex39.14 is a graph of $|\psi(x)|^2$ for an electron.
   a. What is the value of a?
   b. Draw a graph of the wave function $\psi(x)$. (There is more than one acceptable answer.)
   c. What is the probability that the electron is located between $x = 1.0$ nm and $x = 2.0$ nm?

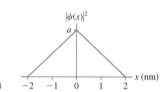

FIGURE EX39.14

15. Figure Ex39.15 is a graph of $|\psi(x)|^2$ for a neutron.
   a. What is the value of a?
   b. Draw a graph of the wave function $\psi(x)$. (There is more than one acceptable answer.)
   c. What is the probability that the neutron is located between $x = -2.0$ mm and $x = 2.0$ mm?

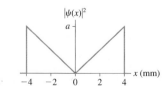

FIGURE EX39.15

16. Figure Ex39.16 shows the wave function of an electron.
    a. What is the value of c?
    b. Draw a graph of $|\psi(x)|^2$.
    c. What is the probability that the electron is located between $x = -1.0$ nm and $x = 1.0$ nm?

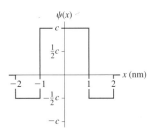

FIGURE EX39.16

17. Figure Ex39.17 shows the wave function of a neutron.
    a. What is the value of c?
    b. Draw a graph of $|\psi(x)|^2$.
    c. What is the probability that the neutron is located between $x = -1.0$ mm and $x = 1.0$ mm?

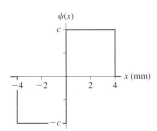

FIGURE EX39.17

Section 39.5 Wave Packets

18. A radar antenna broadcasts electromagnetic waves with a period of 0.100 ns. What range of frequencies would need to be superimposed to create a 1.0-ns-long radar pulse?

19. A radio-frequency amplifier is designed to amplify signals in the frequency range 80 MHz to 120 MHz. What is the smallest duration radio-frequency pulse that can be amplified without distortion?

20. What minimum bandwidth is needed to transmit a pulse that consists of 100 cycles of a 1.00 MHz oscillation?

21. A 1.5-μm-wavelength laser pulse is transmitted through a 2.0-GHz-bandwidth optical fiber. How many oscillations are in the shortest-duration laser pulse that can travel through the fiber?

Section 39.6 The Heisenberg Uncertainty Principle

22. Andrea thinks she's sitting at rest in the middle of her 5.0-m-long dorm room as she does her physics homework. Can Andrea be sure she's at rest? If not, within what range is her velocity likely to be?

23. What is the smallest box in which you can confine an electron if you want to know for certain that the electron's speed is no more than 10 m/s?

24. A thin solid barrier in the xy-plane has a 10-μm-diameter circular hole. An electron traveling in the z-direction with $v_x = 0$ m/s passes through the hole. Afterward, is v_x still zero? If not, within what range is v_x likely to be?

25. A proton is confined within an atomic nucleus of diameter 4.0 fm. Estimate the smallest range of speeds you might find for a proton in the nucleus.

Problems

26. A 1.0-mm-diameter sphere bounces back and forth between two walls at $x = 0$ mm and $x = 100$ mm. The collisions are perfectly elastic, and the sphere repeats this motion over and over with no loss of speed. At a random instant of time, what is the probability that the center of the sphere is
    a. At exactly $x = 50.0$ mm?
    b. Between $x = 49.0$ mm and $x = 51.0$ mm?
    c. At $x \geq 75$ mm?

27. Sound waves of 498 Hz and 502 Hz are superimposed at a temperature where the speed of sound in air is 340 m/s. What is the length Δx of one wave packet?

28. Ultrasound pulses of with a frequency of 1.000 MHz are transmitted into water, where the speed of sound is 1500 m/s. The spatial length of each pulse is 12 mm.
    a. How many complete cycles are contained in one pulse?
    b. What range of frequencies must be superimposed to create each pulse?

29. Figure P39.29 shows a *pulse train*. The period of the pulse train is $T = 2\Delta t$, where Δt is the duration of each pulse. What is the maximum pulse-transmission rate (pulses per second) through an electronics system with a 200 kHz bandwidth? (This is the bandwidth allotted to each FM radio station.)

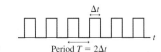

FIGURE P39.29   Period $T = 2\Delta t$

30. Consider a single-slit diffraction experiment using electrons. (Single-slit diffraction was described in Section 22.4.) Using Figure 39.5 as a model, draw
    a. A dot picture showing the arrival positions of the first 40 or 50 electrons.
    b. A graph of $|\psi(x)|^2$ for the electrons on the detection screen.
    c. A graph of $\psi(x)$ for the electrons. Keep in mind that ψ, as a wave-like function, oscillates between positive and negative.

31. An experiment finds electrons to be uniformly distributed over the interval $0 \text{ cm} \leq x \leq 2$ cm, with no electrons falling outside this interval.
    a. Draw a graph of $|\psi(x)|^2$ for these electrons.
    b. What is the probability that an electron will land within the interval 0.79 to 0.81 cm?
    c. If 10^6 electrons are detected, how many will be detected in the interval 0.79 to 0.81 cm?
    d. What is the probability density at $x = 0.80$ cm?

32. In an experiment with 10,000 electrons, which land symmetrically on both sides of $x = 0$, 5000 are detected in the range $-1.0 \text{ cm} \leq x \leq +1.0$ cm, 7500 are detected in the range $-2.0 \text{ cm} \leq x \leq +2.0$ cm, and all 10,000 are detected in the range $-3.0 \text{ cm} \leq x \leq +3.0$ cm. Draw a graph of a probability density that is consistent with these data. (There may be more than one acceptable answer.)

33. Figure P39.33 shows $|\psi(x)|^2$ for the electrons in an experiment.
    a. Is the electron wave function normalized? Explain.
    b. Draw a graph of $\psi(x)$ over this same interval. Provide a numerical scale on both axes. (There may be more than one acceptable answer.)
    c. What is the probability that an electron will be detected in a 0.0010-cm-wide region at $x = 0.00$ cm? At $x = 0.50$ cm? At $x = 0.999$ cm?
    d. If 10^4 electrons are detected, how many are expected to land in the interval -0.30 cm $\leq x \leq 0.30$ cm?

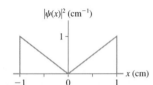

FIGURE P39.33

34. Figure P39.34 shows the wave function of a particle confined between $x = 0$ nm and $x = 1$ nm. The wave function is zero outside this region.
    a. Determine the value of the constant c, as defined in the figure.
    b. Draw a graph of the probability density $P(x) = |\psi(x)|^2$.
    c. Draw a dot picture showing where the first 40 or 50 particles might be found.
    d. Calculate the probability of finding the particle in the interval 0.0 nm $\leq x \leq 0.3$ nm.

FIGURE P39.34

35. Figure P39.35 shows the wave function of a particle confined between $x = -4$ mm and $x = 4$ mm. The wave function is zero outside this region.
    a. Determine the value of the constant c, as defined in the figure.
    b. Draw a graph of the probability density $P(x) = |\psi(x)|^2$.
    c. Draw a dot picture showing where the first 40 or 50 particles might be found.
    d. Calculate the probability of finding the particle in the interval -2.0 mm $\leq x \leq 2.0$ mm.

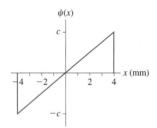

FIGURE P39.35

36. Figure P39.36 shows the probability density for finding a particle at position x.
    a. Determine the value of the constant a, as defined in the figure.

b. At what value of x are you most likely to find the particle? Explain.
c. Within what range of positions centered on your answer to part b are you 75% certain of finding the particle?
d. Interpret your answer to part c by drawing the probability density graph and shading the appropriate region.

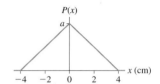

FIGURE P39.36

37. An electron that is confined to $x \geq 0$ nm has the normalized wave function

$$\psi(x) = \begin{cases} 0 & x < 0 \text{ nm} \\ (1.414 \text{ nm}^{-1/2})e^{-x/(1.0 \text{ nm})} & x \geq 0 \text{ nm} \end{cases}$$

where x is in nm.
a. What is the probability of finding the electron in a 0.010-nm-wide region at $x = 1.0$ nm?
b. What is the probability of finding the electron in the interval 0.50 nm $\leq x \leq 1.50$ nm?

38. A particle is described by the wave function

$$\psi(x) = \begin{cases} ce^{x/L} & x \leq 0 \text{ mm} \\ ce^{-x/L} & x \geq 0 \text{ mm} \end{cases}$$

where $L = 2.0$ mm.
a. Sketch graphs of both the wave function and the probability density as functions of x.
b. Determine the normalization constant c.
c. Calculate the probability of finding the particle within 1.0 mm of the origin.
d. Interpret your answer to part b by shading the region representing this probability on the appropriate graph in part a.

39. Consider the electron wave function

$$\psi(x) = \begin{cases} c\sqrt{1 - x^2} & |x| \leq 1 \text{ cm} \\ 0 & |x| \geq 1 \text{ cm} \end{cases}$$

where x is in cm.
a. Determine the normalization constant c.
b. Draw a graph of $\psi(x)$ over the interval -2 cm $\leq x \leq 2$ cm. Provide numerical scales on both axes.
c. Draw a graph of $|\psi(x)|^2$ over the interval -2 cm $\leq x \leq 2$ cm. Provide numerical scales.
d. If 10^4 electrons are detected, how many will be in the interval 0.00 cm $\leq x \leq 0.50$ cm?

40. Consider the electron wave function

$$\psi(x) = \begin{cases} c\sin\left(\dfrac{2\pi x}{L}\right) & 0 \leq x \leq L \\ 0 & x < 0 \text{ or } x > L \end{cases}$$

a. Determine the normalization constant c. Your answer will be in terms of L.
b. Draw a graph of $\psi(x)$ over the interval $-L \leq x \leq 2L$.
c. Draw a graph of $|\psi(x)|^2$ over the interval $-L \leq x \leq 2L$.
d. What is the probability that an electron is in the interval $0 \leq x \leq L/3$?

41. The probability density for finding a particle at position x is

$$P(x) = \begin{cases} \dfrac{a}{(1-x)} & -1 \text{ mm} \le x < 0 \text{ mm} \\ b(1-x) & 0 \text{ mm} \le x \le 1 \text{ mm} \end{cases}$$

and zero elsewhere.
   a. You will learn in Chapter 40 that the wave function must be a *continuous* function. Assuming that to be the case, what can you conclude about the relationship between a and b?
   b. Draw a graph of the probability density over the interval $-2 \text{ mm} \le x \le 2 \text{ mm}$.
   c. Determine values for a and b.
   d. What is the probability that the particle will be found to the left of the origin?

42. A pulse of light is created by the superposition of many waves that span the frequency range $f_0 - \frac{1}{2}\Delta f \le f \le f_0 + \frac{1}{2}\Delta f$, where $f_0 = c/\lambda$ is called the *center frequency* of the pulse. Laser technology can generate a pulse of light that has a wavelength of 600 nm and lasts a mere 6.0 fs (1 fs = 1 femtosecond $= 10^{-15}$ s).
   a. What is the center frequency of this pulse of light?
   b. How many cycles, or oscillations, of the light wave are completed during the 6.0 fs pulse?
   c. What range of frequencies must be superimposed to create this pulse?
   d. What is the spatial length of the laser pulse as it travels through space?
   e. Draw a snapshot graph of this wave packet.

43. A small speck of dust with mass 1.0×10^{-13} g has fallen into the hole shown in Figure P39.43 and appears to be at rest. According to the uncertainty principle, could this particle have enough energy to get out of the hole? If not, what is the deepest hole of this width from which it would have a good chance to escape?

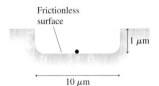

Frictionless surface

1 μm

FIGURE P39.43   10 μm

44. Physicists use laser beams to create an *atom trap* in which atoms are confined within a spherical region of space with a diameter of about 1 mm. The scientists have been able to cool the atoms in an atom trap to a temperature of approximately 1 nK, which is extremely close to absolute zero, but it would be interesting to know if this temperature is close to any limit set by quantum physics. We can explore this issue with a one-dimensional model of a sodium atom in a 1-mm-long box.
   a. Estimate the *smallest* range of speeds you might find for a sodium atom in this box.
   b. Even if we do our best to bring a group of sodium atoms to rest, individual atoms will have speeds within the range you found in part a. Because there's a distribution of speeds, suppose we estimate that the root-mean-square speed v_{rms} of the atoms in the trap is half the value you found in part a. Use this v_{rms} to estimate the temperature of the atoms when they've been cooled to the limit set by the uncertainty principle.

45. You learned in Chapter 37 that, except for hydrogen, the mass of a nucleus with atomic number Z is larger than the mass of the Z protons. The additional mass was ultimately discovered to be due to neutrons, but prior to the discovery of the neutron it was suggested that a nucleus with mass number A might contain A protons and $(A - Z)$ electrons. Such a nucleus would have the mass of A protons, but its net charge would be only Ze.
   a. We know that the diameter of a nucleus is approximately 10 fm. Model the nucleus as a one-dimensional box and find the minimum range of speeds that an electron would have in such a box.
   b. What does your answer imply about the possibility that the nucleus contains electrons? Explain.

46. a. Starting with the expression $\Delta f \Delta t \approx 1$ for a wave packet, find an expression for the product $\Delta E \Delta t$ for a photon.
   b. Interpret your expression. What does it tell you?
   c. The Bohr model of atomic quantization says that an atom in an excited state can jump to a lower-energy state by emitting a photon. The Bohr model says nothing about how long this process takes. You'll learn in Chapter 41 that the time any particular atom spends in the excited state before emitting a photon is unpredictable, but the *average lifetime* Δt of many atoms can be determined. You can think of Δt as being the uncertainty in your knowledge of how long the atom spends in the excited state. A typical value is $\Delta t \approx 10$ ns. Consider an atom that emits a photon with a 500 nm wavelength as it jumps down from an excited state. What is the uncertainty in the energy of the photon? Give your answer in eV.
   d. What is the *fractional uncertainty* $\Delta E/E$ in the photon's energy?

Challenge Problems

47. Figure CP39.47 shows 1.0-μm-diameter dust particles ($m = 1.0 \times 10^{-15}$ kg) in a vacuum chamber. The dust particles are released from rest above a 1.0-μm-diameter hole, fall through the hole (there's just barely room for the particles to go through), and land on a detector at distance d below.
   a. If the particles were purely classical, they would all land in the same 1.0-μm-diameter circle. But quantum effects don't allow this. If $d = 1.0$ m, by how much does the diameter of the circle in which most dust particles land exceed 1.0 μm? Is this increase in diameter likely to be detectable?
   b. Quantum effects would be noticeable if the detection-circle diameter increased by 10% to 1.1 μm. At what distance d would the detector need to be placed to observe this increase in the diameter?

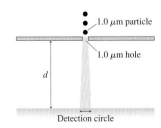

1.0 μm particle

1.0 μm hole

d

FIGURE CP39.47   Detection circle

48. The wave function of a particle is

$$\psi(x) = \sqrt{\frac{b}{\pi(x^2 + b^2)}}$$

   where b is a positive constant. Find the probability that the particle is located in the interval $-b \leq x \leq b$.

49. The wave function of a particle is

$$\psi(x) = \begin{cases} \dfrac{b}{(1 + x^2)} & -1 \text{ mm} \leq x < 0 \text{ mm} \\ c(1 + x)^2 & 0 \text{ mm} \leq x \leq 1 \text{ mm} \end{cases}$$

   and zero elsewhere.

   a. You will learn in Chapter 40 that the wave function must be a *continuous* function. Assuming that to be the case, what can you conclude about the relationship between b and c?
   b. Draw graphs of the wave function and the probability density over the interval -2 mm $\leq x \leq 2$ mm.
   c. What is the probability that the particle will be found to the right of the origin?

50. Consider the electron wave function

$$\psi(x) = \begin{cases} cx & |x| \leq 1 \text{ nm} \\ \dfrac{c}{x} & |x| \geq 1 \text{ nm} \end{cases}$$

   where x is in nm.

   a. Determine the normalization constant c.
   b. Draw a graph of $\psi(x)$ over the interval -5 nm $\leq x \leq 5$ cm. Provide numerical scales on both axes.
   c. Draw a graph of $|\psi(x)|^2$ over the interval -5 nm $\leq x \leq 5$ nm. Provide numerical scales.
   d. If 10^6 electrons are detected, how many will be in the interval -1.0 nm $\leq x \leq 1.0$ nm?

STOP TO THINK ANSWERS

Stop to Think 39.1: 10. The probability of a 1 is $P_1 = \frac{1}{6}$. Similarly, $P_6 = \frac{1}{6}$. The probability of a 1 *or* 6 is $P_{1 \text{ or } 6} = \frac{1}{6} + \frac{1}{6} = \frac{1}{3}$. Thus the expected number is $30\left(\frac{1}{3}\right) = 10$.

Stop to Think 39.2: A > B = D > C. $|A(x)|^2$ is proportional to the density of dots.

Stop to Think 39.3: x_C. The probability is largest at the point where the *square* of $\psi(x)$ is largest.

Stop to Think 39.4: b. The area $\frac{1}{2}a(2 \text{ mm})$ must equal 1.

Stop to Think 39.5: b. $\Delta t = 1.0 \times 10^{-7}$ s. The bandwidth is $\Delta f_B = 1/\Delta t = 1.0 \times 10^7$ Hz = 10 MHz.

Stop to Think 39.6: A. Wave packet A has a smaller spatial extent Δx. The wavelength isn't relevant.

40 One-Dimensional Quantum Mechanics

An example of atomic engineering. Thirty-five xenon atoms have been manipulated into position with the probe tip of a scanning tunneling microscope.

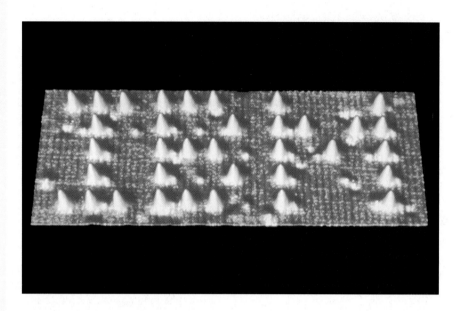

▶ **Looking Ahead**

The goal of Chapter 40 is to understand and apply the essential ideas of quantum mechanics. In this chapter you will learn to:

- Use a strategy for finding and interpreting wave functions.
- Draw wave functions with appropriate shapes.
- Use potential-energy functions to make quantum-mechanical models.
- Understand and use several important quantum-mechanical models.
- Calculate the probability of quantum-mechanical tunneling.

◀ **Looking Back**

Quantum mechanics will be developed around two fundamental ideas: energy diagrams and wave functions. A review of energy diagrams in Chapter 10 is especially important. Please review:

- Section 10.7 Energy diagrams.
- Sections 38.4–38.5 Matter waves and the Bohr model of quantization.
- Sections 39.3–39.4 Wave functions and normalization.

Quantum mechanics is not just for physicists any more. It is now an essential tool in the design of semiconductor devices such as diode lasers. Whole new classes of devices, called *quantum-well devices,* have been designed and built to exploit the quantized energy levels. We will look at some examples in this chapter.

Also at the cutting edge of engineering science is the design and manufacture of *nanostructures*—small machines or other devices only a few hundred nanometers in size. Many scientists and engineers envision a day in the near future when nanostructures will be constructed literally atom by atom. Quantum effects will be important in devices this small. This photograph, of a curious symbolic structure that scientists at IBM's research laboratory built by moving xenon atoms around on a metal surface, shows an early example of "atomic engineering."

Our goal for this chapter is to introduce the essential ideas of quantum mechanics. Although the real world is three-dimensional, we will limit our study of quantum mechanics to one dimension. This will allow us to focus on the fundamental concepts of quantum physics without becoming overwhelmed by mathematical complications. We will discuss some of the aspects of finding and using wave functions, then look at several applications of quantum mechanics. We'll conclude this chapter with a look at a phenomenon called *quantum-mechanical tunneling,* one of the more startling aspects of quantum physics.

40.1 Schrödinger's Equation: The Law of Psi

In the fall of 1925, just before Christmas, the Austrian physicist Erwin Schrödinger gathered together a few books and headed off to a villa in the Swiss Alps. He had recently learned of de Broglie's 1924 suggestion that matter has wave-like properties, and he wanted some time free from distractions to think about it. Before the trip was done, Schrödinger had discovered the law of quantum mechanics.

Erwin Schrödinger.

Schrödinger's goal was to predict the outcome of atomic experiments, a goal that had eluded classical physics. The mathematical equation that he developed is now called the **Schrödinger equation.** It is the law of quantum mechanics in the same way that Newton's laws are the laws of classical mechanics. It would make sense to call it Schrödinger's law, but by tradition it is called simply the Schrödinger equation.

You learned in Chapter 39 that a matter particle is characterized in quantum physics by its wave function $\psi(x)$. If you know a particle's wave function, you can predict the probability of detecting it in some region of space. That's all well and good, but Chapter 39 didn't provide any method for determining wave functions. The Schrödinger equation is the missing piece of the puzzle. It is an equation for finding a particle's wave function $\psi(x)$ along the x-axis.

Consider an atomic particle with mass m and mechanical energy E whose interactions with the environment can be characterized by a one-dimensional potential-energy function $U(x)$. The Schrödinger equation for the particle's wave function is

$$\frac{d^2\psi}{dx^2} = -\frac{2m}{\hbar^2}[E - U(x)]\psi(x) \qquad \text{(the Schrödinger equation)} \qquad (40.1)$$

This is a differential equation whose solution is the wave function $\psi(x)$ that we seek. Our first goal is to learn what this equation means and how it is used.

Justifying the Schrödinger Equation

The Schrödinger equation can be neither derived nor proved. It is not an outgrowth of any previous theory. Its success depended on its ability to explain the various phenomena that had refused to yield to a classical-physics analysis and to make new predictions that were subsequently verified.

Although the Schrödinger equation cannot be derived, the reasoning behind it can at least be made *plausible*. De Broglie had postulated a wave-like nature for matter in which a particle of mass m, velocity v, and momentum $p = mv$ has a wavelength

$$\lambda = \frac{h}{p} = \frac{h}{mv} \qquad (40.2)$$

Schrödinger's goal was to find a *wave equation* for which the solution would be a wave function having the de Broglie wavelength.

An oscillatory wave-like function with wavelength λ is

$$\psi(x) = \psi_0 \sin\left(\frac{2\pi x}{\lambda}\right) \qquad (40.3)$$

where ψ_0 is the amplitude of the wave function. Suppose we take a second derivative of $\psi(x)$:

$$\frac{d\psi}{dx} = \frac{2\pi}{\lambda}\psi_0 \cos\left(\frac{2\pi x}{\lambda}\right)$$

$$\frac{d^2\psi}{dx^2} = \frac{d}{dx}\frac{d\psi}{dx} = -\frac{(2\pi)^2}{\lambda^2}\psi_0 \sin\left(\frac{2\pi x}{\lambda}\right)$$

We can use the definition of $\psi(x)$, from Equation 40.3, to write the second derivative as

$$\frac{d^2\psi}{dx^2} = -\frac{(2\pi)^2}{\lambda^2}\psi(x) \qquad (40.4)$$

Equation 40.4 relates the wavelength λ to a combination of the wave function $\psi(x)$ and its second derivative.

NOTE ▶ These manipulations are not specific to quantum mechanics. Equation 40.4, which is well known for classical waves, applies equally well to sound waves or waves on a string. ◀

Schrödinger's insight was to identify λ with the de Broglie wavelength of a particle. We can write the de Broglie wavelength in terms of the particle's kinetic energy K as

$$\lambda = \frac{h}{mv} = \frac{h}{\sqrt{2m(\frac{1}{2}mv^2)}} = \frac{h}{\sqrt{2mK}} \qquad (40.5)$$

Notice that **the de Broglie wavelength increases as the particle's kinetic energy decreases.** This observation will play a key role as we develop an understanding of wave functions.

If we square this expression for λ and substitute it into Equation 40.4, we find

$$\frac{d^2\psi}{dx^2} = -\frac{(2\pi)^2 2mK}{h^2}\psi(x) = -\frac{2m}{\hbar^2}K\psi(x) \qquad (40.6)$$

where $\hbar = h/2\pi$. Equation 40.6 is a differential equation for the function $\psi(x)$. The solution to this differential equation is the sinusoidal wave function of Equation 40.3, where λ is the de Broglie wavelength for a particle with kinetic energy K.

Our derivation of Equation 40.6 assumed that the particle's kinetic energy K is constant. The energy diagram of Figure 40.1a reminds you that a particle's kinetic energy remains constant as it moves along the x-axis only if its potential energy U is constant. In this case, the de Broglie wavelength is the same at all positions.

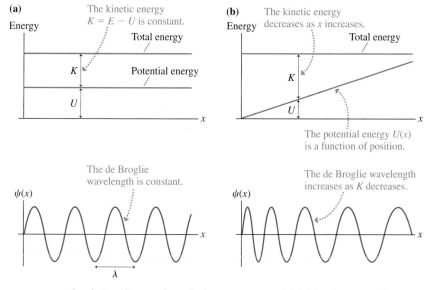

FIGURE 40.1 The de Broglie wavelength changes as a particle's kinetic energy changes.

In contrast, Figure 40.1b shows the energy diagram for a particle whose potential energy changes with x and whose kinetic energy is *not* constant. This particle speeds up and slows down as it moves along the x-axis, transforming potential energy to kinetic energy or vice versa. Consequently, its de Broglie wavelength changes with position.

Suppose a particle's potential energy—gravitational or electric or any other kind of potential energy—is described by the function $U(x)$ or $U(y)$. That is, the potential energy is a *function of position* along the axis of motion. For example, the gravitational potential energy near the earth's surface is the function $U(y) = mgy$.

If E is the particle's total mechanical energy, its kinetic energy at position x is

$$K = E - U(x) \qquad (40.7)$$

If we use this expression for K in Equation 40.6, that equation becomes

$$\frac{d^2\psi}{dx^2} = -\frac{2m}{\hbar^2}[E - U(x)]\psi(x)$$

This is Equation 40.1, the Schrödinger equation for the particle's wave function $\psi(x)$.

> **NOTE** ▶ This has not been a derivation of the Schrödinger equation. We've made a *plausibility argument,* based on de Broglie's hypothesis about matter waves, but only experimental evidence will show if this equation has merit. ◀

STOP TO THINK 40.1 Three de Broglie waves are shown for particles of equal mass. Rank in order, from largest to smallest, the speeds of particles a, b, and c.

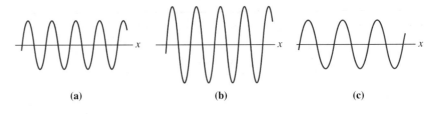

(a)   (b)   (c)

Quantum-Mechanical Models

Long ago, in your study of Newtonian mechanics, you learned of the importance of *models*. To understand the motion of an object, we made many simplifying assumptions: that the object could be represented by a particle, that friction could be described in a simple way, that air resistance could be neglected, and so on. In other words, we constructed a model that gave a good, though not perfect, description of reality. Models allowed us to understand the primary features of an object's motion without getting lost in the details.

The same holds true in quantum mechanics. The exact description of a microscopic atom or a solid is extremely complicated. Our only hope for using quantum mechanics effectively is to make a number of simplifying assumptions—that is, to make a **quantum-mechanical model** of the situation. Much of this chapter will be about building and using quantum mechanical models.

The test of a model's success is its agreement with experimental measurement. Laboratory experiments cannot measure $\psi(x)$, and they rarely make direct measurements of probabilities. Thus it will be important to tie our models to measurable quantities such as wavelengths, charges, currents, times, and temperatures.

There's one very important difference between models in classical mechanics and quantum mechanics. Classical models are described in terms of *forces,* and Newton's laws are a connection between force and motion. The Schrödinger equation for the wave function is written in terms of *energies.* Consequently, quantum-mechanical modeling involves finding a potential-energy function $U(x)$ that describes a particle's interactions with its environment

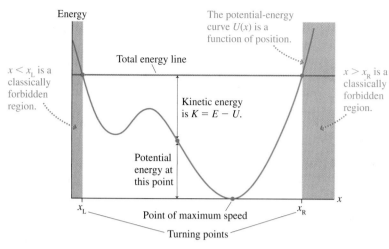

FIGURE 40.2 Interpreting an energy diagram.

Figure 40.2 reminds you how to interpret an energy diagram. We will use energy diagrams extensively in this and the remaining chapters to portray quantum-mechanical models. A review of Section 10.7, where energy diagrams were introduced, is highly recommended.

The primary limitation to our quantum-mechanical models is the fact that the real world is three-dimensional. Unfortunately, solving the Schrödinger equation in three dimensions raises the mathematical level beyond what is appropriate for this text. Although the one-dimensional models that we will use are oversimplifications, there are many situations for which a one-dimensional model is a reasonably good approximation. Our major goal is to learn how quantum-mechanical concepts are used, and one-dimensional quantum mechanics will be sufficient for this purpose.

Activ
Physics ONLINE   20.1

40.2 Solving the Schrödinger Equation

The Schrödinger equation is a second-order differential equation, meaning that it is a differential equation for $\psi(x)$ involving its second derivative. However, this textbook does not assume that you know how to solve differential equations. As we did with Newton's laws, we will restrict ourselves to situations where the mathematical skills are those you have been developing in calculus.

The solution to an algebraic equation is simply a number. For example, $x = 3$ is the solution to the equation $2x = 6$. In contrast, the solution to a differential equation is a *function*. You saw this idea in the previous section, where Equation 40.6 was constructed so that the function $\psi(x) = \psi_0 \sin(2\pi x/\lambda)$ would be a solution.

The Schrödinger equation can't be solved until the potential-energy function $U(x)$ has been specified. Different potential-energy functions result in different wave functions, just as different forces lead to different trajectories in classical mechanics. Once $U(x)$ has been specified, the solution of the differential equation is a *function* $\psi(x)$ that is defined at all values of x. We will usually display the solution as a graph of $\psi(x)$ versus x.

Restrictions and Boundary Conditions

Not all functions $\psi(x)$ make *acceptable* solutions to the Schrödinger equation. That is, there may be functions that satisfy the Schrödinger equation but that are not physically meaningful. We have previously encountered restrictions in our solutions of algebraic equations. We insist, for physical reasons, that masses be positive rather than negative numbers, that positions be real rather than imaginary numbers, and so on. Mathematical solutions not meeting these restrictions are rejected as being unphysical.

Because we want to interpret $|\psi(x)|^2$ as a probability density, we have to insist that the function $\psi(x)$ be one for which this interpretation is possible. The conditions or restrictions on acceptable solutions are called the **boundary conditions.** You will see, in later examples, how the boundary conditions help us to choose the correct solution for $\psi(x)$. The primary conditions that the wave function must obey are

1. $\psi(x)$ is a continuous function.
2. $\psi(x) = 0$ if x is in a region where it is physically impossible for the particle to be.
3. $\psi(x) \rightarrow 0$ as $x \rightarrow +\infty$ and $x \rightarrow -\infty$.
4. $\psi(x)$ is a normalized function.

The last is not, strictly speaking, a boundary condition but is an auxiliary condition we require for the wave function to have a useful interpretation. Boundary condition 3 is needed to enable the normalization integral $\int |\psi(x)|^2 dx$ to converge.

Once boundary conditions have been established, there are three general approaches to solving the Schrödinger equation: Use general techniques for solving second-order differential equations, use a numerical technique to solve the equation on a computer, or guess.

More advanced courses make extensive use of the first and second approaches. However, we are not assuming a knowledge of differential equations, so you will not be asked to use these methods. The third, although it sounds almost like cheating, is widely used in simple situations where we can use physical arguments to infer the functional form of the wave function. The upcoming examples will illustrate this third approach.

A quadratic algebraic equation has two different solutions. Similarly, a second-order differential equation has two independent solutions $\psi_1(x)$ and $\psi_2(x)$. By "independent solutions" we mean that $\psi_2(x)$ is not just a constant multiple of $\psi_1(x)$, such as $3\psi_1(x)$, but that $\psi_1(x)$ and $\psi_2(x)$ are totally different functions.

Suppose that $\psi_1(x)$ and $\psi_2(x)$ are known to be two independent solutions of the Schrödinger equation. A theorem you will learn in differential equations states that a *general solution* of the equation can be written

$$\psi(x) = A\psi_1(x) + B\psi_2(x) \tag{40.8}$$

where A and B are constants whose values are determined by the boundary conditions. In other words, the general solution is a *superposition,* one of the primary characteristics of waves. Equation 40.8 is a powerful statement, although one that will make more sense after you see it applied in upcoming examples. The main point is that **if we can find two solutions $\psi_1(x)$ and $\psi_2(x)$ by guessing, then Equation 40.8 is** *the* **general solution to the Schrödinger equation.**

Quantization

We've asserted that the Schrödinger equation is the law of quantum mechanics, but thus far we've not said anything about quantization. Although the particle's total energy E appears in the Schrödinger equation, it is treated in the equation as an unspecified constant. However, it will turn out that there are *no* acceptable solutions for most values of E. That is, there are no functions $\psi(x)$ that satisfy both the Schrödinger equation *and* the boundary conditions. Acceptable solutions exist only for *discrete* values of E. The energies for which solutions exist are the quantized energies of the system. Thus, as you'll see, the Schrödinger equation has quantization as a built-in feature.

Problem Solving in Quantum Mechanics

Our problem-solving strategy for classical mechanics focused on identifying and using forces. In quantum mechanics we're interested in *energy* rather than forces. The critical step in solving a problem in quantum mechanics is to determine the particle's potential-energy function $U(x)$. Identifying the interactions that cause a potential energy is the *physics* of the problem. Once the potential-energy function is known, it is "just mathematics" to solve for the wave function.

 PROBLEM-SOLVING STRATEGY 40.1 Quantum-mechanics problems

MODEL Determine a potential-energy function that describes the particle's interactions. Make simplifying assumptions.

VISUALIZE The potential-energy curve is the pictorial representation.

- Draw the potential-energy curve.
- Identify known information.
- Establish the boundary conditions that the wave function must satisfy.

SOLVE The Schrödinger equation is the mathematical representation.

- Utilize the boundary conditions.
- Normalize the wave functions.
- Draw graphs of $\psi(x)$ and $|\psi(x)|^2$.
- Determine the allowed energy levels.
- Calculate probabilities, wavelengths, or other specific quantities.

ASSESS Check that your result has the correct units, is reasonable, and answers the question.

The solutions to the Schrödinger equation are the stationary states of the system. Bohr had postulated the existence of stationary states, but he didn't know how to find them. Now we have a strategy for finding them.

Bohr's idea of transitions, or quantum jumps, between stationary states remains very important in Schrödinger's quantum mechanics. The system can jump from one stationary state, characterized by wave function $\psi_i(x)$ and energy E_i, to another state, characterized by $\psi_f(x)$ and E_f, by emitting or absorbing a photon of frequency

$$f = \frac{\Delta E}{h} = \frac{|E_f - E_i|}{h}$$

Thus the solutions to the Schrödinger equation will allow us to predict the emission and absorption spectra of a quantum system. These predictions will test the validity of Schrödinger's theory.

40.3 A Particle in a Rigid Box: Energies and Wave Functions

Figure 40.3 shows a particle of mass m confined in a rigid, one-dimensional box of length L. The walls of the box are assumed to be perfectly rigid, and the particle undergoes perfectly elastic reflections from the ends. This situation is known as a "particle in a box."

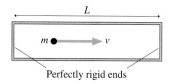

FIGURE 40.3 A particle in a rigid box of length L.

20.2 Activ
Physics

A classical particle bounces back and forth between the walls of the box. There are no restrictions on the speed or kinetic energy of a classical particle. In contrast, a wave-like particle characterized by a de Broglie wavelength sets up a standing wave as it reflects back and forth. In Chapters 24 and 38, we found that a standing de Broglie wave automatically leads to energy quantization. That is, only certain discrete energies are allowed. However, our hypothesis of a de Broglie standing wave was just a guess, with no real justification, because we had no theory as to how a wave-like particle ought to behave.

We will now revisit this problem from the new perspective of quantum mechanics. The basic questions we want to answer in this, and any quantum-mechanics problem, are

- What are the allowed energies of the particle?
- What is the wave function associated with each energy?
- In which part of the box is the particle most likely to be found?

We can answer these questions by following the steps of the problem-solving strategy.

Model: Identify a Potential-Energy Function

By a *rigid box* we mean a box whose walls are so sturdy that they can confine a particle no matter how fast the particle moves. Furthermore, the walls are so stiff that they do not flex or give as the particle bounces. No real container has these attributes, so the rigid box is a *model* of a situation in which a particle is extremely well confined. Our first task is to characterize the rigid box in terms of a potential-energy function.

Let's establish a coordinate axis with the boundaries of the box at $x = 0$ and $x = L$. The rigid box has three important characteristics:

1. The particle can move freely between 0 and L at constant speed and thus with constant kinetic energy.
2. No matter how much kinetic energy the particle has, its turning points are at $x = 0$ and $x = L$.
3. The regions $x < 0$ and $x > L$ are forbidden. The particle cannot leave the box.

A potential-energy function that describes the particle in this situation is

$$U_{\text{rigid box}}(x) = \begin{cases} 0 & 0 \le x \le L \\ \infty & x < 0 \quad \text{or} \quad x > L \end{cases} \tag{40.9}$$

Inside the box, the particle has only kinetic energy. The infinitely high potential-energy barriers prevent the particle from ever having $x < 0$ or $x > L$ no matter how much kinetic energy it may have. It is this potential energy for which we want to solve the Schrödinger equation.

Visualize: Establish Boundary Conditions

Figure 40.4 is the energy diagram of a particle in the rigid box. You can see that $U = 0$ and $E = K$ inside the box. The upward arrows labeled ∞ indicate that the potential energy becomes infinitely large at the walls of the box ($x = 0$ and $x = L$).

NOTE ▶ Figure 40.4 is not a picture of the box. It is a graphical representation of the particle's kinetic and potential energy. ◀

Next, we need to establish the boundary conditions that the solution must satisfy. Because it is physically impossible for the particle to be outside the box, we require

$$\psi(x) = 0 \quad \text{for } x < 0 \quad \text{or} \quad x > L \tag{40.10}$$

That is, there is zero probability of finding the particle outside the box.

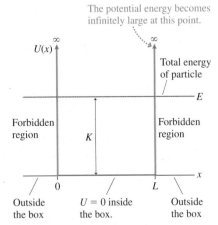

The potential energy becomes infinitely large at this point.

FIGURE 40.4 The energy diagram of a particle in a rigid box of length L.

Furthermore, the wave function must be a *continuous* function. That is, there can be no break in the wave function at any point. Because the solution is zero everywhere outside the box, continuity requires that the wave function inside the box obey the two conditions

$$\psi(\text{at } x = 0) = 0 \quad \text{and} \quad \psi(\text{at } x = L) = 0 \qquad (40.11)$$

In other words, as Figure 40.5 shows, the oscillating wave function inside the box must go to zero at the boundaries to be continuous with the wave function outside the box. This requirement of the wave function is equivalent to saying that a standing wave on a string must have a node at the ends because it is fixed at those points and cannot move.

Solve I: Find the Wave Functions

At all points *inside* the box the potential energy is $U(x) = 0$. Thus the Schrödinger equation inside the box is

$$\frac{d^2\psi}{dx^2} = -\frac{2mE}{\hbar^2}\psi(x) \qquad (40.12)$$

There are two aspects to solving this equation:

1. For what values of E does Equation 40.12 have physically meaningful solutions?
2. What are the solutions $\psi(x)$ for those values of E?

To begin, let's simplify the notation by defining $\beta^2 = 2mE/\hbar^2$. Equation 40.12 is then

$$\frac{d^2\psi}{dx^2} = -\beta^2\psi(x) \qquad (40.13)$$

We're going to solve this differential equation by guessing! Can you think of any functions whose second derivative is a *negative* constant times the function itself? Two such functions are

$$\psi_1(x) = \sin\beta x \quad \text{and} \quad \psi_2(x) = \cos\beta x \qquad (40.14)$$

Both are solutions to Equation 40.13 because

$$\frac{d^2\psi_1}{dx^2} = \frac{d^2}{dx^2}(\sin\beta x) = -\beta^2\sin\beta x = -\beta^2\psi_1(x)$$

$$\frac{d^2\psi_2}{dx^2} = \frac{d^2}{dx^2}(\cos\beta x) = -\beta^2\cos\beta x = -\beta^2\psi_2(x)$$

Furthermore, these are *independent* solutions because $\psi_2(x)$ is not a multiple or a rearrangement of $\psi_1(x)$. Consequently, according to Equation 40.8, the general solution to the Schrödinger equation for the particle in a rigid box is

$$\psi(x) = A\sin\beta x + B\cos\beta x \qquad (40.15)$$

where

$$\beta = \frac{\sqrt{2mE}}{\hbar} \qquad (40.16)$$

The constants A and B must be determined by using the boundary conditions of Equation 40.11. First, the wave function must go to zero at $x = 0$. That is,

$$\psi(\text{at } x = 0) = A\cdot 0 + B\cdot 1 = 0 \qquad (40.17)$$

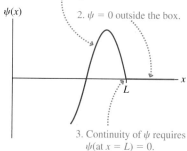

1. Inside the box, ψ is oscillating in some way still to be determined.

2. $\psi = 0$ outside the box.

3. Continuity of ψ requires $\psi(\text{at } x = L) = 0$.

FIGURE 40.5 Applying boundary conditions to the wave function of a particle in a box.

This boundary condition can be satisfied only if $B = 0$. The $\cos \beta x$ term may satisfy the differential equation in a mathematical sense, but it is not a physically meaningful solution for this problem because it does not satisfy the boundary conditions. Thus the physically meaningful solution is

$$\psi(x) = A \sin \beta x$$

The wave function must also go to zero at $x = L$. That is,

$$\psi(\text{at } x = L) = A \sin \beta L = 0 \tag{40.18}$$

This condition could be satisfied by $A = 0$, but then we wouldn't have a wave function at all! Fortunately, that isn't necessary. This boundary condition is also satisfied if

$$\beta L = n\pi \quad \text{or} \quad \beta = \frac{n\pi}{L} \qquad n = 1, 2, 3, \ldots \tag{40.19}$$

Notice that n starts with 1, not 0. The value $n = 0$ would give $\beta = 0$ and make $\psi = 0$ at all points, a physically meaningless solution.

Thus the solutions to the Schrödinger equation for a particle in a rigid box are

$$\psi_n(x) = A \sin \beta_n x = A \sin\left(\frac{n\pi x}{L}\right) \qquad n = 1, 2, 3, \ldots \tag{40.20}$$

We've found a whole *family* of solutions, each corresponding to a different value of the integer n. These wave functions represent the stationary states of the particle in the box. The constant A remains to be determined.

Solve II: Find the Allowed Energies

Equation 40.16 defined β. Equation 40.19 then placed restrictions on the possible values of β:

$$\beta_n = \frac{\sqrt{2mE_n}}{\hbar} = \frac{n\pi}{L} \qquad n = 1, 2, 3, \ldots \tag{40.21}$$

where the value of β and the energy associated with the integer n have been labeled β_n and E_n. We can solve for E_n by squaring both sides:

$$E_n = n^2 \frac{\pi^2 \hbar^2}{2mL^2} = n^2 \frac{h^2}{8mL^2} \qquad n = 1, 2, 3, \ldots \tag{40.22}$$

where, in the last step, we used the definition $\hbar = h/2\pi$. For a particle in a box, **these energies are the only values of E for which there are physically meaningful solutions to the Schrödinger equation.**

We have found that the particle's energy is quantized! It is useful to write the energies of the stationary states as

$$E_n = n^2 E_1 \tag{40.23}$$

where E_n is the energy of the stationary state with *quantum number n*. The smallest possible energy $E_1 = h^2/8mL^2$ is the energy of the $n = 1$ *ground state*. These allowed energies are shown in the *energy-level diagram* of Figure 40.6, where you can see that the allowed energies increase with the square of the quantum number. Recall, from Chapter 38, that an energy-level diagram is not a graph (the horizontal axis doesn't represent anything) but a "ladder" of allowed energies.

Equation 40.22 is identical to the energies we found in Chapter 38 by requiring the de Broglie wave of a particle in a box to form a standing wave. Only now we have a theory that tells not only the energies but also the wave functions.

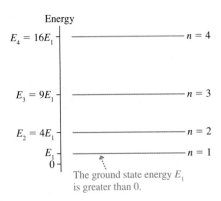

FIGURE 40.6 The energy-level diagram for a particle in a box.

EXAMPLE 40.1 **An electron in a box**

An electron is confined to a rigid box. What is the length of the box if the energy difference between the first and second states is 3.0 eV?

MODEL Model the electron as a particle in a rigid one-dimensional box of length L.

SOLVE The first two quantum states, with $n = 1$ and $n = 2$, have energies E_1 and $E_2 = 4E_1$. Thus the energy difference between the states is

$$\Delta E = 3E_1 = \frac{3h^2}{8mL^2} = 3.0\ \text{eV} = 4.8 \times 10^{-19}\ \text{J}$$

The length of the box for which $\Delta E = 3.0$ eV is

$$L = \sqrt{\frac{3h^2}{8m\,\Delta E}} = 6.14 \times 10^{-10}\ \text{m} = 0.614\ \text{nm}$$

ASSESS The expression for E_1 is in SI units, so energies must be in J, not eV.

Solve III: Normalize the Wave Functions

We can determine the constant A by requiring the wave functions to be normalized. The normalization condition, which we found in Chapter 39, is

$$\int_{-\infty}^{\infty} |\psi(x)|^2 dx = 1$$

This is the mathematical statement that the particle must be *somewhere* on the x-axis. The integration limits extend to $\pm\infty$, but here we need to integrate only from 0 to L because the wave function is zero outside the box. Thus normalization requires

$$\int_0^L |\psi_n(x)|^2 dx = A_n^2 \int_0^L \sin^2\left(\frac{n\pi x}{L}\right) dx = 1 \tag{40.24}$$

or

$$A_n = \left[\int_0^L \sin^2\left(\frac{n\pi x}{L}\right) dx\right]^{-1/2} \tag{40.25}$$

We placed a subscript n on A_n because it is possible that the normalization constant is different for each wave function in the family. This is a standard integral. We will leave it as a homework problem for you to show that its value, for any n, is

$$A_n = \sqrt{\frac{2}{L}} \qquad n = 1, 2, 3, \ldots \tag{40.26}$$

We now have a complete solution to the problem. The normalized wave function for the particle in quantum state n is

$$\psi_n(x) = \begin{cases} \sqrt{\dfrac{2}{L}} \sin\left(\dfrac{n\pi x}{L}\right) & 0 \leq x \leq L \\ 0 & x < 0 \text{ and } x > L \end{cases} \tag{40.27}$$

40.4 A Particle in a Rigid Box: Interpreting the Solution

Our solution to the quantum-mechanical problem of a particle in a box tells us that

1. The particle must have energy $E_n = n^2 E_1$ where $n = 1, 2, 3, \ldots$ is the quantum number and where $E_1 = h^2/8mL^2$ is the energy of the $n = 1$ ground state.

2. The wave function for a particle in quantum state n is

$$\psi_n(x) = \begin{cases} \sqrt{\dfrac{2}{L}}\sin\left(\dfrac{n\pi x}{L}\right) & 0 \le x \le L \\ 0 & x < 0 \text{ and } x > L \end{cases}$$

These are the stationary states of the system.

3. The probability density for finding the particle at position x inside the box is

$$P_n(x) = \left|\psi_n(x)\right|^2 = \frac{2}{L}\sin^2\left(\frac{n\pi x}{L}\right) \tag{40.28}$$

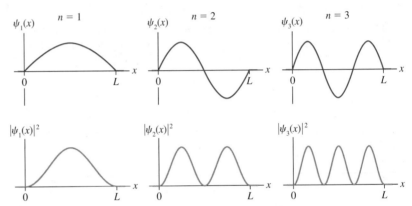

FIGURE 40.7 Wave functions and probability densities for a particle in a rigid box of length L.

A graphical presentation will make these results more meaningful. Figure 40.7 shows the wave functions $\psi(x)$ and the probability densities $P(x) = \left|\psi(x)\right|^2$ for quantum states $n = 1$ to 3. Notice that the wave functions go to zero at the boundaries and thus are continuous with $\psi = 0$ outside the box.

The wave functions $\psi(x)$ for a particle in a rigid box are analogous to standing waves on a string that is tied at both ends. You can see that $\psi_n(x)$ has $n - 1$ nodes (zeros), excluding the ends, and n antinodes (maxima and minima). This is a general result for any wave function, not just for a particle in a rigid box.

Figure 40.8 shows another way in which energies and wave functions are shown graphically in quantum mechanics. First, the graph shows the potential-energy function $U(x)$ of the particle. Second, the allowed energies are shown as horizontal lines (total energy lines) across the potential-energy graph. These are labeled with the quantum number n and the energy E_n. Third—and this is a bit tricky—the wave function for each n is drawn *as if* the energy line were the x-axis. That is, the graph of $\psi_n(x)$ is drawn on top of the E_n energy line. This allows energies and wave functions to be displayed simultaneously, but it does *not* imply that ψ_2 is in any sense "above" ψ_1. Both oscillate sinusoidally about zero, as Figure 40.7 shows.

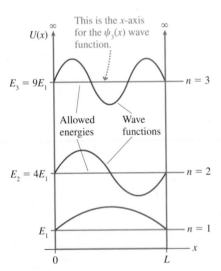

FIGURE 40.8 An alternative way to show the potential-energy diagram, the energies, and the wave functions.

EXAMPLE 40.2 Energy levels and quantum jumps

A semiconductor device known as a *quantum-well device* is designed to "trap" electrons in a 1.0-nm-wide region. Treat this as a one-dimensional problem:

a. What are the energies of the first three quantum states?
b. What wavelengths of light can these electrons absorb?

MODEL Model an electron in a quantum-well device as a particle confined in a rigid box of length $L = 1.0$ nm.

VISUALIZE Figure 40.9 shows the first three energy levels and the transitions by which an electron in the ground state can absorb a photon.

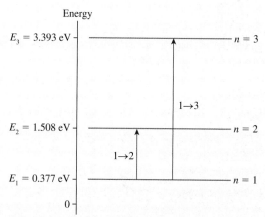

FIGURE 40.9 Energy levels and quantum jumps for an electron in a quantum-well device.

SOLVE

a. The particle's mass is $m = m_e = 9.11 \times 10^{-31}$ kg. The allowed energies, in both J and eV, are

$$E_1 = \frac{h^2}{8mL^2} = 6.03 \times 10^{-20} \text{ J} = 0.377 \text{ eV}$$

$$E_2 = 4E_1 = 1.508 \text{ eV}$$

$$E_3 = 9E_1 = 3.393 \text{ eV}$$

b. An electron spends most of its time in the $n = 1$ ground state. According to Bohr's model of stationary states, the electron can absorb a photon of light and undergo a transition, or quantum jump, to $n = 2$ or $n = 3$ if the light has frequency $f = \Delta E/h$. The wavelengths, given by $\lambda = c/f = hc/\Delta E$, are

$$\lambda_{1\to2} = \frac{hc}{E_2 - E_1} = 1098 \text{ nm}$$

$$\lambda_{1\to3} = \frac{hc}{E_3 - E_1} = 411 \text{ nm}$$

ASSESS In practice, various complications usually make the $1 \to 3$ transition unobservable. But quantum-well devices do indeed exhibit strong absorption and emission at the $\lambda_{1\to2}$ wavelength. In this example, which is typical of quantum-well devices, the wavelength is in the near-infrared portion of the spectrum. Devices such as these are used to construct the semiconductor lasers used in CD players and laser printers.

NOTE ▶ The wavelengths of light emitted or absorbed by a quantum system are determined by the *difference* between two allowed energies. Quantum jumps involve two stationary states. ◀

Zero-Point Motion

The lowest energy state in Example 40.2, the $n = 1$ ground state, has $E_1 = 0.38$ eV. There is no stationary state having $E = 0$. Unlike a classical particle, **a quantum particle in a box cannot be at rest!** No matter how much its energy is reduced, such as by cooling it toward absolute zero, it cannot have energy less than E_1.

The particle motion associated with energy E_1, called the **zero-point motion,** is a consequence of Heisenberg's uncertainty principle. Because the particle is somewhere in the box, its position uncertainty is $\Delta x = L$. If the particle were at rest in the box, we would know that its velocity and momentum are exactly zero with *no* uncertainty: $\Delta p_x = 0$. But then $\Delta x \Delta p_x = 0$ would violate the Heisenberg uncertainty principle. One of the conclusions that follows from the uncertainty principle is that **a confined particle cannot be at rest.**

Although the particle's position and velocity are uncertain, the particle's energy in each state can be calculated with a high degree of precision. This distinction between a precise energy and an uncertain position and velocity seems rather strange, but it is just our old friend the standing wave. In order to *have* a stationary state at all, the de Broglie waves have to form standing waves. Only for very precise frequencies, and thus precise energies, can the standing-wave pattern appear.

EXAMPLE 40.3 **Nuclear energies**

Protons and neutrons are tightly bound within the nucleus of an atom. If we use a one-dimensional model of a nucleus, what are the first three energy levels of a neutron in a 10-fm-diameter nucleus (1 fm = 10^{-15} m)?

MODEL Model the nucleus as a one-dimensional box of length $L = 10$ fm. The neutron is confined within the box.

SOLVE The energy levels, with $L = 10$ fm and $m = m_n = 1.67 \times 10^{-27}$ kg, are

$$E_1 = \frac{h^2}{8mL^2} = 3.29 \times 10^{-13}\,\text{J} = 2.06\,\text{MeV}$$

$$E_2 = 4E_1 = 8.24\,\text{MeV}$$

$$E_3 = 9E_1 = 18.54\,\text{MeV}$$

ASSESS An electron confined in an atom-size space has energies of a few eV. A neutron confined in a nucleus-size space has energies of a few *million* eV.

EXAMPLE 40.4 **The probabilities of locating the particle**

A particle in a rigid box of length L is in its ground state.

a. Where is the particle most likely to be found?
b. What are the probabilities of finding the particle in an interval of width $0.01L$ at $x = 0, 0.25L,$ and $0.50L$?
c. What is the probability of finding the particle in the center half of the box?

MODEL The wave functions for a particle in a rigid box have been determined.

VISUALIZE Figure 40.10 shows the probability density $P_1(x) = |\psi_1(x)|^2$ in the ground state.

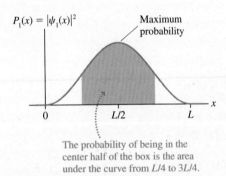

$P_1(x) = |\psi_1(x)|^2$ Maximum probability

The probability of being in the center half of the box is the area under the curve from $L/4$ to $3L/4$.

FIGURE 40.10 Probability density for a particle in the ground state.

SOLVE

a. The particle is most likely to be found at the point where the probability density $P(x)$ is a maximum. You can see from Figure 40.10 that the point of maximum probability for $n = 1$ is $x = L/2$.

b. For a *small* width δx, the probability of finding the particle in δx at position x is

$$\text{Prob(in } \delta x \text{ at } x) = P_1(x)\delta x = |\psi_1(x)|^2\delta x = \frac{2}{L}\sin^2\left(\frac{\pi x}{L}\right)\delta x$$

The interval $\delta x = 0.01L$ is sufficiently small for this to be valid. The probabilities of finding the particle are

Prob(in $0.01L$ at $x = 0.00L$) = 0.000 = 0.0%

Prob(in $0.01L$ at $x = 0.25L$) = 0.010 = 1.0%

Prob(in $0.01L$ at $x = 0.50L$) = 0.020 = 2.0%

c. The center half of the box stretches from $x = L/4$ to $x = 3L/4$. The probability that the particle is in this interval is the area under the probability-density curve:

$$\text{Prob}\left(\text{in interval } \frac{1}{4}L \text{ to } \frac{3}{4}L\right) = \int_{L/4}^{3L/4} P_1(x)\,dx$$

$$= \frac{2}{L}\int_{L/4}^{3L/4}\sin^2\left(\frac{\pi x}{L}\right)dx$$

$$= \left[\frac{x}{L} - \frac{1}{\pi}\sin\left(\frac{\pi x}{L}\right)\cos\left(\frac{\pi x}{L}\right)\right]_{L/4}^{3L/4}$$

$$= \frac{1}{2} + \frac{1}{\pi} = 0.818$$

ASSESS If a particle in a box is in the $n = 1$ ground state, there is an 81.8% chance of finding it in the center half of the box. The probability is greater than 50% because, as you can see in Figure 40.10, the probability density $P_1(x)$ is larger near the center of the box than near the boundaries.

This has been a lengthy presentation of the particle-in-a-box problem. However, it was important that we explore the method of solution completely. Future examples will now go more quickly because many of the issues discussed here will not need to be repeated.

STOP TO THINK 40.2 A particle in a rigid box in the $n = 2$ stationary state is most likely to be found

    a. In the center of the box.
    b. One-third of the way from either end.
    c. One-quarter of the way from either end.
    d. It is equally likely to be found at any point in the box.

40.5 The Correspondence Principle

Suppose we confine an electron in a microscopic box, then allow the box to get bigger and bigger. What started out as a quantum-mechanical situation should, when the box becomes macroscopic in size, eventually look like a classical-physics situation. Similarly, a classical situation such as two charged particles revolving about each other should begin to exhibit quantum behavior as the size becomes smaller and smaller.

These examples suggest that there should be some in-between size, or energy, for which the quantum-mechanical solution corresponds in some way to the solution of classical mechanics. Niels Bohr put forward the idea that the *average* behavior of a quantum system should begin to look like the classical solution in the limit that the quantum number becomes very large—that is, as $n \to \infty$. Because the radius of the Bohr hydrogen atom is $r = n^2 a_B$, the atom becomes a macroscopic object as n becomes very large. Bohr's idea, that the quantum world should blend smoothly into the classical world for high quantum numbers, is today known as the **correspondence principle.**

Our quantum knowledge of a particle in a box is given by its probability density

$$P_{quant}(x) = |\psi_n(x)|^2 = \frac{2}{L}\sin^2\left(\frac{n\pi x}{L}\right) \qquad (40.29)$$

To what classical quantity can the probability density be compared as $n \to \infty$?

Interestingly, we can also define a classical probability density $P_{class}(x)$. A classical particle follows a well-defined trajectory, but suppose we observe the particle at random times. For example, suppose the box containing a classical particle has a viewing window. The window is normally closed, but at random times, selected by a random-number generator, the window opens for a brief interval of time δt and you can measure the particle's position. When the window opens, what is the probability that the particle will be in a narrow interval δx at position x?

The probability of finding a classical particle within a small interval δx is equal to the *fraction of its time* that it spends passing through δx. That is, you're more likely to find the particle in those intervals δx where it spends lots of time, less likely to find it in a δx where it spends very little time.

If the particle oscillates between two turning points with period T, the time it spends moving from one turning point to the other is $\frac{1}{2}T$. As it moves between the turning points, it passes once through the interval δx at position x, taking time δt to do so. Consequently, the probability of finding the particle within this interval is

$$Prob_{class}(\text{in } \delta x \text{ at } x) = \text{fraction of time spent in } \delta x = \frac{\delta t}{\frac{1}{2}T} \qquad (40.30)$$

The amount of time needed to pass through δx is $\delta t = \delta x / v(x)$, where $v(x)$ is the particle's velocity at position x. Thus the probability of finding the particle in the interval δx at position x is

$$\text{Prob}_{\text{class}}(\text{in } \delta x \text{ at } x) = \frac{\delta x / v(x)}{\frac{1}{2}T} = \frac{2}{Tv(x)}\delta x \qquad (40.31)$$

(a) Uniform speed

Particle in an empty box

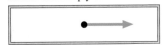

Motion diagram

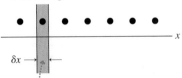

The probability of finding the particle in δx is the fraction of time the particle spends in δx.

(b) Nonuniform speed

Particle on a spring

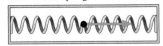

Motion diagram

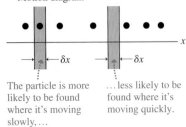

The particle is more likely to be found where it's moving slowly,...   ...less likely to be found where it's moving quickly.

FIGURE 40.11 The classical probability density is indicated by the density of dots in a motion diagram.

You learned in Chapter 39 that the probability is related to the probability density by

$$\text{Prob}_{\text{class}}(\text{in } \delta x \text{ at } x) = P_{\text{class}}(x)\,\delta x$$

Thus the classical probability density for finding a particle at position x is

$$P_{\text{class}}(x) = \frac{2}{Tv(x)} \qquad (40.32)$$

where the velocity $v(x)$ is expressed as a function of x. **Classically, a particle is more likely to be found where it is moving slowly, less likely to be found where it is moving quickly.**

NOTE ▶ Our derivation of Equation 40.32 made no assumptions about the particle's motion other than the requirement that it be periodic. This is the classical probability density for any oscillatory motion. ◀

Figure 40.11a shows a motion diagram of a classical particle in a rigid box of length L. The particle's speed is a *constant* $v(x) = v_0$ as it bounces back and forth between the walls. The particle travels distance $2L$ during one round trip, so the period is $T = 2L/v_0$. Consequently, the classical probability density for a particle in a box is

$$P_{\text{class}}(x) = \frac{2}{(2L/v_0)v_0} = \frac{1}{L} \qquad (40.33)$$

$P_{\text{class}}(x)$ is independent of x, telling us that the particle is equally likely to be found *anywhere* in the box.

In contrast, Figure 40.11b shows a particle with nonuniform speed. A mass on a spring slows down near the turning points, so it spends more time near the ends of the box than in the middle. Consequently the classical probability density for this particle is a maximum at the edges and a minimum at the center. We'll look at this classical probability density again later in the chapter.

EXAMPLE 40.5 The classical probability of locating the particle

A classical particle is in a rigid 10-cm-long box. What is the probability that, at a random instant of time, the particle is in a 1.0-mm-wide interval at the center of the box?

SOLVE The particle's probability density is

$$P_{\text{class}}(x) = \frac{1}{L} = \frac{1}{10 \text{ cm}} = 0.10 \text{ cm}^{-1}$$

The probability that the particle is in an interval of width $\delta x = 1.0 \text{ mm} = 0.10 \text{ cm}$ is

$$\text{Prob}(\text{in } \delta x \text{ at } x = 5 \text{ cm}) = P(x)\delta x = (0.10 \text{ cm}^{-1})(0.10 \text{ cm})$$

$$= 0.010 = 1.0\%$$

ASSESS The classical probability is 1.0% because 1.0 mm is 1% of the 10 cm length.

Figure 40.12 shows the quantum and the classical probability densities for the $n = 1$ and the $n = 20$ quantum states of a particle in a rigid box. Notice that

■ The quantum probability density oscillates between a minimum of 0 and a maximum of $2/L$, so it oscillates around the classical probability density $1/L$.

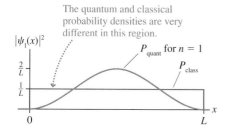

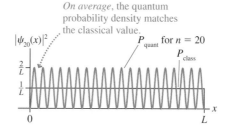

FIGURE 40.12 The quantum and classical probability densities for a particle in a box.

- For $n = 1$, the quantum and classical probability densities are quite different. The ground state of the quantum system will be very nonclassical.
- For $n = 20$, you can see that *on average* the quantum particle's behavior looks very much like that of the classical particle.

As n gets even bigger, and the number of oscillations increases, the probability of finding the particle in an interval δx will be the same for both the quantum and the classical particle as long as δx is large enough to include several oscillations of the wave function. As Bohr predicted, the quantum mechanical solution "corresponds" to the classical solution in the limit $n \to \infty$.

40.6 Finite Potential Wells

Figure 40.4, the potential-energy diagram for a particle in a rigid box, is an example of a **potential well,** so named because the graph of the potential-energy "hole" looks like a well from which you might draw water. The rigid box was an *infinite* potential well. There was no chance that a particle inside could escape the infinitely high walls.

No box is infinitely strong. A more realistic model of a confined particle is the *finite* potential well shown in Figure 40.13a. A particle with total energy $E < U_0$ is confined within the well, bouncing back and forth between turning points at $x = 0$ and $x = L$. The regions $x < 0$ and $x > L$ are **classically forbidden regions** for a particle with $E < U_0$. However, the particle will escape the well if it somehow manages to acquire energy $E > U_0$.

Recall that the zero of energy is arbitrary. Figure 40.13a defined $U = 0$ as the potential energy inside the well. Figure 40.13b has repositioned the zero of energy at the level of the "energy plateau" on both sides of the well. **Figures 40.13a and 40.13b are the same potential well.** Both have width L and depth U_0, and both have the same wave functions and the same allowed energies (relative to our choice of $E = 0$). Which we use is a matter of convenience for the situation we are modeling.

We've made no mention of the *force* that is responsible for this potential well. An electron confined within a semiconductor by an electric force has a potential energy that can be modeled as a finite potential well. So does a proton confined within the nucleus by the nuclear force. The Schrödinger equation depends on the *shape* of the potential-energy function, not the cause. Hence *any* situation in which a force confines a particle to a well-defined region can be modeled as a finite potential well.

Although it is possible to solve the Schrödinger equation exactly for the finite potential well, the result is cumbersome and not especially illuminating. Instead, we'll present the results of numerical calculations. The derivation of the wave functions and energy levels is not as important as understanding and interpreting the results.

(a) $U = 0$ inside the well.

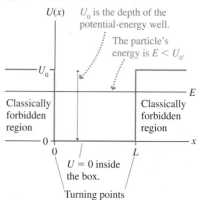

(b) $U = 0$ outside the well.

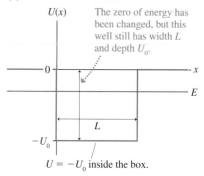

FIGURE 40.13 A finite potential well of width L and depth U_0.

As a first example, consider an electron in a 2.0-nm-wide potential well of depth $U_0 = 1.0$ eV. These are reasonable parameters for an electron in a semiconductor device. Figure 40.14a is a graphical presentation of the allowed energies and wave functions. For comparison, Figure 40.14b shows the first three energy levels and wave functions for a rigid box ($U_0 \rightarrow \infty$) with the same 2.0 nm width.

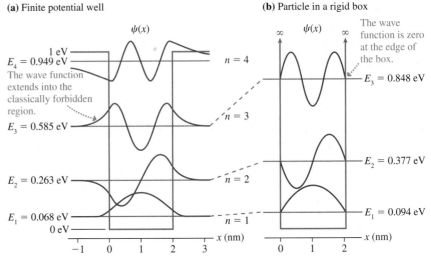

FIGURE 40.14 Energy levels and wave functions for a finite potential well. For comparison, the energies and wave functions are shown for a rigid box of equal width.

 20.3

The quantum-mechanical solution for a particle in a finite potential well has several important properties:

- The particle's energy is quantized. A particle in the potential well *must* be in one of the stationary states with quantum numbers $n = 1, 2, 3, \ldots$
- There are only a finite number of **bound states**—four in this example, although the number will be different in other examples. These wave functions represent electrons confined to, or bound in, the potential well. There are no stationary states with $E > U_0$ because such a particle would not remain in the well.
- The wave functions are qualitatively similar to those of a particle in a rigid box, but the energies are somewhat lower. This is because the wave functions are slightly more spread out. A slightly larger de Broglie wavelength corresponds to a lower velocity and thus a lower energy.
- Most interesting, perhaps, is that the wave functions of Figure 40.14a extend into the classically forbidden regions. It is as though a tennis ball penetrated partly *through* the racket's strings before bouncing back, but without breaking the strings.

EXAMPLE 40.6 Absorption spectrum of an electron
What wavelengths of light are absorbed by a semiconductor device in which electrons are confined in a 2.0-nm-wide region with a potential-energy depth of 1.0 eV?

MODEL The electron is in the finite potential well whose energies and wave functions were shown in Figure 40.14a.

SOLVE Photons of light can be absorbed if a photon's energy $E_{\text{photon}} = hf$ exactly matches the energy difference ΔE between two energy levels of the system. Because most electrons will be in the $n = 1$ ground state, the absorption transitions will be $1 \rightarrow 2$, $1 \rightarrow 3$, and $1 \rightarrow 4$.

The absorption wavelengths $\lambda = c/f$ are

$$\lambda_{n \rightarrow m} = \frac{hc}{\Delta E} = \frac{hc}{|E_n - E_m|}$$

For this example, we find

$$\Delta E_{1-2} = 0.195 \text{ eV} \qquad \lambda_{1 \rightarrow 2} = 6.37 \text{ } \mu\text{m}$$
$$\Delta E_{1-3} = 0.517 \text{ eV} \qquad \lambda_{1 \rightarrow 3} = 2.40 \text{ } \mu\text{m}$$
$$\Delta E_{1-4} = 0.881 \text{ eV} \qquad \lambda_{1 \rightarrow 4} = 1.41 \text{ } \mu\text{m}$$

ASSESS These transitions are all in the infrared portion of the spectrum.

STOP TO THINK 40.3   This is a wave function for a particle in a finite quantum well. What is the particle's quantum number?

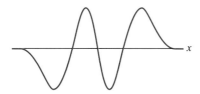

The Classically Forbidden Region

The extension of a particle's wave functions into the classically forbidden region is an important difference between classical and quantum physics. Let's take a closer look at the wave function in the region $x \geq L$. The potential energy in the classically forbidden region is U_0, thus the Schrödinger equation for $x \geq L$ is

$$\frac{d^2\psi}{dx^2} = -\frac{2m}{\hbar^2}(E - U_0)\psi(x)$$

We're assuming a confined particle, with E less than U_0, so $E - U_0$ is negative. It will be useful to reverse the order of these and write

$$\frac{d^2\psi}{dx^2} = \frac{2m}{\hbar^2}(U_0 - E)\psi(x) = \frac{1}{\eta^2}\psi(x) \tag{40.34}$$

where

$$\eta^2 = \frac{\hbar^2}{2m(U_0 - E)} \tag{40.35}$$

is a *positive* constant. As a homework problem, you can show that the units of η are m.

The Schrödinger equation of Equation 40.34 is one we can solve by guessing. We simply need to think of two functions whose second derivatives are a positive constant times the functions themselves. Two such functions, as you can quickly confirm, are $e^{x/\eta}$ and $e^{-x/\eta}$. Thus, according to Equation 40.8, the general solution of the Schrödinger equation for $x \geq L$ is

$$\psi(x) = Ae^{x/\eta} + Be^{-x/\eta} \quad \text{for } x \geq L \tag{40.36}$$

One requirement of the wave function is that $\psi \to 0$ as $x \to \infty$. The function $e^{x/\eta}$ diverges as $x \to \infty$, so the only way to satisfy this requirement is to set $A = 0$. This leaves

$$\psi(x) = Be^{-x/\eta} \quad \text{for } x \geq L \tag{40.37}$$

This is an exponentially decaying function. Notice that all the wave functions in Figure 40.14a look like exponential decays for $x > L$.

The wave function must also be continuous. Suppose the oscillating wave function within the potential well ($x \leq L$) has the value ψ_{edge} when it reaches the classical boundary at $x = L$. To be continuous, the wave function of Equation 40.37 has to match this value at $x = L$. That is,

$$\psi\,(\text{at } x = L) = Be^{-L/\eta} = \psi_{\text{edge}} \tag{40.38}$$

This boundary condition at $x = L$ is sufficient to determine that the constant B is

$$B = \psi_{\text{edge}}e^{L/\eta} \tag{40.39}$$

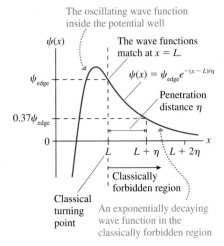

FIGURE 40.15 The wave function in the classically forbidden region.

If we use the Equation 40.39 result for B in Equation 40.37, we find that the wave function in the classically forbidden region of a finite potential well is

$$\psi(x) = \psi_{edge}e^{-(x-L)/\eta} \quad \text{for } x \geq L \qquad (40.40)$$

In other words, **the wave function oscillates until it reaches the classical turning point at $x = L$, then it decays exponentially within the classically forbidden region.**

Figure 40.15 shows the wave function in the classically forbidden region. You can see that the wave function at $x = L + \eta$ has decreased to

$$\psi(\text{at } x = L + \eta) = e^{-1}\psi_{edge} = 0.37\psi_{edge}$$

Although an exponential decay does not have a sharp ending point, the parameter η measures "about how far" the wave function extends past the classical turning point before the probability of finding the particle has decreased nearly to zero. This distance is called the **penetration distance:**

$$\text{penetration distance } \eta = \frac{\hbar}{\sqrt{2m(U_0 - E)}} \qquad (40.41)$$

A classical particle reverses direction at the $x = L$ turning point. But atomic particles are not classical. Because of wave-particle duality, an atomic particle is "fuzzy" and without a well-defined edge. Thus an atomic particle can spread a distance of roughly η into the classically forbidden region.

The penetration distance is unimaginably small for any macroscopic mass, but it can be significant for atomic particles. Notice that the penetration distance depends inversely on the quantity $U_0 - E$, which is the distance of the energy level below the top of the potential well. You can see in Figure 40.14a that η is much larger for the $n = 4$ state, near the top of the potential well, than for the $n = 1$ state.

NOTE ▶ In making use of Equation 40.41 you *must* use SI units of J s for $\hbar$ and J for the energies. The penetration distance η is then in m. ◀

EXAMPLE 40.7 Penetration distance of an electron
An electron is confined in a 2.0-nm-wide region with a potential-energy depth of 1.00 eV. What are the penetration distances into the classically forbidden region for an electron in the $n = 1$ and $n = 4$ states?

MODEL The electron is in the finite potential well whose energies and wave functions were shown in Figure 40.14a.

SOLVE The ground state has $U_0 - E_1 = 1.000$ eV $- 0.068$ eV $= 0.932$ eV. Similarly, $U_0 - E_4 = 0.051$ eV in the $n = 4$ state. We can use Equation 40.41 to calculate

$$\eta = \frac{\hbar}{\sqrt{2m(U_0 - E)}} = \begin{cases} 0.20 \text{ nm} & n = 1 \\ 0.86 \text{ nm} & n = 4 \end{cases}$$

ASSESS These values are consistent with the penetration distances that you can estimate visually in Figure 40.14a.

Quantum-Well Devices

In Part VI we developed a model of electrical conductivity in which the valence electrons of a metal form a loosely bound "sea of electrons." The typical speed of an electron is the rms speed

$$v_{rms} = \sqrt{\frac{3k_B T}{m}}$$

where k_B is Boltzmann's constant. Hence at room temperature, where $v_{rms} \approx 1 \times 10^5$ m/s, the de Broglie wavelength of a typical conduction electron is

$$\lambda \approx \frac{h}{mv_{rms}} \approx 6 \text{ nm}$$

There is a range of wavelengths, because the electrons have a range of speeds, but this is a typical value.

You've now seen many times that wave effects are significant only when the sizes of physical structures are comparable to or smaller than the wavelength. This is why the interference and diffraction of light are hard to observe and why the wave-like nature of matter becomes important only on microscopic scales. Because the de Broglie wavelength of conduction electrons is only a few nm, quantum effects are insignificant in electronic devices whose features are larger than about 100 nm. The electrons in macroscopic devices can be treated as classical particles, which is how we analyzed electric current in Chapter 28.

However, devices smaller than about 100 nm do exhibit quantum effects. Some semiconductor devices, such as the semiconductor lasers used in fiber-optic communications, now incorporate features only a few nm in size. Quantum effects play an important role in these devices.

Figure 40.16a shows the construction of a *semiconductor diode laser.* Although the operating principles of diodes are beyond the scope of this textbook, we can note that a current travels through this device from left to right. In the center is a very thin layer of the semiconductor gallium arsenide (GaAs). It is surrounded on either side by layers of gallium aluminum arsenide (GaAlAs), and these in turn are embedded within the larger structure of the diode. The electrons within the central GaAs layer begin to emit laser light when the current through the diode exceeds some *threshold current.*

You can learn in a solid-state physics or materials engineering course that the electric potential energy of an electron is slightly lower in GaAs than in GaAlAs. This makes the GaAs layer a potential well for electrons, with higher-potential-energy GaAlAs "walls" on either side. As a result, the electrons become trapped within the thin GaAs layer. Such a device is called a **quantum-well laser.**

As an example, Figure 40.16b shows a quantum-well device with a 1.0-nm-thick GaAs layer in which the electron's potential energy is 0.30 eV less than in the surrounding GaAlAs layers. A numerical solution of the Schrödinger equation finds that this potential well has only a *single* quantum state, $n = 1$ with $E_1 = 0.125$ eV. Every electron trapped in this quantum well has the *same* energy—a very nonclassical result! The fact that the electron energies are so well defined, in contrast to the range of electron energies in bulk material, is what makes this a useful device. You can also see from the probability density $|\psi|^2$ that the electrons are more likely to be found in the center of the layer than at the edges. This concentration of electrons makes it easier for the device to begin laser action.

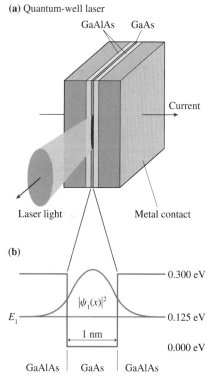

(a) Quantum-well laser

GaAlAs   GaAs

Current

Laser light   Metal contact

(b)

0.300 eV

$|\psi_1(x)|^2$

E_1 0.125 eV

1 nm

0.000 eV

GaAlAs | GaAs | GaAlAs

FIGURE 40.16 A semiconductor diode laser with a single quantum well.

Nuclear Physics

The nucleus of an atom consists of an incredibly dense assembly of protons and neutrons. The positively charged protons exert extremely strong electric repulsive forces on each other, so you might wonder how the nucleus keeps from exploding. During the 1930s, physicists found that protons and neutrons also exert an *attractive* force on each other. This force, one of the fundamental forces of nature, is called the *strong force.* It is the force that holds the nucleus together.

The primary characteristic of the strong force, other than its strength, is that it is a *short range* force. The attractive strong force between two *nucleons* (a nucleon is either a proton or a neutron; the strong force does not distinguish between them) rapidly decreases to zero if they are separated by more than about 2 fm. This is in sharp contrast to the long-range nature of the electric force.

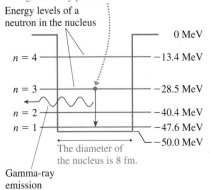

A radioactive decay has left the neutron in the $n = 3$ excited state. The neutron jumps to the $n = 1$ ground state, emitting a gamma-ray photon.

Energy levels of a neutron in the nucleus

0 MeV

$n = 4$ ———— -13.4 MeV

$n = 3$ ———— -28.5 MeV

$n = 2$ ———— -40.4 MeV
$n = 1$ ———— -47.6 MeV
———— -50.0 MeV

The diameter of the nucleus is 8 fm.

Gamma-ray emission

FIGURE 40.17 There are four allowed energy levels for a neutron in this nuclear potential well.

A reasonable model of the nucleus is to think of the protons and neutrons as particles in a nuclear potential well that is created by the strong force. The diameter of the potential well is equal to the diameter of the nucleus (this varies with atomic mass), and nuclear physics experiments have found that the depth of the potential well is ≈ 50 MeV.

The real potential well is three-dimensional, but let's make a simplified model of the nucleus as a one-dimensional potential well. Figure 40.17 shows the potential energy of a neutron along an x-axis passing through the center of the nucleus. Notice that the zero of energy has been chosen such that a "free" neutron, one outside the nucleus, has $E = 0$. Thus the potential energy inside the nucleus is -50 MeV. The 8 fm diameter shown is appropriate for a nucleus having atomic mass number $A \approx 40$, such as argon or potassium. Lighter nuclei will be a little smaller, heavier nuclei somewhat larger. (The potential-energy diagram for a proton is similar, but is complicated a bit by the electric potential energy.)

A numerical solution of the Schrödinger equation finds the four stationary states shown in Figure 40.17. The wave functions have been omitted, but they look essentially identical to the wave functions in Figure 40.14a. The major point to note is that the allowed energies differ by several *million* electron volts! These are enormous energies compared to those of an electron in an atom or a semiconductor. But recall that the energies of a particle in a rigid box, $E_n = n^2h^2/8mL^2$, are proportional to $1/L^2$. Our previous examples, with nanometer-size boxes, found energies in the eV range. When the box size is reduced to femtometers, the energies jump up into the MeV range.

It often happens that the nuclear decay of a radioactive atom leaves a neutron in an excited state. For example, Figure 40.17 shows a neutron that has been left in the $n = 3$ state by a previous radioactive decay. This neutron can now undergo a quantum jump to the $n = 1$ ground state by emitting a photon with energy

$$E_{photon} = E_3 - E_1 = 19.1 \text{ MeV}$$

and wavelength

$$\lambda_{photon} = \frac{c}{f} = \frac{hc}{E_{photon}} = 6.50 \times 10^{-5} \text{ nm}$$

This photon is $\approx 10^7$ times more energetic, and its wavelength $\approx 10^7$ times smaller, than the photons of visible light! These extremely high-energy photons are called **gamma rays.** Gamma-ray emission is, indeed, one of the primary processes in the decay of radioactive elements.

Our one-dimensional model cannot be expected to give accurate results for the energy levels or gamma-ray energies of any specific nucleus. Nonetheless, this model does provide a reasonable understanding of the energy-level structure in nuclei and correctly predicts that nuclei can emit photons having energies of several million electron volts. This model, when extended to three dimensions, becomes the basis of the *shell model* of the nucleus in which the protons and neutrons are grouped in various shells analogous to the electron shells around an atom that you remember from chemistry. You can learn more about nuclear physics and the shell model in Chapter 42.

40.7 Wave-Function Shapes

Bound-state wave functions are standing de Broglie waves. In addition to boundary conditions, two other factors govern the shapes of wave functions:

1. The de Broglie wavelength is inversely dependent on the particle's speed. Consequently, the node spacing is smaller (shorter wavelength) where the

kinetic energy is larger, and it is larger (longer wavelength) where the kinetic energy is smaller.

2. A classical particle is more likely to be found where it is moving more slowly. In quantum mechanics, the probability of finding the particle increases as the wave-function amplitude increases. Consequently, the wave-function amplitude is larger where the kinetic energy is smaller, and it is smaller where the kinetic energy is larger.

We can use this information to draw reasonably accurate wave functions for the different allowed energies in a potential-energy well.

TACTICS BOX 40.1 Drawing wave functions

❶ Draw a graph of the potential energy $U(x)$. Show the allowed energy E as a horizontal line. Locate the classical turning points.

❷ Draw the wave function as a continuous, oscillatory function between the turning points. The wave function for quantum state n has n antinodes and $n - 1$ nodes (excluding the ends).

❸ Make the wavelength longer (larger node spacing) and the amplitude higher in regions where the kinetic energy is smaller. Make the wavelength shorter and the amplitude lower in regions where the kinetic energy is larger.

❹ Bring the wave function to zero at the edge of an infinitely high potential-energy "wall."

❺ Let the wave function decay exponentially inside a classically forbidden region where $E < U$. The penetration distance η increases as E gets closer to the top of the potential-energy well.

EXAMPLE 40.8 Sketching wave functions
Figure 40.18a shows a potential-energy well and the allowed energies for the $n = 1$ and $n = 4$ quantum states. Sketch the $n = 1$ and $n = 4$ wave functions.

VISUALIZE The steps of Tactics Box 40.1 have been followed to sketch the wave functions shown in Figure 40.18b.

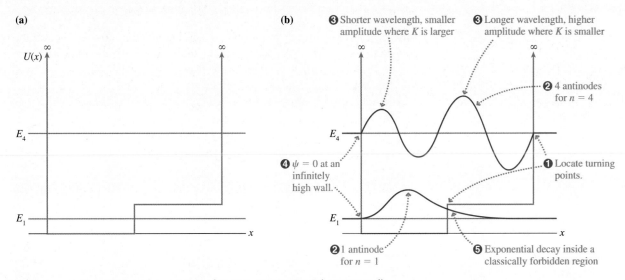

FIGURE 40.18 The $n = 1$ and $n = 4$ wave functions in a potential-energy well.

STOP TO THINK 40.4   For which potential energy is this an appropriate $n = 4$ wave function?

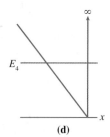

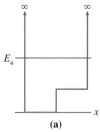

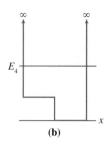

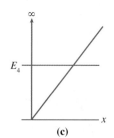

(a)   (b)   (c)   (d)

40.8 The Quantum Harmonic Oscillator

Simple harmonic motion is exceptionally important in classical physics, where it serves as a prototype for more complex oscillations. As you might expect, a microscopic oscillator—the **quantum harmonic oscillator**—is equally important as a model of oscillations at the atomic level.

The defining characteristic of simple harmonic motion is a linear restoring force: $F = -kx$, where k is the spring constant. The corresponding potential-energy function, as you learned in Chapter 10, is

$$U(x) = \frac{1}{2}kx^2 \tag{40.42}$$

where we'll assume that the equilibrium position is $x_e = 0$. The potential energy of a harmonic oscillator is shown in Figure 40.19. It is a potential-energy well with curved sides.

A classical particle of mass m oscillates with angular frequency

$$\omega = \sqrt{\frac{k}{m}} \tag{40.43}$$

between the two turning points where the energy line crosses the parabolic potential-energy curve. As you've learned, this classical description fails if m represents an atomic particle, such as an electron or an atom. In that case, we need to solve the Schrödinger equation to find the wave functions.

The Schrödinger equation for a quantum harmonic oscillator with $U(x) = \frac{1}{2}kx^2$ is

$$\frac{d^2\psi}{dx^2} = -\frac{2m}{\hbar^2}\left(E - \frac{1}{2}kx^2\right)\psi(x) \tag{40.44}$$

We will assert, without deriving them, that the wave functions of the first three states are

$$\psi_1(x) = A_1 e^{-x^2/2b^2}$$

$$\psi_2(x) = A_2\frac{x}{b}e^{-x^2/2b^2} \tag{40.45}$$

$$\psi_3(x) = A_3\left(1 - \frac{2x^2}{b^2}\right)e^{-x^2/2b^2}$$

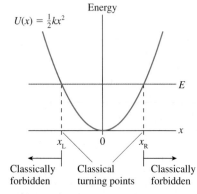

FIGURE 40.19 The potential energy of a harmonic oscillator.

$U(x) = \frac{1}{2}kx^2$

Energy

x_L 0 x_R

Classically forbidden   Classical turning points   Classically forbidden

where b is

$$b = \sqrt{\frac{\hbar}{m\omega}} \qquad (40.46)$$

The constant b has dimensions of length. We will leave it as a homework problem for you to show that b is the classical turning point of an oscillator in the $n = 1$ ground state. The constants A_1, A_2, and A_3 are normalization constants. For example, A_1 can be found by requiring

$$\int_{-\infty}^{\infty} |\psi_1(x)|^2 dx = A_1^2 \int_{-\infty}^{\infty} e^{-x^2/b^2} dx = 1 \qquad (40.47)$$

The completion of this calculation will also be left as a homework problem.

As expected, stationary states of a quantum harmonic oscillator exist only for certain discrete energy levels, the quantum states of the oscillator. The allowed energies are given by the very simple equation

$$E_n = \left(n - \frac{1}{2}\right)\hbar\omega \qquad n = 1, 2, 3, \ldots \qquad (40.48)$$

where ω is the classical angular frequency of Equation 40.43 and n is the quantum number.

NOTE ▶ The ground-state energy of the quantum harmonic oscillator is $E_1 = \frac{1}{2}\hbar\omega$. An atomic mass on a spring can *not* be brought to rest. This is a consequence of the uncertainty principle. ◀

Figure 40.20 shows the first three energy levels and wave functions of a quantum harmonic oscillator. Notice that the energy levels are equally spaced by $\Delta E = \hbar\omega$. This result differs from the particle in a box, where the energy levels get increasingly farther apart. Also notice that the wave functions, like those of the finite potential well, extend beyond the turning points into the classically forbidden region.

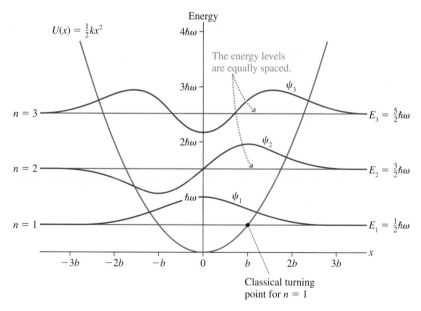

FIGURE 40.20 The first three energy levels and wave functions of a quantum harmonic oscillator.

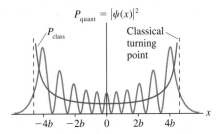

FIGURE 40.21 The quantum and classical probability densities for the $n = 11$ state of a quantum harmonic oscillator.

Figure 40.21 shows the probability density $|\psi(x)|^2$ for the $n = 11$ state of a quantum harmonic oscillator. Notice how the node spacing and the amplitude both increase as the particle moves away from the equilibrium position at $x = 0$. This is consistent with rule 3 of Tactics Box 40.1. The particle slows down as it moves away from the origin, causing its de Broglie wavelength *and* the probability of finding it to increase.

Section 40.5 introduced the classical probability density $P_{\text{class}}(x)$ and noted that a classical particle is most likely to be found where it is moving the slowest. Figure 40.21 shows $P_{\text{class}}(x)$ for a classical particle with the same total energy as the $n = 11$ quantum state. You can see that *on average* the quantum probability density $|\psi(x)|^2$ mimics the classical probability density. This is just what the correspondence principle leads us to expect.

EXAMPLE 40.9 Light emission by an oscillating electron
An electron in a harmonic-oscillator potential well emits light of wavelength 600 nm as it jumps from one level to the next. What is the spring constant of the restoring force?

MODEL The electron is a quantum harmonic oscillator.

SOLVE A photon is emitted as the electron undergoes the quantum jump $n \rightarrow n - 1$. We can use Equation 40.48 for the energy levels to find that the electron loses energy

$$\Delta E = E_n - E_{n-1} = \left(n - \frac{1}{2}\right)\hbar\omega_e - \left(n - 1 - \frac{1}{2}\right)\hbar\omega_e = \hbar\omega_e$$

$\Delta E = \hbar\omega_e$ for *all* transitions, independent of n, because the energy levels of the quantum harmonic oscillator are equally spaced. We need to distinguish the oscillations of the electron from the oscillations of the light wave, hence the subscript e on ω_e.

The emitted photon has energy $E_{\text{photon}} = hf_{\text{ph}} = \Delta E$. Thus

$$\hbar\omega_e = \frac{h}{2\pi}\omega_e = hf_{\text{ph}} = \frac{hc}{\lambda}$$

The wavelength of the light is $\lambda = 600$ nm, hence the classical angular frequency of the oscillating electron is

$$\omega_e = 2\pi\frac{c}{\lambda} = 3.14 \times 10^{15} \text{ rad/s}$$

The electron's angular frequency is related to the spring constant of the restoring force by

$$\omega_e = \sqrt{\frac{k}{m}}$$

Thus $k = m\omega_e^2 = 9.0$ N/m.

Molecular Vibrations

We've made many uses of the idea that atoms are held together by spring-like molecular bonds. We've always assumed that the bonds could be modeled as classical springs. The classical model is acceptable for some purposes, but it fails to explain some important features of molecular vibrations. Not surprisingly, the quantum harmonic oscillator is a better model of a molecular bond.

Figure 40.22 is the potential energy of two atoms connected by a molecular bond. Nearby atoms attract each other through a polarization force, much as a charged rod picks up small pieces of paper. If the atoms get too close, a *repulsive* force between the negative electrons pushes them away. The equilibrium separation at which the attractive and repulsive forces are balanced is r_0, and two classical atoms would be at rest at this separation. But quantum particles, even in their lowest energy state, have $E > 0$. Consequently, the molecule *vibrates* as the two atoms oscillate back and forth along the bond.

U_{dissoc} is the energy at which the molecule will *dissociate* and the two atoms will fly apart. Dissociation can occur at very high temperatures or after the molecule has absorbed a high-energy (ultraviolet) photon, but under typical conditions a molecule has energy $E \ll U_{\text{dissoc}}$. In other words, the molecule is in an energy level near the bottom of the potential well.

You can see that the lower portion of the potential well is very nearly a parabola. Consequently, we can model a molecular bond as a quantum harmonic

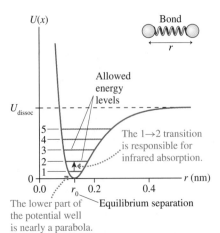

FIGURE 40.22 The potential energy of a molecular bond and a few of the allowed energies.

oscillator. The energy associated with the molecular vibration is quantized and can have *only* the values

$$E_{\text{vib}} \approx \left(n - \frac{1}{2}\right)\hbar\omega \qquad n = 1, 2, 3, \ldots \qquad (40.49)$$

where ω is the angular frequency with which the atoms would vibrate if the bond were a classical spring. The molecular potential-energy curve is not exactly that of a harmonic oscillator, hence the $\approx$ sign; but the model is very good for low values of the quantum number n. The energy levels calculated by Equation 40.49 are called the oscillatory function **vibrational energy levels** of the molecule. The first few vibrational energy levels are shown in Figure 40.22.

At room temperature, most molecules are in the $n = 1$ vibrational ground state. Their vibrational motion can be excited by absorbing photons of frequency $f = \Delta E/h$. This frequency is usually in the infrared region of the spectrum, and these *vibrational transitions* give each molecule a unique and distinctive infrared absorption spectrum.

As an example, Figure 40.23 shows the infrared absorption spectrum of acetone. The vertical axis is the percentage of the light intensity passing all the way through the sample. The sample is essentially transparent at most wavelengths (transmission $\approx 100\%$), but there are two prominent absorption features. The transmission drops to $\approx 75\%$ at $\lambda = 3.3$ μm and to a mere 7% at $\lambda = 5.8$ μm. The 3.3 μm absorption is due to the $n = 1$ to $n = 2$ transition in the vibration of a $C-CH_3$ carbon-methyl bond. The 5.8 μm absorption is the $1 \rightarrow 2$ transition of a vibrating $C = O$ carbon-oxygen double bond.

Absorption spectra such as this are known for thousands of molecules, and chemists routinely use absorption spectroscopy to identify the chemicals in a sample. A specific bond has the same absorption wavelength regardless of the larger molecule in which it is embedded, thus the presence of that absorption wavelength is a "signature" that the bond is present within a molecule.

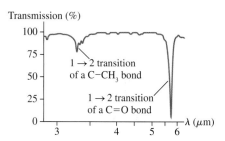

FIGURE 40.23 The absorption spectrum of acetone.

STOP TO THINK 40.5   Which probability density represents a quantum harmonic oscillator with $E = \frac{5}{2}\hbar\omega$?

     (a)   (b)   (c)   (d)

40.9 More Quantum Models

In this section we'll look at two more examples of quantum-mechanical models.

A Particle in a Capacitor

Many semiconductor devices are designed to confine electrons within a layer only a few nanometers thick. If a potential difference is applied across the layer, the electrons act very much as if they are trapped within a microscopic capacitor.

Figure 40.24a on the next page shows two capacitor plates separated by distance L. The left plate is positive, so the electric field points to the right with strength $E = \Delta V_0/L$. Because of its negative charge, an electron launched from the left plate is slowed by a *retarding* force. The electron makes it across to the right plate if it starts with sufficient kinetic energy; otherwise it reaches a turning point and then is pushed back to the positive plate.

(a)

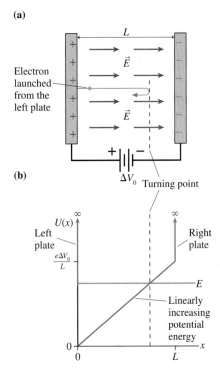

(b)

FIGURE 40.24 An electron in a capacitor.

This classical analysis is a valid model of a macroscopic capacitor. But if L becomes sufficiently small, comparable to the de Broglie wavelength of an electron, then the wave-like properties of the electron cannot be ignored. We need a quantum-mechanical model.

Let's establish a coordinate system with $x = 0$ at the left plate and $x = L$ at the right plate. Define the electric potential to be zero at the positive plate. The potential *decreases* in the direction of the field, so the potential inside the capacitor (see Section 29.5) is

$$V(x) = -Ex = -\frac{\Delta V_0}{L}x$$

The electron, with charge $q = -e$, has potential energy

$$U(x) = qV(x) = +\frac{e\,\Delta V_0}{L}x \qquad 0 < x < L \qquad (40.50)$$

This potential energy increases linearly for $0 < x < L$. If we assume that the capacitor plates act like the walls of a rigid box, then $U(x) \to \infty$ at $x = 0$ and $x = L$.

Figure 40.24b shows the electron's potential-energy function. It is the particle-in-a-rigid-box potential with a sloping "floor" due to the electric field. The figure also shows the total energy line E of an electron in the capacitor. The energy is purely kinetic at $x = 0$, where $K = E$, but it is converted to potential energy as the electron moves to the right. The right turning point occurs where the energy line E crosses the potential-energy curve $U(x)$. If the electron is a classical particle, it must reverse position at this point.

> **NOTE** ▶ This is also the potential energy for a microscopic bouncing ball that is trapped between a floor at $y = 0$ and a ceiling at $y = L$. ◀

It is physically impossible for an electron to be outside the capacitor, so the wave function must be zero for $x < 0$ and $x > L$. The continuity of ψ requires the same boundary conditions as for a particle in a rigid box: $\psi = 0$ at $x = 0$ and at $x = L$. The wave functions inside the capacitor are too complicated to find by guessing, so we have solved the Schrödinger equation numerically and will present the results graphically.

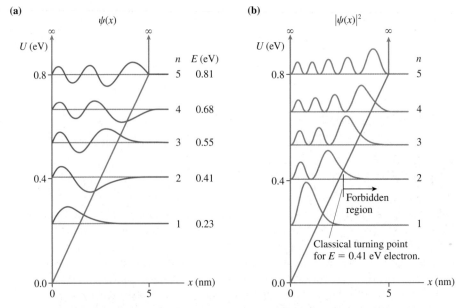

FIGURE 40.25 Energy levels, wave functions, and probability densities for an electron in a 5.0-nm-wide capacitor with a 0.80 V potential difference.

Figure 40.25 shows the wave functions and probability densities for the first five quantum states of an electron confined in a 5.0-nm-thick layer that has a 0.80 V potential difference across it. Each allowed energy is represented as a horizontal line, with the numerical values shown on the right. They range from $E_1 = 0.23$ eV up to $E_5 = 0.81$ eV. An electron *must* have one of the allowed energies shown in the figure. An electron cannot have $E = 0.30$ eV in this capacitor because no de Broglie wave with that energy can match the necessary boundary conditions.

NOTE ▶ Remember that each wave function is graphed as if its energy line is the x-axis. ◀

We can make several observations about the Schrödinger equation solutions:

1. The energies E_n become more closely spaced as n increases. This behavior is in contrast to the particle in a box, for which E_n became more widely spaced.
2. The spacing between the nodes of a wave function is not constant but increases toward the right. This is because an electron on the right side of the capacitor has less kinetic energy and thus a slower speed and a larger de Broglie wavelength.
3. The height of the probability density $|\psi|^2$ increases toward the right. That is, we are more likely to find the electron on the right side of the capacitor than on the left. But this also makes sense if, classically, the electron is moving more slowly when on the right side and thus spending more time there than on the left side.
4. The electron penetrates *beyond* the classical turning point into the classically forbidden region.

EXAMPLE 40.10 The emission spectrum of an electron in a capacitor

What are the frequencies of photons emitted by electrons in the $n = 4$ state of Figure 40.25?

SOLVE Photon emission occurs as the electrons make $4 \to 3$, $4 \to 2$, and $4 \to 1$ quantum jumps. In each case, the photon frequency is $f = \Delta E/h$ and the wavelength is

$$\lambda = \frac{c}{f} = \frac{hc}{\Delta E}$$

The energies of the quantum jumps, which can be read from Figure 40.25a, are $\Delta E_{4\to3} = 0.13$ eV, $\Delta E_{4\to2} = 0.27$ eV, and $\Delta E_{4\to1} = 0.45$ eV. Thus

$$\lambda_{4\to3} = 9500 \text{ nm} = 9.5 \text{ }\mu\text{m}$$
$$\lambda_{4\to2} = 4600 \text{ nm} = 4.6 \text{ }\mu\text{m}$$
$$\lambda_{4\to1} = 2800 \text{ nm} = 2.8 \text{ }\mu\text{m}$$

ASSESS The $n = 4$ electrons in this device emit three distinct infrared wavelengths.

The Covalent Bond

You probably recall from chemistry that a **covalent molecular bond,** such as the bond between the two atoms in molecules such as H_2 and O_2, is a bond in which the electrons are shared between the atoms. The basic idea of covalent bonding can be understood with a one-dimensional quantum-mechanical model.

The simplest molecule, the hydrogen molecular ion H_2^+, consists of two protons and one electron. Although it seems surprising that such a system could be stable, the two protons form a molecular bond with one electron. This is the simplest covalent bond.

How can we model the H_2^+ ion? To begin, Figure 40.26a on the next page shows a one-dimensional model of a hydrogen *atom* in which the electron's Coulomb potential energy, with its $1/r$ dependence, has been approximated by a finite potential well of width 0.10 nm ($\approx 2a_B$) and depth 24.2 eV. You learned in Chapter 38 that an electron in the ground state of the Bohr hydrogen atom orbits

(a) Simple one-dimensional model of an electron in a hydrogen atom

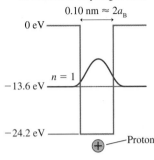

(b) An H_2^+ molecule modeled as an electron with two protons separated by 0.12 nm

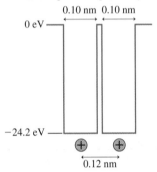

FIGURE 40.26 A molecule can be modeled as two closely spaced potential wells, one representing each atom.

the proton with radius $r_1 = a_B$ (the Bohr radius) and energy $E_1 = -13.6$ eV. A numerical solution of the Schrödinger equation finds that the ground-state energy of this finite potential well is $E_1 = -13.6$ eV. Clearly this model of a hydrogen atom is oversimplified, but it does have the correct size and the correct ground-state energy.

We can model H_2^+ by bringing two of these potential wells close together. The molecular bond length of H_2^+ is known to be ≈ 0.12 nm, so Figure 40.26b shows potential wells with 0.12 nm between their centers. This is a model of H_2^+, not a complete H_2 molecule, because this is the potential energy of a single electron. (Modeling H_2 is more complex because we would need to consider the repulsion between the two electrons.)

Figure 40.27 shows the allowed energies and wave functions for an electron with this potential energy. The $n = 1$ wave function has a high probability of being found within the classically forbidden region *between* the two protons. In other words, an electron in this quantum state really is "shared" by the protons and spends most of its time between them.

In contrast, an electron in the $n = 2$ energy level has zero probability of being found between the two protons because the $n = 2$ wave function has a node at the center. The probability density shows that an $n = 2$ electron is "owned" by one proton or the other rather than being shared.

To learn the consequences of these wave functions we need to calculate the total energy of the molecule: $E_{mol} = E_{p-p} + E_{elec}$. The $n = 1$ and $n = 2$ energies shown in Figure 40.27 are the energies E_{elec} of the electron. At the same time, the protons repel each other and have electric potential energy E_{p-p}. It's not hard to calculate that $E_{p-p} = 12.0$ eV for two protons separated by 0.12 nm. Thus

$$E_{mol} = E_{p-p} + E_{elec} = \begin{cases} 12.0 \text{ eV} - 17.5 \text{ eV} = -5.5 \text{ eV} & n = 1 \\ 12.0 \text{ eV} - 9.0 \text{ eV} = +3.0 \text{ eV} & n = 2 \end{cases}$$

The $n = 1$ molecular energy is less than zero, showing that this is a *bound state*. The $n = 1$ wave function is called a **bonding molecular orbital.** Although

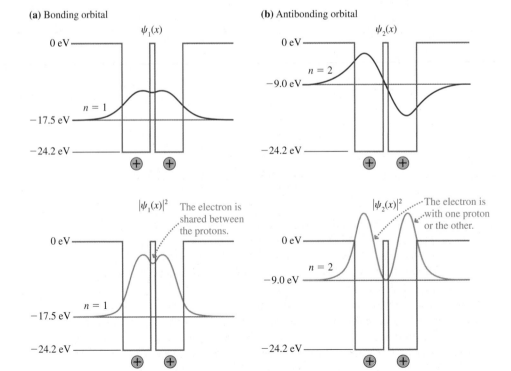

FIGURE 40.27 The wave functions and probability densities of the electron in H_2^+.

the protons repel each other, the shared electron provides sufficient "glue" to hold the system together. The $n = 2$ molecular energy is positive, so this is *not* a bound state. The system would be more stable as a hydrogen atom and a distant proton. The $n = 2$ wave function is called an **antibonding molecular orbital.**

Both E_{elec} and E_{p-p} depend on the separation between the protons, which we assumed to be 0.12 nm in this calculation. If we were to calculate and graph E_{mol} for many different values of the proton separation, the graph would look like the molecular-bond energy curve shown in Figure 40.22. In other words, a molecular bond has an equilibrium length where the bond energy is a minimum *because* of the interplay between E_{p-p} and E_{elec}.

Although real molecular wave functions are more complex than this one-dimensional model, the $n = 1$ wave function captures the essential idea of a covalent bond. Notice that a "classical" molecule cannot have a covalent bond because the electron would not be able to exist in the classically forbidden region. Covalent bonds can be understood only within the context of quantum mechanics. In fact, the explanation of molecular bonds was one of the earliest successes of quantum mechanics.

40.10 Quantum-Mechanical Tunneling

Figure 40.28a shows a ball rolling toward a hill. A ball with sufficient kinetic energy can go over the top of the hill, slowing down as it ascends and speeding up as it rolls down the other side. A ball with insufficient energy rolls partway up the hill, then reverses direction and rolls back down.

Activ Physics 20.4

(a)

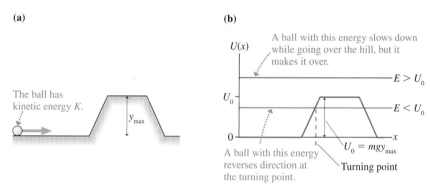

The ball has kinetic energy K.

y_{max}

(b)

$U(x)$

A ball with this energy slows down while going over the hill, but it makes it over.

$E > U_0$

U_0

$E < U_0$

0

A ball with this energy reverses direction at the turning point.

$U_0 = mgy_{max}$

Turning point

x

FIGURE 40.28 A hill is an energy barrier to a rolling ball.

We can think of the hill as an "energy barrier" of height $U_0 = mgy_{max}$. As Figure 40.28b shows, a ball incident from the left with energy $E > U_0$ can go over the barrier (i.e., roll over the hill), but a ball with $E < U_0$ will reflect from the energy barrier at the turning point. According to the laws of classical physics, a ball that is incident on the energy barrier from the left with $E < U_0$ will never be found on the right side of the barrier.

NOTE ▶ Figure 40.28b is not a "picture" of the energy barrier. And when we say that a ball with energy $E > U_0$ can go "over" the barrier, we don't mean that the ball is thrown from a higher elevation in order to go over the top of the hill. The ball rolls *on the ground* the entire time, as Figure 40.28a shows, and Figure 40.28b describes the kinetic and potential energy of the ball as it rolls. A higher total energy line means a larger initial kinetic energy, not a higher elevation. ◄

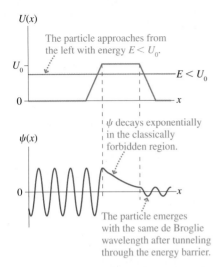

The particle approaches from the left with energy $E < U_0$.

ψ decays exponentially in the classically forbidden region.

The particle emerges with the same de Broglie wavelength after tunneling through the energy barrier.

FIGURE 40.29 A quantum particle can penetrate through the energy barrier.

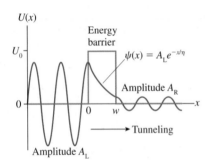

FIGURE 40.30 Tunneling through an idealized energy barrier.

Figure 40.29 shows the situation from the perspective of quantum mechanics. As you've learned, quantum particles can penetrate with an exponentially decreasing wave function into the classically forbidden region of an energy barrier. Suppose that the barrier is very narrow. Although the wave function decreases within the barrier, starting at the classical turning point, it hasn't vanished when it reaches the other side. In other words, there is some probability that a quantum particle will pass *through* the barrier and emerge on the other side!

It is very much as if the ball of Figure 40.28a gets to the turning point and then, instead of reversing direction and rolling back down, tunnels its way *through* the hill and emerges on the other side. Although this feat is strictly forbidden in classical mechanics, it is apparently acceptable behavior for quantum particles. The process is called **quantum-mechanical tunneling.**

The process of tunneling through a potential-energy barrier is one of the strangest and most unexpected predictions of quantum mechanics. Yet it does happen, and you will see that it even has many practical applications.

NOTE ▶ The word "tunneling" is used as a metaphor. If a classical particle really did tunnel, it would expend energy doing so and emerge on the other side with less energy. Quantum-mechanical tunneling requires no expenditure of energy. The total energy line is at the same height on both sides of the barrier. A particle that tunnels through a barrier emerges with *no* loss of energy. That is why the de Broglie wavelength is the same on both sides of the potential barrier in Figure 40.29. ◀

To simplify our analysis of tunneling, Figure 40.30 shows an idealized energy barrier of height U_0 and width w. We've superimposed the wave function on top of the energy diagram so that you can see how it aligns with the potential energy. The wave function to the left of the barrier is a sinusoidal oscillation with amplitude A_L. The wave function *within* the barrier is the decaying exponential we found in Equation 40.40:

$$\psi_{\text{in}}(0 \le x \le w) = \psi_{\text{edge}}e^{-x/\eta} = A_L e^{-x/\eta} \quad (40.51)$$

where we've assumed $\psi_{\text{edge}} = A_L$. The penetration distance η was given in Equation 40.41 as

$$\eta = \frac{\hbar}{\sqrt{2m(U_0 - E)}}$$

NOTE ▶ You *must* use SI units when calculating values of η. Energies must be in J and $\hbar$ in J s. The penetration distance η has units of m. ◀

The wave function decreases exponentially within the barrier, but before it can decay to zero it emerges again on the right side ($x > w$) as an oscillation with amplitude

$$A_R = \psi_{\text{in}}(\text{at } x = w) = A_L e^{-w/\eta} \quad (40.52)$$

The probability that the particle is to the left of the barrier is proportional to $|A_L|^2$ and the probability of finding it to the right of the barrier is proportional to $|A_R|^2$. Thus the probability that a particle striking the barrier from the left will emerge on the right is

$$P_{\text{tunnel}} = \frac{|A_R|^2}{|A_L|^2} = (e^{-w/\eta})^2 = e^{-2w/\eta} \quad (40.53)$$

This is the probability that a particle will tunnel through the energy barrier.

Now, our analysis, we have to say, has not been terribly rigorous. For example, we assumed that the oscillatory wave functions on the left and the right were exactly at a maximum where they reached the barrier at $x = 0$ and $x = w$. There is no reason this has to be the case. We have taken other liberties, which experts will spot, but—fortunately—it really makes no difference. Our result, Equation 40.53, turns out to be perfectly adequate for most applications of tunneling.

Because the tunneling probability is an exponential function, it is *very* sensitive to the values of w and η. The tunneling probability can be substantially reduced by even a small increase in the thickness of the barrier. The parameter η, which measures how far the particle can penetrate into the barrier, depends both on the particle's mass and on $U_0 - E$. A particle with E only slightly less than U_0 will have a larger value of η and thus a larger tunneling probability than will an identical particle with less energy.

EXAMPLE 40.11 Electron tunneling

a. Find the probability that an electron will tunnel through a 1.0-nm-wide energy barrier if the electron's energy is 0.10 eV less than the height of the barrier.

b. Find the tunneling probability if the barrier in part a is widened to 3.0 nm.

c. Find the tunneling probability if the electron in part a is replaced by a proton with the same energy.

SOLVE

a. An electron with energy 0.10 eV less than the height of the barrier has $U_0 - E = 0.10 \text{ eV} = 1.60 \times 10^{-20}$ J. Thus its penetration distance is

$$\eta = \frac{\hbar}{\sqrt{2m(U_0 - E)}}$$

$$= \frac{1.05 \times 10^{-34} \text{ J s}}{\sqrt{2(9.11 \times 10^{-31} \text{ kg})(1.60 \times 10^{-20} \text{ J})}}$$

$$= 6.18 \times 10^{-10} \text{ m} = 0.618 \text{ nm}$$

The probability that this electron will tunnel through a barrier of width $w = 1.0$ nm is

$$P_{\text{tunnel}} = e^{-2w/\eta} = e^{-2(1.0 \text{ nm})/(0.618 \text{ nm})} = 0.039 = 3.9\%$$

b. Changing the width to $w = 3.0$ nm has no effect on η. The new tunneling probability is

$$P_{\text{tunnel}} = e^{-2w/\eta} = e^{-2(3.0 \text{ nm})/(0.618 \text{ nm})} = 6.0 \times 10^{-5}$$

$$= 0.006\%$$

Increasing the width by a factor of 3 decreases the tunneling probability by a factor of 660!

c. A proton is more massive than an electron. Thus a proton with $U_0 - E = 0.10$ eV has $\eta = 0.014$ nm. Its probability of tunneling through a 1.0-nm-wide barrier is

$$P_{\text{tunnel}} = e^{-2w/\eta} = e^{-2(1.0 \text{ nm})/(0.014 \text{ nm})} \approx 1 \times 10^{-64}$$

For practical purposes, the probability that a proton will tunnel through this barrier is zero.

ASSESS If the probability of a proton tunneling through a mere 1 nm is only 10^{-64}, you can see that a macroscopic object will "never" tunnel through a macroscopic distance!

Quantum-mechanical tunneling seems so obscure that it is hard to imagine practical applications. Surprisingly, there are many. We will look at two: the scanning tunneling microscope and resonant tunneling diodes.

The Scanning Tunneling Microscope

Diffraction limits the view of an optical microscope to objects no smaller than about a wavelength of light—roughly 500 nm. This is more than 1000 times the size of an atom, so there is no hope of resolving atoms or molecules via optical microscopy. Electron microscopes are similarly limited by the de Broglie wavelength of the electrons. Their resolution is much better than an optical microscope, but still not quite at the level of resolving individual atoms.

This situation changed dramatically in 1981 with the invention of the **scanning tunneling microscope,** or STM as it is usually called. The STM allowed scientists, for the first time, to "see" surfaces literally atom by atom. Figure 40.31 on the next page shows two pictures taken with an STM. In one you can see individual atoms of carbon on the surface of graphite. The other shows a

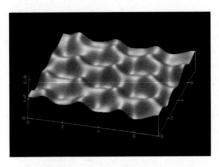

Individual atoms of carbon on
the surface of graphite

The surface of silicon

FIGURE 40.31 Two pictures made with a scanning tunneling microscope.

(a)
Sea of electrons

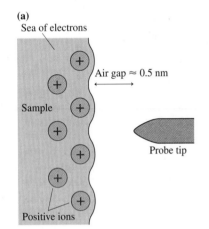

Air gap ≈ 0.5 nm

Sample

Probe tip

Positive ions

(b)

Energy level of
an electron in the
sample or the probe

$U(x)$

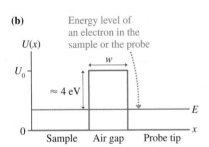

U_0

≈ 4 eV

w

E

x

0

Sample   Air gap   Probe tip

FIGURE 40.32 A scanning tunneling
microscope.

somewhat less magnified surface of silicon. These pictures, and many others you have likely seen (but may not have known where they came from) are stupendous, but how are they made?

Figure 40.32a shows how the scanning tunneling microscope works. A conducting probe with a *very* sharp tip, just a few atoms wide, is brought to within a few tenths of a nanometer of a surface. Preparing the tips and controlling the spacing are both difficult technical challenges, but scientists have learned how to do both. Once positioned, the probe can be mechanically scanned back and forth across the surface.

When we analyzed the photoelectric effect, you learned that electrons are bound inside metals by an amount of energy called the *work function* E_0. A typical work function is 4 or 5 eV. This is the energy that must be supplied—by a photon or otherwise—to lift an electron out of the metal. In other words, the electron's energy in the metal is E_0 less than its energy outside the metal.

This fact is the basis for the potential-energy diagram of Figure 40.32b. The small air gap between the sample and the probe tip is a potential-energy barrier. The energy of an electron in the metal of the sample or the probe tip is less than the energy of an electron in the air by ≈4 eV, the work function. The absorption of a photon with $E_{photon} > 4$ eV would lift the electron *over* the barrier, from the sample to the probe. This is just the photoelectric effect. Alternatively, electrons can tunnel *through* the barrier if it is sufficiently narrow. This creates a *tunneling current* from the sample into the probe.

In operation, the tunneling current is recorded as the probe tip scans across the surface. You saw above that the tunneling current is extremely sensitive to the barrier thickness. As the tip scans over the position of an atom, the gap decreases by ≈0.1 nm and the current increases. The gap is larger when the tip is between atoms, so the current drops. Today's STMs can sense changes in the gap of as little as 0.001 nm, or about 1% of an atomic diameter! The images you see, such as those in Figure 40.31, are computer-generated from the current measurements at each position.

The STM has revolutionized the science and engineering of microscopic objects. They are now used to study everything from how surfaces corrode and oxidize, a topic of great practical importance, to how biological molecules are structured. Another example of quantum mechanics working for you!

The Resonant Tunneling Diode

The semiconductor diode laser that we examined in Section 40.6 had a narrow GaAs layer surrounded by wide layers of GaAlAs. Because an electron's potential energy is ≈0.3 eV less in GaAs than in GaAlAs, this structure provides a quantum well in which electrons are confined in a single energy level.

Suppose we manufacture a device in which a thin layer of GaAs is surrounded by still thinner layers of GaAlAs, only a few nanometers thick. Figure 40.33a is the potential-energy diagram of an electron in such a device. Because the GaAlAs layers are very thin, an electron inside the quantum well can tunnel through to the outside.

Conversely, an electron coming from the outside and impinging on the GaAlAs barrier might tunnel *into* the quantum well. However, tunneling into the well from the outside is hindered by a serious energy mismatch. An electron inside the quantum well *must* have one of the allowed energies. Typically there is a single allowed quantum state with $E_1 \approx 0.15$ eV. Electrons on the outside have thermal energy

$$E_{th} \approx \frac{3}{2} k_B T = 6.0 \times 10^{-21} \text{ J} = 0.040 \text{ eV}$$

at room temperature. Tunneling may be a strange phenomenon, but energy does still have to be conserved. An electron approaching the barrier with $E \approx 0.04$ eV cannot tunnel inside unless there is a quantum state with this allowed energy.

Figure 40.33b shows the effect of placing a potential difference ΔV across the three layers of the device, with the left side more positive. Because electrons are negative, ΔV *lowers* the potential energy on the left side. As the potential difference is increased, it will reach a value ΔV_{res} at which the energy level inside the quantum well matches the energy of an electron approaching from the right. We then have a *resonance,* much as when an external driving frequency matches the natural frequency of an oscillator.

Once the energies match, electrons approaching from the right can easily tunnel into the quantum well. They then tunnel through the opposite barrier and emerge on the left with kinetic energy $K \approx e\Delta V$. In other words, there is a current through the device when the potential difference is ΔV_{res}. This device is called a **resonant tunneling diode.**

Applying too much voltage destroys the resonance. As Figure 40.33c shows, a large ΔV drops the energy level in the quantum well too low, so again electrons from the right side have no matching energy level into which they can tunnel. Charge flows through a resonant tunneling diode only for a small range of voltages near ΔV_{res}.

Figure 40.34 is an experimental current-voltage graph for a device having a 4 nm GaAs quantum well surrounded by 10-nm-wide GaAlAs barriers. There is a small range of voltages around 0.25 volts for which the current shoots up by a factor of 10. This is ΔV_{res}, and the current is due to electrons tunneling through the diode. The current then drops back to near zero by the time $\Delta V = 0.40$ V. (The current increase for $\Delta V > 0.7$ V is "normal" diode behavior. A resonant tunneling diode would not be operated with voltages that large.)

The ability to drastically change current with just a small change in voltage makes tunneling diodes very useful in the digital circuits of high-speed computers. These diodes can also be used as very-high-speed oscillators, creating oscillating voltages with frequencies as high as 500 GHz.

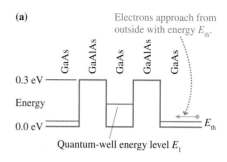

(a)

Electrons approach from outside with energy E_{th}.

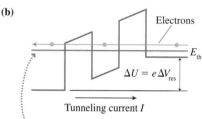

(b)

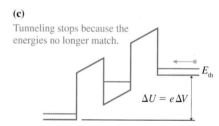

The quantum-well energy matches the electron energy, allowing the electrons to tunnel through.

(c)

Tunneling stops because the energies no longer match.

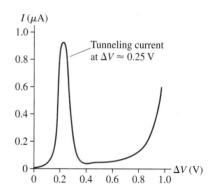

FIGURE 40.33 Electron potential energy in a resonant tunneling diode.

FIGURE 40.34 Experimental measurement of the current-voltage characteristics of a resonant tunneling diode.

STOP TO THINK 40.6 A particle with energy E approaches an energy barrier with height $U_0 > E$. If U_0 is slowly decreased, the probability that the particle reflects from the barrier

a. Increases.

b. Decreases.

c. Does not change.

SUMMARY

The goal of Chapter 40 has been to understand and apply the essential ideas of quantum mechanics.

GENERAL PRINCIPLES

The Schrödinger Equation (The "law of psi")

$$\frac{d^2\psi}{dx^2} = -\frac{2m}{\hbar^2}[E - U(x)]\psi(x)$$

This equation determines the wave function $\psi(x)$ and, through $\psi(x)$, the probabilities of finding a particle of mass m with potential energy $U(x)$.

Boundary conditions

- $\psi(x)$ is a continuous function.
- $\psi(x) \rightarrow 0$ as $x \rightarrow \pm\infty$.
- $\psi(x) = 0$ in a region where it is physically impossible for the particle to be.
- $\psi(x)$ is normalized.

Shapes of wave functions

- The wave function oscillates between the classical turning points.
- State n has n antinodes.
- Node spacing and amplitude increase as kinetic energy K decreases.
- $\psi(x)$ decays exponentially in a classically forbidden region.

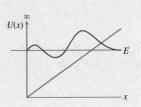

Quantum-mechanical models are characterized by the particle's potential-energy function $U(x)$.

- Wave-function solutions exist only for certain values of E. Thus energy is quantized.
- Photons are emitted or absorbed in quantum jumps.

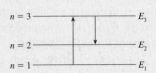

IMPORTANT CONCEPTS

Quantum-mechanical tunneling

A wave function can penetrate into a classically forbidden region with

$$\psi(x) = \psi_{\text{edge}}e^{-(x-L)/\eta}$$

where the **penetration distance** is

$$\eta = \frac{\hbar}{\sqrt{2m(U_0 - E)}}$$

The probability of tunneling through a barrier of width w is

$$P_{\text{tunnel}} = e^{-2w/\eta}$$

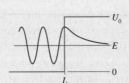
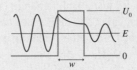

The correspondence principle says that the quantum world blends smoothly into the classical world for high quantum numbers. This is seen by comparing $|\psi(x)|^2$ to the classical probability density

$$P_{\text{class}} = \frac{2}{Tv(x)}$$

P_{class} expresses the idea that a classical particle is more likely to be found where it is moving slowly.

APPLICATIONS

Particle in a rigid box $E_n = n^2\dfrac{h^2}{8mL^2}$ $n = 1, 2, 3, \ldots$

Quantum harmonic oscillator $E_n = (n - \frac{1}{2})\hbar\omega$ $n = 1, 2, 3, \ldots$

Other applications were studied through numerical solution of the Schrödinger equation.

TERMS AND NOTATION

Schrödinger equation	bound state	bonding molecular orbital
quantum-mechanical model	penetration distance, η	antibonding molecular orbital
boundary conditions	quantum-well laser	quantum-mechanical tunneling
zero-point motion	gamma rays	scanning tunneling microscope (STM)
correspondence principle	quantum harmonic oscillator	resonant tunneling diode
potential well	vibrational energy levels	
classically forbidden regions	covalent molecular bond	

EXERCISES AND PROBLEMS

Exercises

Section 40.3 A Particle in a Rigid Box: Energies and Wave Functions

Section 40.4 A Particle in a Rigid Box: Interpreting the Solution

1. An electron in a rigid box absorbs light with a wavelength of 600 nm. How long is the box?

2. The electrons in a rigid box emit photons of wavelength 1484 nm during the $3 \rightarrow 2$ transition.
   a. What kind of photons are they—infrared, visible, or ultraviolet?
   b. How long is the box in which the electrons are confined?

3. Figure Ex40.3 shows the wave function of an electron in a rigid box. The electron energy is 6.0 eV. How long is the box?

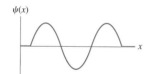

$\psi(x)$

FIGURE EX40.3

4. Figure Ex40.4 shows the wave function of an electron in a rigid box. The electron energy is 12.0 eV. What is the energy of the electron's ground state?

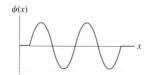

$\psi(x)$

FIGURE EX40.4

Section 40.6 Finite Potential Wells

5. Show that the penetration distance η has units of m.

6. a. Sketch graphs of the probability density $|\psi(x)|^2$ for the four states in the finite potential well of Figure 40.14a. Stack them vertically, similar to the Figure 40.14a graphs of $\psi(x)$.

   b. What is the probability that a particle in the $n = 2$ state of the finite potential well will be found at the center of the well? Explain.

   c. Is your answer to part b consistent with what you know about waves? Explain.

7. For a particle in a finite potential well of depth U_0, what is the ratio of the probability Prob(in δx at $x = L + \eta$) to the probability Prob(in δx at $x = L$)?

8. A finite potential well has depth $U_0 = 2.00$ eV. What is the penetration distance for an electron with energy (a) 0.50 eV, (b) 1.00 eV, and (c) 1.50 eV?

9. An electron in a finite potential well has a 1.0 nm penetration distance into the classically forbidden region. How far below U_0 is the electron's energy?

10. A helium atom is in a finite potential well. The atom's energy is 1.0 eV below U_0. What is the atom's penetration distance into the classically forbidden region?

Section 40.7 Wave-Function Shapes

11. Sketch the $n = 6$ wave function for the potential energy shown in Figure Ex40.11.

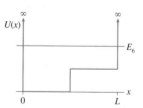

$U(x)$ ∞ ∞ E_6

FIGURE EX40.11 0 L

12. Sketch the $n = 8$ wave function for the potential energy shown in Figure Ex40.12.

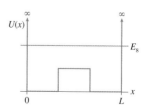

$U(x)$ ∞ ∞ E_8

FIGURE EX40.12 0 L

13. The graph in Figure Ex40.13 shows the potential-energy function $U(x)$ of a particle. Solution of the Schrödinger equation finds that the $n = 3$ level has $E_3 = 0.5$ eV and that the $n = 6$ level has $E_6 = 2.0$ eV.
    a. Redraw this figure and add to it the energy lines for the $n = 3$ and $n = 6$ states.
    b. Sketch the $n = 3$ and $n = 6$ wave functions. Show them as oscillating about the appropriate energy line.

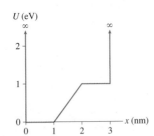

FIGURE EX40.13

14. Sketch the $n = 1$ and $n = 7$ wave functions for the potential energy shown in Figure Ex40.14.

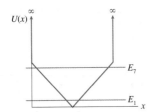

FIGURE EX40.14

Section 40.8 The Quantum Harmonic Oscillator

15. An electron in a harmonic potential well emits a photon with a wavelength of 300 nm as it undergoes a $3 \rightarrow 2$ quantum jump. What wavelength photon is emitted in a $3 \rightarrow 1$ quantum jump?

16. Consider a quantum harmonic oscillator.
    a. What happens to the spacing between the nodes of the wave function as $|x|$ increases? Why?
    b. What happens to the heights of the antinodes of the wave function as $|x|$ increases? Why?
    c. Sketch a reasonably accurate graph of the $n = 8$ wave function of a quantum harmonic oscillator.

17. An electron is confined in a harmonic potential well that has a spring constant of 2.0 N/m.
    a. What are the first three energy levels of the electron?
    b. What wavelength photon is emitted if the electron undergoes a $3 \rightarrow 1$ quantum jump?

18. An electron is confined in a harmonic potential well that has a spring constant of 12.0 N/m. What is the longest wavelength of light that the electron can absorb?

19. Two adjacent energy levels of an electron in a harmonic potential well are known to be 2.0 eV and 2.8 eV. What is the spring constant of the potential well?

Section 40.10 Quantum-Mechanical Tunneling

20. What is the probability that an electron will tunnel through a 0.45 nm gap from a metal to a STM probe if the work function is 4.0 eV?

21. A proton's energy is 1.0 MeV below the top of 10-fm-wide energy barrier. What is the probability that the proton will tunnel through the barrier?

Problems

22. A 2.0-μm-diameter water droplet is moving with a speed of 1.0 μm/s in a 20-μm-long box.
    a. Estimate the particle's quantum number.
    b. Use the correspondence principle to determine whether quantum mechanics is needed to understand the particle's motion or if it is "safe" to use classical physics.

23. Suppose that $\psi_1(x)$ and $\psi_2(x)$ are both solutions to the Schrödinger equation for the same potential energy $U(x)$. Prove that the superposition $\psi(x) = A\psi_1(x) + B\psi_2(x)$ is also a solution to the Schrödinger equation.

24. Figure 40.26a modeled a hydrogen atom as a finite potential well with rectangular edges. A more realistic model of a hydrogen atom, although still a one-dimensional model, would be the electron + proton electrostatic potential energy in one dimension:

$$U(x) = -\frac{e^2}{4\pi\epsilon_0|x|}$$

    a. Draw a graph of $U(x)$ versus x. Center your graph at $x = 0$.
    b. Despite the divergence at $x = 0$, the Schrödinger equation can be solved to find energy levels and wave functions for the electron in this potential. Draw a horizontal line across your graph of part a about one-third of the way from the bottom to the top. Label this line E_2, then, on this line, sketch a plausible graph of the $n = 2$ wave function.
    c. Redraw your graph of part a and add a horizontal line about two-thirds of the way from the bottom to the top. Label this line E_3, then, on this line, sketch a plausible graph of the $n = 3$ wave function.

25. a. Derive an expression for $\lambda_{2\rightarrow1}$, the wavelength of light emitted by a particle in a rigid box during a quantum jump from $n = 2$ to $n = 1$.
    b. In what length rigid box will an electron undergoing a $2 \rightarrow 1$ transition emit light with a wavelength of 694 nm? This is the wavelength of a ruby laser.

26. Model an atom as an electron in a rigid box of length 0.10 nm, roughly twice the Bohr radius.
    a. What are the four lowest energy levels of the electron?
    b. Calculate all the wavelengths that would be seen in the emission spectrum of this atom due to quantum jumps between these four energy levels. Give each wavelength a label $\lambda_{n\rightarrow m}$ to indicate the transition.
    c. Are these wavelengths in the infrared, visible, or ultraviolet portion of the spectrum?
    d. The stationary states of the Bohr hydrogen atom have negative energies. The stationary states of this model of the atom have positive energies. Is this a physically significant difference? Explain.
    e. Compare this model of an atom to the Bohr hydrogen atom. In what ways are the two models similar? Other than the signs of the energy levels, in what ways are they different?

27. Show that the normalization constant A_n for the wave functions of a particle in a rigid box has the value given in Equation 40.26.
28. A particle confined in a rigid one-dimensional box of length 10 fm has an energy level $E_n = 32.9$ MeV and an adjacent energy level $E_{n+1} = 51.4$ MeV.
    a. Determine the values of n and $n + 1$.
    b. Draw an energy-level diagram showing all energy levels from 1 through $n + 1$. Label each level and write the energy beside it.
    c. Sketch the $n + 1$ wave function on the $n + 1$ energy level.
    d. What is the wavelength of a photon emitted in the $n + 1 \rightarrow n$ transition? Compare this to a typical visible-light wavelength.
    e. What is the mass of the particle? Can you identify it?
29. Consider a particle in a rigid box of length L. For each of the states $n = 1$, $n = 2$, and $n = 3$:
    a. Sketch graphs of $|\psi(x)|^2$. Label the points $x = 0$ and $x = L$.
    b. Where, in terms of L, are the positions at which the particle is *most* likely to be found?
    c. Where, in terms of L, are the positions at which the particle is *least* likely to be found?
    d. Determine, by examining your $|\psi(x)|^2$ graphs, if the probability of finding the particle in the left one-third of the box is less than, equal to, or greater than $\frac{1}{3}$. Explain your reasoning.
    e. *Calculate* the probability that the particle will be found in the left one-third of the box.
30. For the quantum-well laser of Figure 40.16, *estimate* the probability that an electron will be found within one of the GaAlAs layers rather than in the GaAs layer. Explain your reasoning.
31. In a nuclear physics experiment, a proton is fired toward a $Z = 13$ nucleus with the diameter and neutron energy levels shown in Figure 40.17. The nucleus, which was initially in its ground state, subsequently emits a gamma ray with wavelength 1.73×10^{-4} nm. What was the *minimum* initial speed of the proton?
    Hint: Don't neglect the proton-nucleus collision.
32. Use the data from Figure 40.23 to calculate the first three vibrational energy levels of a C=O carbon-oxygen double bond.
33. Verify that the $n = 1$ wave function $\psi_1(x)$ of the quantum harmonic oscillator really is a solution of the Schrödinger equation. That is, show that the right and left sides of the Schrödinger equation are equal if you use the $\psi_1(x)$ wave function.
34. Show that the constant b used in the quantum-harmonic-oscillator wave functions (a) has units of length and (b) is the classical turning point of an oscillator in the $n = 1$ ground state.
35. a. Determine the normalization constant A_1 for the $n = 1$ ground-state wave function of the quantum harmonic oscillator. Your answer will be in terms of b.
    b. Write an expression for the probability that a quantum harmonic oscillator in its $n = 1$ ground state will be found in the classically forbidden region.
    c. (Optional) Use a numerical integration program to evaluate your probability expression of part b.
    Hint: It helps to simplify the integral by making a change of variables to $u = x/b$.

36. a. Derive an expression for the classical probability density $P_{class}(x)$ for a simple harmonic oscillator with amplitude A.
    b. Graph your expression between $x = -A$ and $x = +A$.
    c. Interpret your graph. Why is it shaped as it is?
37. a. Derive an expression for the classical probability density $P_{class}(y)$ for a ball that bounces between the ground and height h. The collisions with the ground are perfectly elastic.
    b. Graph your expression between $y = 0$ and $y = h$.
    c. Interpret your graph. Why is it shaped as it is?
38. Even the smoothest mirror finishes are "rough" when viewed at a scale of 100 nm. When two very smooth metals are placed in contact with each other, the actual distance between the surfaces varies from 0 nm at a few points of real contact to ≈ 100 nm. The average distance between the surfaces is ≈ 50 nm. The work function of aluminum is 4.3 eV. What is the probability that an electron will tunnel between two pieces of aluminum that are 50 nm apart?
39. An electron approaches a 1.0-nm-wide potential-energy barrier of height 5.0 eV. What energy electron has a tunneling probability of (a) 10%, (b) 1.0%, and (c) 0.10%?

Challenge Problems

40. A typical electron in a piece of metallic sodium has energy $-E_0$ compared to a free electron, where E_0 is the 2.7 eV work function of sodium.
    a. At what distance *beyond* the surface of the metal is the electron's probability density 10% of its value *at* the surface?
    b. How does this distance compare to the size of an atom?
41. Consider a particle in a rigid box of length L with walls at $x = -L/2$ and $x = +L/2$.
    a. What is the wave function $\psi(x)$ for $x < -L/2$ and $x > L/2$? Explain.
    b. Write the Schrödinger equation in the region $-L/2 \le x \le L/2$ for a particle with energy E.
    c. Write down a general solution to the Schrödinger equation that is valid in the region $-L/2 \le x \le L/2$.
    d. What are the boundary conditions this wave function must satisfy?
    e. Apply the boundary conditions to determine the allowed energy levels. Note that there are two different ways to satisfy the boundary conditions, each giving a different set of wave functions and energy levels.
    f. Compare your results to the rigid box that was analyzed in this chapter. In what ways are the results the same and in what ways are they different? Are any differences physically meaningful?
42. A particle of mass m has the wave function $\psi(x) = Ax\exp(-x^2/a^2)$ when it is in an allowed energy level with $E = 0$.
    a. Draw a graph of $\psi(x)$ versus x.
    b. At what value or values of x is the particle most likely to be found?
    c. Find and graph the potential-energy function $U(x)$.

43. In most metals, the atomic ions form a regular arrangement called a *crystal lattice*. The conduction electrons in the sea of electrons move through this lattice. Figure CP40.43 is a one-dimensional model of a crystal lattice. The ions have mass m, charge e, and an equilibrium separation b.

    a. Suppose the middle charge is displaced a very small distance $(x \ll b)$ from its equilibrium position while the outer charges remain fixed. Show that the net electric force on the middle charge is given approximately by

    $$F = -\frac{e^2}{b^3 \pi \epsilon_0} x$$

    In other words, the charge experiences a linear restoring force.

    b. Suppose this crystal consists of aluminum ions with an equilibrium spacing of 0.30 nm. What are the energies of the four lowest vibrational states of these ions?

    c. What wavelength photons are emitted during quantum jumps between *adjacent* energy levels? Is this wavelength in the infrared, visible, or ultraviolet portion of the spectrum?

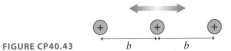

FIGURE CP40.43

44. a. What is the probability that an electron will tunnel through a 0.50 nm air gap from a metal to a STM probe if the work function is 4.0 eV?

    b. The probe passes over an atom that is 0.050 nm "tall." By what factor does the tunneling current increase?

    c. If a 10% current change is reliably detectable, what is the smallest height change the STM can detect?

45. Tennis balls traveling at greater than 100 mph routinely bounce off tennis rackets. At some sufficiently high speed, however, the ball will break through the strings and keep going. The racket is a potential-energy barrier whose height is the energy of the slowest string-breaking ball. Suppose that a 100 g tennis ball traveling at 200 mph is just sufficient to break the 2.0-mm-thick strings. Estimate the probability that a 120 mph ball will tunnel through the racket without breaking the strings. Give your answer as a power of 10 rather than a power of e.

STOP TO THINK ANSWERS

Stop to Think 40.1: $v_a = v_b > v_c$. The de Broglie wavelength is $\lambda = h/mv$, so slower particles have longer wavelengths. The wave amplitude is not relevant.

Stop to Think 40.2: c. The $n = 2$ state has a node in the middle of the box. The antinodes are centered in the left and right halves of the box.

Stop to Think 40.3: $n = 4$. There are four antinodes, three nodes (excluding the ends).

Stop to Think 40.4: d. The wave function reaches zero abruptly on the right, indicating an infinitely high potential-energy wall. The exponential decay on the left shows that the left wall of the poten-tial energy is *not* infinitely high. The node spacing and the amplitude increase steadily in going from right to left, indicating a *steadily* decreasing kinetic energy and thus a steadily increasing potential energy.

Stop to Think 40.5: c. $E = (n - \frac{1}{2})\hbar\omega$, so $\frac{5}{2}\hbar\omega$ is the energy of the $n = 3$ state. An $n = 3$ state has 3 antinodes.

Stop to Think 40.6: b. The probability of tunneling through the barrier increases as the difference between E and U_0 decreases. If the tunneling probability increases, the reflection probability must decrease.

41 Atomic Physics

Lasers are one of the most important applications of the quantum-mechanical properties of atoms and light.

▶ **Looking Ahead**

The goal of Chapter 41 is to understand the structure and properties of atoms. In this chapter you will learn to:

- Use a quantum-mechanical model of the hydrogen atom.
- Understand the idea of electron spin.
- Apply Schrödinger's quantum theory to multielectron atoms.
- Interpret atomic spectra.
- Understand how lasers work.

◀ **Looking Back**

The material in this chapter depends on an understanding of the Bohr model of atomic quantization and one-dimensional quantum mechanics. Please review:

- Sections 38.5–38.7 Bohr's model of quantization and the hydrogen atom.
- Sections 39.3–39.4 Interpreting and using wave functions.
- Sections 40.1–40.2 The basic ideas of quantum mechanics.

The problem of discovering the structure of atoms is one that we have continued to revisit. The first model of an atom we looked at, Rutherford's solar-system model, was purely classical. This model incorporated Rutherford's discovery of a very small nucleus, but otherwise it had almost no agreement with the experimental evidence about atoms. It could not explain their discrete spectra, nor could it explain why atoms are stable!

The Bohr model of the hydrogen atom was a big step forward. The concept of stationary states provided a means of understanding both the stability of atoms and the quantum jumps that lead to discrete spectra. And Bohr's ability to derive the Balmer formula for the hydrogen spectrum indicated that he was on the right track. Yet, as we have seen, the Bohr model was not successful for any atom other than hydrogen.

Now it's Schrödinger's turn. Is Schrödinger's theory of quantum mechanics better at explaining atomic structure than other models? The answer, as you can probably anticipate, is a decisive yes. This chapter is an overview of how quantum mechanics finally provides us with an understanding of atomic structure and atomic properties.

41.1 The Hydrogen Atom: Angular Momentum and Energy

Let's begin with a quantum-mechanical model of the hydrogen atom. Figure 41.1 shows an electron at distance r from a proton. The proton is much more massive than the electron, so we will assume that the proton remains at rest at the origin.

As you learned in Chapter 40, the problem-solving procedure in quantum mechanics consists of two basic steps:

1. Specify a potential-energy function.
2. Solve the Schrödinger equation to find the wave functions, allowed energy levels, and other quantum properties.

The first step is easy. The proton and electron are charged particles with $q = \pm e$, so the potential energy of a hydrogen atom as a function of the electron distance r is

$$U(r) = -\frac{1}{4\pi\epsilon_0}\frac{e^2}{r} \tag{41.1}$$

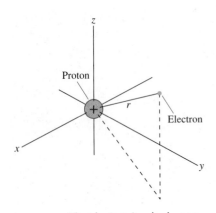

FIGURE 41.1 The electron in a hydrogen atom is distance r from the proton.

The difficulty arises with the second step. The Schrödinger equation of Chapter 40 was for one-dimensional problems. Atoms are three-dimensional, and the three-dimensional Schrödinger equation turns out to be a partial differential equation whose solution is outside the scope of this textbook. Consequently, we'll present results that can be derived only in a more advanced course in quantum mechanics. The good news is that you have learned enough quantum mechanics to interpret and use the results.

Stationary States of Hydrogen

In one dimension, energy quantization appeared as a consequence of *boundary conditions* on the wave function. That is, there were solutions to the Schrödinger equation that satisfied the boundary conditions only for certain discrete energies, characterized by the quantum number n. In three dimensions, the wave function must satisfy *three* different boundary conditions. Consequently, solutions to the three-dimensional Schrödinger equation have *three* quantum numbers and *three* quantized parameters.

Solutions to the Schrödinger equation for the hydrogen atom potential energy exist only if three conditions are satisfied:

1. The atom's energy must be one of the values

$$E_n = -\frac{1}{n^2}\left(\frac{1}{4\pi\epsilon_0}\frac{e^2}{2a_B}\right) = -\frac{13.60\text{ eV}}{n^2} \qquad n = 1, 2, 3, \ldots \tag{41.2}$$

where $a_B = 4\pi\epsilon_0\hbar^2/me^2 = 0.0529$ nm is the Bohr radius. The integer n is called the **principal quantum number.** These energies are the same as those in the Bohr hydrogen atom.

2. The angular momentum L of the electron's orbit must be one of the values

$$L = \sqrt{l(l + 1)}\hbar \qquad l = 0, 1, 2, 3, \ldots, n - 1 \tag{41.3}$$

The integer l is called the **orbital quantum number.**

3. The z-component of the angular momentum L_z must be one of the values

$$L_z = m\hbar \qquad m = -l, -l + 1, \ldots, 0, \ldots, l - 1, l \qquad (41.4)$$

The integer m is called the **magnetic quantum number.**

In other words, each stationary state of the hydrogen atom is identified by a triplet of quantum numbers (n, l, m). Each quantum number is associated with a physical property of the atom.

NOTE ▶ The energy of the stationary state depends only on the principal quantum number n, not on l or m. ◀

EXAMPLE 41.1 Listing quantum numbers
List all possible states of a hydrogen atom that have energy $E = -3.40$ eV.

SOLVE Energy depends only on the principal quantum number n. States with $E = -3.40$ eV have

$$n = \sqrt{\frac{-13.60 \text{ eV}}{-3.40 \text{ eV}}} = 2$$

An atom with principal quantum number $n = 2$ could have either $l = 0$ or $l = 1$, but $l \geq 2$ is ruled out. If $l = 0$, the only

possible value for the magnetic quantum number m is $m = 0$. If $l = 1$, then the atom could have $m = -1$, $m = 0$, or $m = +1$. Thus the possible quantum numbers are

n	l	m
2	0	0
2	1	1
2	1	0
2	1	-1

These four states all have the same energy.

Hydrogen turns out to be unique. For all other elements, the allowed energies depend on both n *and* l (but not m). Consequently, it is useful to label the stationary states by their values of n and l. The lowercase letters shown in Table 41.1 are customarily used to represent the various values of quantum number l. These symbols come from spectroscopic notation used in prequantum-mechanics days, when some spectral lines were classified as sharp, others as principal, and so on.

Using these symbols, the ground state of the hydrogen atom, with $n = 1$ and $l = 0$, is called the 1s state. The 3d state has $n = 3$, $l = 2$. In Example 41.1, we found one 2s state (with $l = 0$) and three 2p states (with $l = 1$), all with the same energy.

TABLE 41.1 Symbols used to represent quantum number l

l	Symbol
0	s
1	p
2	d
3	f

Angular Momentum Is Quantized

If the hydrogen atom were classical, the electron's orbit, like that of a planet in the solar system, would be an ellipse. Furthermore, the orbit need not lie in the xy-plane. Figure 41.2 shows a possible classical orbit tilted at angle θ below the xy-plane.

We introduced the angular momentum vector $\vec{L}$ in Chapter 13. It will be useful to call $\vec{L}$ the *orbital* angular momentum in order to distinguish it later from *spin* angular momentum. Figure 41.2 reminds you that the vector $\vec{L}$ is perpendicular to the plane of the electron's orbit. The angular momentum vector has a z-component $L_z = L\cos\theta$ along the z-axis.

Classically, L and L_z can have any values. Not so in quantum mechanics. Quantum conditions 2 and 3 tell us that **the electron's orbital angular momentum is quantized.** The magnitude of the orbital angular momentum must be one of the discrete values

$$L = \sqrt{l(l + 1)}\,\hbar = 0, \sqrt{2}\hbar, \sqrt{6}\hbar, \sqrt{12}\hbar, \ldots$$

where l is an integer. Simultaneously, the z-component L_z must have one of the values $L_z = m\hbar$, where m is an integer between $-l$ and l. No other values of L or L_z allow the wave function to satisfy the boundary conditions.

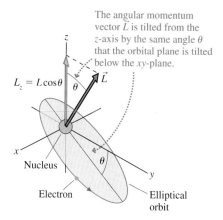

The angular momentum vector $\vec{L}$ is tilted from the z-axis by the same angle θ that the orbital plane is tilted below the xy-plane.

$L_z = L\cos\theta$

FIGURE 41.2 The angular momentum of an elliptical orbit.

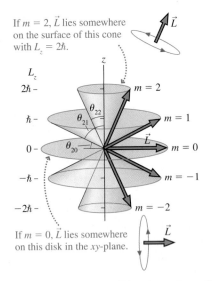

If $m = 2$, $\vec{L}$ lies somewhere on the surface of this cone with $L_z = 2\hbar$.

If $m = 0$, $\vec{L}$ lies somewhere on this disk in the xy-plane.

FIGURE 41.3 Five possible orientations of the angular momentum vector for $l = 2$. The angular momentum vectors all have length $L = \sqrt{6}\hbar = 2.45\hbar$.

The quantization of angular momentum places restrictions on the shape and orientation of the electron's orbit. To see this, consider a hydrogen atom with orbital quantum number $l = 2$. In this state, the *magnitude* of the electron's angular momentum must be $L = \sqrt{6}\hbar = 2.45\hbar$. Furthermore, the angular momentum vector must point in a *direction* such that $L_z = m\hbar$, where m is one of only five integers in the range $-2 \le m \le 2$.

The combination of these two requirements allows $\vec{L}$ to point only in certain directions in space, as shown in Figure 41.3. This is a rather unusual figure that requires a little thought to understand. Suppose $m = 0$ and thus $L_z = 0$. With no z-component, the angular momentum vector $\vec{L}$ must lie somewhere in the xy-plane. Furthermore, because the length of $\vec{L}$ is constrained to be $2.45\hbar$, the tip of $\vec{L}$ must lie somewhere on the circle labeled $m = 0$. These values of $\vec{L}$ correspond to classical orbits tipped into a vertical plane.

Similarly, $m = 2$ requires $\vec{L}$ to lie along the cone whose height is $2\hbar$ and whose side has length $2.45\hbar$. These values of $\vec{L}$ correspond to classical orbits tilted slightly out of the xy-plane. Notice that $\vec{L}$ **cannot point directly along the z-axis.** The maximum possible value of L_z, when $m = l$, is $(L_z)_{\max} = l\hbar$. But $l < \sqrt{l(l+1)}$, so $(L_z)_{\max} < L$. The angular momentum vector *must* have either an x- or a y-component (or both). In other words, the corresponding classical orbit cannot lie in the xy-plane.

An angular momentum vector $\vec{L}$ tilted at angle θ from the z-axis corresponds to an orbit tilted at angle θ out of the xy-plane. The quantization of angular momentum restricts the orbital planes to only a few discrete angles. For quantum state (n, l, m), the angle of the angular momentum vector is

$$\theta_{lm} = \cos^{-1}\left(\frac{L_z}{L}\right) = \cos^{-1}\left(\frac{m\hbar}{\sqrt{l(l+1)}\hbar}\right) = \cos^{-1}\left(\frac{m}{\sqrt{l(l+1)}}\right) \qquad (41.5)$$

Angles θ_{22}, θ_{21}, and θ_{20} are labeled in Figure 41.3. Orbital planes at other angles are not allowed because they don't satisfy the quantization conditions for angular momentum.

EXAMPLE 41.2 The angle of the angular momentum vector

What is the angle between $\vec{L}$ and the z-axis for a hydrogen atom in the stationary state $(n, l, m) = (4, 2, 1)$?

SOLVE The angle θ_{21} is labeled in Figure 41.3. The state $(4, 2, 1)$ has $l = 2$ and $m = 1$, thus

$$\theta_{21} = \cos^{-1}\left(\frac{1}{\sqrt{6}}\right) = 65.9°$$

ASSESS This quantum state corresponds to a classical orbit tilted 65.9° away from the xy-plane.

NOTE ▶ The ground state of hydrogen, with $l = 0$, has *no* angular momentum. A classical particle cannot orbit unless it has angular momentum, but apparently a quantum particle does not have this requirement. We will examine this issue in the next section. ◀

Energy Levels of the Hydrogen Atom

The energy of the hydrogen atom is quantized. Only those energies given by Equation 41.2 allow the wave function to satisfy the boundary conditions. The allowed energies of hydrogen depend only on the principal quantum number n, but for other atoms the energies will depend on both n and l. In anticipation of using both quantum numbers, Figure 41.4 is an *energy-level diagram* for the hydrogen atom in which the rows are labeled by n and the columns by l. The left column contains all of the $l = 0$ s states, the next column is the $l = 1$ p states, and so on.

Because the quantum condition of Equation 41.3 requires $n > l$, the s states begin with $n = 1$, the p states begin with $n = 2$, and the d states with $n = 3$. That is, the lowest-energy d state is $3d$ because states with $n = 1$ or $n = 2$ cannot have $l = 2$. For hydrogen, where the energy levels do not depend on l, the energy-level diagram shows that the $3s$, $3p$, and $3d$ states have equal energy. Figure 41.4 shows only the first few energy levels for each value of l, but there really are an infinite number of levels, as $n \to \infty$, crowding together beneath $E = 0$. The dotted line at $E = 0$ is the atom's *ionization limit*, the energy of a hydrogen atom in which the electron has been moved infinitely far away to form an H^+ ion.

The lowest energy state, the $1s$ state with $E_1 = -13.60$ eV, is the *ground state* of hydrogen. The value $|E_1| = 13.60$ eV is the **ionization energy,** the *minimum* energy that would be needed to form a hydrogen ion by removing the electron from the ground state. All of the states with $n > 1$ are *excited states*.

Quantum number l	0	1	2	3
Symbol	s	p	d	f

n	$E = 0$ eV		Ionization limit	
4	-0.85 eV	$4s$	$4p$ $4d$ $4f$	
3	-1.51 eV	$3s$	$3p$ $3d$	
2	-3.40 eV	$2s$	$2p$	
1	-13.60 eV	$1s$ — Ground state		

FIGURE 41.4 Energy-level diagram for the hydrogen atom.

STOP TO THINK 41.1 What are the quantum numbers n and l for a hydrogen atom with $E = -(13.60/9)$ eV and $L = \sqrt{2}\hbar$?

41.2 The Hydrogen Atom: Wave Functions and Probabilities

You learned in Chapter 40 that the probability of finding a particle in a small interval of width δx at the position x is given by

$$\text{Prob(in } \delta x \text{ at } x) = |\psi(x)|^2 \delta x = P(x)\,\delta x$$

where

$$P(x) = |\psi(x)|^2$$

is the probability density. This interpretation of $|\psi(x)|^2$ as a probability density lies at the heart of quantum mechanics. However, $P(x)$ was for a one-dimensional wave function. Because we're now looking at a three-dimensional atom, we need to consider the probability of finding a particle in a small *volume* of space δV at the position described by the three coordinates (x, y, z). This probability is

$$\text{Prob(in } \delta V \text{ at } x, y, z) = |\psi(x, y, z)|^2 \delta V \qquad (41.6)$$

We can still interpret $|\psi(x, y, z)|^2$ as a probability density.

In one-dimensional quantum mechanics we could simply graph $P(x)$ versus x. Portraying the probability density of a three-dimensional wave function is more of a challenge. One way to do so, shown in Figure 41.5 on the next page, is to use denser shading to indicate regions of larger probability density. That is, the amplitude of ψ is larger, and the electron is more likely to be found in regions where the shading is darker. These figures show the probability densities of the $1s$, $2s$, and $2p$ states of hydrogen. As you can see, the probability density in three dimensions creates what is often called an **electron cloud** around the nucleus.

These figures contain a lot of information. For example, notice how the p electrons have directional properties. These directional properties allow p electrons to "reach out" toward nearby atoms, forming molecular bonds. The quantum mechanics of bonding goes beyond what we can study in this text, but the electron-cloud pictures of the p electrons begin to suggest how bonds could form.

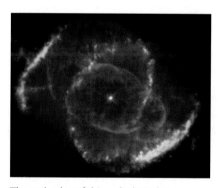

The red color of this nebula is due to the emission of light from hydrogen atoms. The atoms are excited by intense ultraviolet light from the star in the center. They then emit red light ($\lambda = 656$ nm) in a $3 \to 2$ transition, part of the Balmer series of spectral lines emitted by hydrogen.

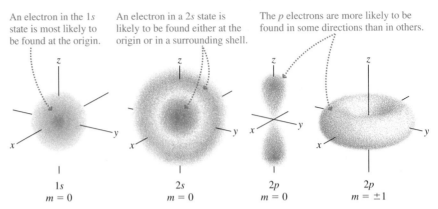

An electron in the 1s state is most likely to be found at the origin.

An electron in a 2s state is likely to be found either at the origin or in a surrounding shell.

The p electrons are more likely to be found in some directions than in others.

1s
m = 0

2s
m = 0

2p
m = 0

2p
m = ±1

FIGURE 41.5 The probability densities of the electron in the 1s, 2s, and 2p states of hydrogen.

Radial Wave Functions

Figures such as Figure 41.5 are useful for "seeing" the electron clouds, but these figures are hard to use. Often, we would simply like to know the probability of finding the electron at a certain *distance* from the nucleus. That is, what is the probability that the electron is to be found within the small range of distances δr at the distance r?

It turns out that the solutions to the three-dimensional Schrödinger equation can be written in a form that focuses on the electron's radial distance r from the proton. The portion of the wave function that depends only on r is called the **radial wave function**. These functions, which depend on the quantum numbers n and l, are designated $R_{nl}(r)$. The first three radial wave functions are

$$R_{1s}(r) = \frac{1}{\sqrt{\pi a_B^3}} e^{-r/a_B}$$

$$R_{2s}(r) = \frac{1}{\sqrt{8\pi a_B^3}}\left(1 - \frac{r}{2a_B}\right)e^{-r/2a_B} \qquad (41.7)$$

$$R_{2p}(r) = \frac{1}{\sqrt{24\pi a_B^3}}\left(\frac{r}{2a_B}\right)e^{-r/2a_B}$$

where a_B is the Bohr radius.

The radial wave functions may seem mysterious, because we haven't shown where they come from, but they are essentially the same as the one-dimensional wave functions $\psi(x)$ you learned to work with in Chapter 40. In fact, these radial wave functions are similar to the one-dimensional wave functions of the simple harmonic oscillator. One important difference, however, is that r ranges from 0 to ∞. For one-dimensional wave functions, x ranged from $-\infty$ to ∞.

Figure 41.6 shows the radial wave functions for the 1s and 2s states. Notice that the radial wave function is nonzero at $r = 0$, the position of the nucleus. This is surprising, but it is consistent with our observation in Figure 41.5 that the 1s and 2s electrons have a strong probability of being found at the origin.

We can gain some understanding of the s-state wave functions by considering the angular momentum. A classical particle, for which $L = mvr$, can have $L = 0$ only if the radius of its orbit shrinks to zero. This is impossible for a classical particle, but zero angular momentum *is* achievable for quantum particles because the uncertainty principle prevents a quantum particle from being localized at a single

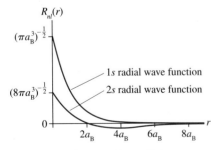

FIGURE 41.6 The 1s and 2s radial wave functions of hydrogen.

point. The *s*-state wave functions of Figure 41.6, with their maximum values at $r = 0$, are the quantum analogs of a classical particle orbiting with $r = 0$.

Our purpose for introducing the radial wave functions was to determine the probability of finding the electron a certain *distance* from the nucleus. Figure 41.7 shows a shell of radius r and thickness δr centered on the nucleus. The probability of finding the electron at distance r from the nucleus is equivalent to the probability that the electron is located somewhere within this shell. The volume of a thin shell is its surface area multiplied by its thickness δr. The surface area of a sphere is $4\pi r^2$, so the volume of this thin shell is

$$\delta V = 4\pi r^2 \delta r \tag{41.8}$$

We will assert, without proof, that the probability of finding the electron within this shell is

$$\text{Prob(in } \delta r \text{ at } r) = |R_{nl}(r)|^2 \delta V = 4\pi r^2 |R_{nl}(r)|^2 \delta r = P_r(r)\,\delta r \tag{41.9}$$

where

$$P_r(r) = 4\pi r^2 |R_{nl}(r)|^2 \tag{41.10}$$

is called the **radial probability density** for the state *nl*.

The radial probability density tells us the relative likelihood of finding the electron at distance r from the nucleus. The volume factor $4\pi r^2$ reflects the fact that more space is available in a shell of larger r, and this additional space increases the probability of finding the electron at that distance.

The probability of finding the electron between $r_{\min}$ and $r_{\max}$ is

$$\text{Prob}(r_{\min} \leq r \leq r_{\max}) = \int_{r_{\min}}^{r_{\max}} P_r(r)\,dr = 4\pi \int_{r_{\min}}^{r_{\max}} r^2 |R_{nl}(r)|^2 dr \tag{41.11}$$

The electron must be *somewhere* between $r = 0$ and $r = \infty$, so the integral of $P_r(r)$ between 0 and ∞ must equal 1. This normalization condition was used to determine the constants in front of the radial wave functions of Equation 41.7.

Figure 41.8 shows the radial probability densities for the $n = 1$, 2, and 3 states of the hydrogen atom, all drawn to the same scale so that you can compare them to each other. The horizontal scale is in units of the Bohr radius a_B.

You can see that the 1*s*, 2*p*, and 3*d* states, with maxima at a_B, $4a_B$ and $9a_B$, respectively, are following the pattern $r_{\text{peak}} = n^2 a_B$. These are exactly the radii of the orbits in the Bohr hydrogen atom. There we simply bent a one-dimensional de Broglie wave into a circle of that radius. Now we have a three-dimensional wave function for which the electron is *most likely* to be this distance from the nucleus, although it *could* be found at other values of r. The physical situation is very different in quantum mechanics, but it is good to see that various aspects of the Bohr atom can be reproduced.

But why is it the 3*d* state that agrees with the Bohr atom rather than 3*s* or 3*p*? All states with the same value of n form a collection of "orbits" having the same energy. As Figure 41.9 shows, the state with $l = n - 1$ has the largest angular momentum of the group. Consequently, the maximum-*l* state corresponds to a circular classical orbit and matches the circular orbits of the Bohr atom. Notice that the radial probability densities for the 2*p* and 3*d* states have a single peak, corresponding to a classical orbit at a constant distance.

States with smaller *l* correspond to elliptical orbits. You can see in Figure 41.8 that the radial probability density of a 3*s* electron has a peak in close to the nucleus. The 3*s* electron also has a good chance of being found *farther* from the nucleus than a 3*d* electron, suggesting an orbit that alternately swings in near the nucleus, then moves out past the circular orbit with the same energy. This distinction

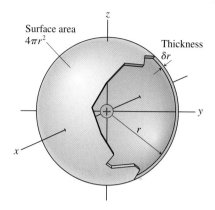

FIGURE 41.7 The radial probability density gives the probability of finding the electron in a spherical shell of thickness δr at radius r.

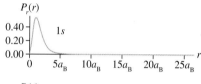
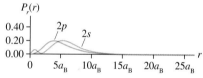
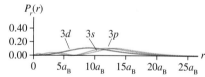

FIGURE 41.8 The radial probability densities for $n = 1$, 2, and 3.

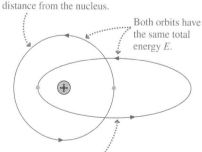

The circular orbit has the largest angular momentum. The electron stays at a constant distance from the nucleus.

Both orbits have the same total energy *E*.

The elliptical orbit has a smaller angular momentum. Compared to the circular orbit, the electron gets both closer to and farther from the nucleus.

FIGURE 41.9 More circular orbits have larger angular momenta.

between circular and elliptical orbits will be important when we discuss the energy levels in multielectron atoms.

NOTE ▶ In quantum mechanics, nothing is really orbiting. However, the probability densities for the electron to be, or not to be, at any given distance from the nucleus mimic certain aspects of classical orbits and provide us with a useful analogy. ◀

You can see in Figure 41.8 that the most likely distance from the nucleus of an $n = 1$ electron is approximately a_B. The distance of an $n = 2$ electron is most likely to be between about $3a_B$ and $7a_B$. An $n = 3$ electron is most likely to be found between about $8a_B$ and $15a_B$. In other words, the radial probability densities give the clear impression that each value of n has a fairly well defined range of radii where the electron is most likely to be found. This is the basis of the **shell model** of the atom that is used in chemistry.

However, there's one significant puzzle. In Figure 41.5, the fuzzy sphere representing the $1s$ ground state is densest at the center, where the electron is most likely to be found. This maximum density at $r = 0$ agrees with the $1s$ radial wave function of Figure 41.6, which is a maximum at $r = 0$, but it seems to be in sharp disagreement with the $1s$ graph of Figure 41.8, which is *zero* at the nucleus and peaks at $r = a_B$.

Resolving this puzzle requires distinguishing between the probability density $|\psi(x, y, z)|^2$ and the *radial* probability density $P_r(r)$. The $1s$ wave function, and thus the $1s$ probability density, really does peak at the nucleus. But $|\psi(x, y, z)|^2$ is the probability of being in a small volume δV, such as a small box with sides δx, δy, and δz, whereas $P_r(r)$ is the probability of being in a spherical shell of thickness δr. Compared to $r = 0$, the probability density $|\psi(x, y, z)|^2$ is smaller at any *one* point having $r = a_B$. But the volume of *all* points with $r \approx a_B$ (i.e., the volume of the spherical shell at $r = a_B$) is so large that the radial probability density P_r peaks at this distance.

To use a mass analogy, consider a fuzzy ball that is densest at the center. Even though the density away from the center has decreased, a spherical shell of modest radius r can have *more total mass* than a small-radius spherical shell of the same thickness simply because it has so much more volume.

EXAMPLE 41.3 Maximum probability

Show that an electron in the $2p$ state is most likely to be found at $r = 4a_B$.

SOLVE We can use the $2p$ radial wave function from Equation 41.7 to write the radial probability density

$$P_r(r) = 4\pi r^2 |R_{2p}(r)|^2 = 4\pi r^2 \left[\frac{1}{\sqrt{24\pi a_B^3}} \left(\frac{r}{2a_B} \right) e^{-r/2a_B} \right]^2$$

$$= Cr^4 e^{-r/a_B}$$

where $C = (24a_B)^{-5}$ is a constant. This expression for $P_r(r)$ was graphed in Figure 41.8.

The most probable value of r occurs at the point where the derivative of $P_r(r)$ is zero:

$$\frac{dP_r}{dr} = C(4r^3)(e^{-r/a_B}) + C(r^4)\left(\frac{-1}{a_B} e^{-r/a_B} \right)$$

$$= Cr^3\left(4 - \frac{r}{a_B} \right)e^{-r/a_B} = 0$$

This expression is zero only if $r = 4a_B$, so $P_r(r)$ is maximum at $r = 4a_B$. An electron in the $2p$ state is most likely to be found at this distance from the nucleus.

STOP TO THINK 41.2   How many maxima will there be in a graph of the radial probability density for the $4s$ state of hydrogen?

41.3 The Electron's Spin

Recall, from Chapter 32, that an electron orbiting a nucleus is a microscopic *magnetic moment* $\vec{\mu}$. Figure 41.10 reminds you that a magnetic moment, like a compass needle, has a north and south pole. Consequently, a magnetic moment in an external magnetic field experiences forces and torques. In the early 1920s, the German physicists Otto Stern and Walter Gerlach developed a technique to measure the magnetic moments of atoms. Their apparatus, shown in Figure 41.11, prepares an *atomic beam* by evaporating atoms out of a hole in an "oven." These atoms, traveling in a vacuum, pass through a *nonuniform* magnetic field. The field is stronger toward the top of the magnet, weaker toward the bottom.

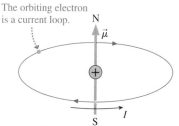

The orbiting electron is a current loop.

A current loop generates a magnetic moment with a north and south magnetic pole.

FIGURE 41.10 An orbiting electron generates a magnetic moment.

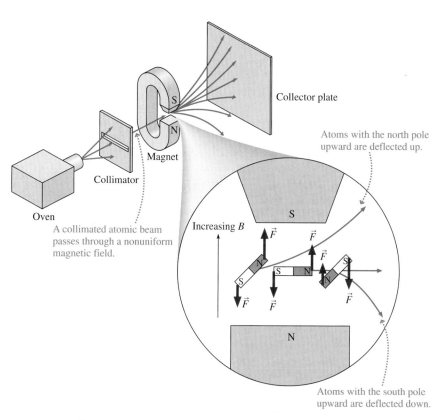

Collector plate

Magnet

Collimator

Oven

A collimated atomic beam passes through a nonuniform magnetic field.

Atoms with the north pole upward are deflected up.

Increasing B

Atoms with the south pole upward are deflected down.

FIGURE 41.11 The Stern-Gerlach experiment.

A nonuniform field exerts forces of *different* strengths on the north and south poles of a magnetic moment, causing a *net force*. An atom whose magnetic moment vector $\vec{\mu}$ is tilted upward ($\mu_z > 0$) has an upward force on its north pole that is larger than the downward force on its south pole. As the figure shows, this atom is deflected upward as it passes through the magnet. A downward-tilted magnetic moment ($\mu_z < 0$) experiences a net downward force and is deflected downward. A magnetic moment perpendicular to the field ($\mu_z = 0$) feels no net force and passes through the magnet without deflection. In other words, an atom's deflection as it passes through the magnet is proportional to μ_z, the z-component of its magnetic moment.

It's not hard to show, although we will omit the proof, that an atom's magnetic moment is proportional to the electron's orbital angular momentum: $\vec{\mu} \propto \vec{L}$. Because the deflection of an atom depends on μ_z, measuring the deflections in a nonuniform field provides information about the L_z values of the atoms in the atomic beam. The measurements are made by allowing the atoms to stick on a collector plate at the end of the apparatus. After the experiment has been run for several hours, the collector plate is removed and examined to learn how the atoms are being deflected.

With the magnet turned off, the atoms pass through without deflection and land along a narrow line at the center, as shown in Figure 41.12a. If the orbiting electrons are classical particles, they should have a continuous range of angular momenta. Turning on the magnet should produce a continuous range of vertical deflections, and the distribution of atoms collected on the plate should look like Figure 41.12b. But if angular momentum is *quantized,* as Bohr had suggested several years earlier, the atoms should be deflected to discrete positions on the collector plate.

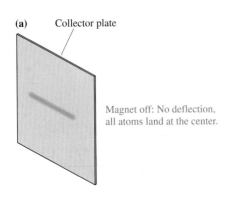

(a) Collector plate

Magnet off: No deflection, all atoms land at the center.

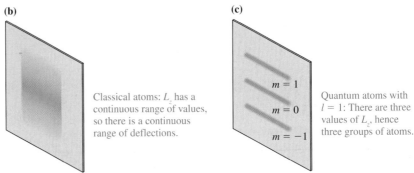

(b)

Classical atoms: L_z has a continuous range of values, so there is a continuous range of deflections.

(c)

$m = 1$
$m = 0$
$m = -1$

Quantum atoms with $l = 1$: There are three values of L_z, hence three groups of atoms.

FIGURE 41.12 Distribution of the atoms on the collector plate.

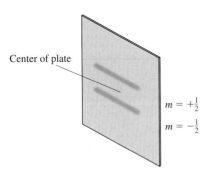

Center of plate

$m = +\frac{1}{2}$
$m = -\frac{1}{2}$

FIGURE 41.13 The outcome of the Stern-Gerlach experiment for hydrogen atoms.

For example, an atom with $l = 1$ has three distinct values of L_z corresponding to quantum numbers $m = -1, 0,$ and 1. This leads to a prediction of the three distinct groups of atoms shown in Figure 41.12c. There should always be an *odd* number of groups because there are $2l + 1$ values of L_z.

In 1927, with Schrödinger's quantum theory brand new, the Stern-Gerlach technique was used to measure the magnetic moment of hydrogen atoms. The ground state of hydrogen is 1s, with $l = 0$, so the atoms should have *no* magnetic moment and there should be *no* deflection at all. Instead, the experiment produced the two-peaked distribution shown in Figure 41.13.

Because the hydrogen atoms were deflected, they *must* have a magnetic moment. But where does it come from if $L = 0$? Even stranger was the deflection into two groupings, rather than an odd number. The deflection is proportional to L_z, and $L_z = m\hbar$ where m ranges in integer steps from $-l$ to $+l$. The experimental results would make sense only if $l = \frac{1}{2}$, allowing m to take on the two possible values $-\frac{1}{2}$ or $+\frac{1}{2}$. But according to Schrödinger's theory, the quantum numbers l and m must be integers.

An explanation for these observations was soon suggested, then confirmed: The electron has an *inherent* magnetic moment. After all, the electron has an inherent gravitational character, its mass m_e, and an inherent electric character, its charge $q_e = -e$. These are simply part of what an electron is. Thus it is plausible that an electron should also have an inherent magnetic character described by a built-in magnetic moment $\vec{\mu}_e$. A classical electron, if thought of as a little ball of charge, could spin on its axis as it orbits the nucleus. A spinning ball of charge would have a magnetic moment associated with its angular momentum. This inherent magnetic moment of the electron is what caused the unexpected deflection in the Stern-Gerlach experiment.

If the electron has an inherent magnetic moment, it must have an inherent angular momentum. This angular momentum is called the electron's **spin,** which is designated $\vec{S}$. The outcome of the Stern-Gerlach experiment tells us that the z-component of this spin angular momentum is

$$S_z = m_s \hbar \quad \text{where } m_s = +\frac{1}{2} \quad \text{or} \quad -\frac{1}{2} \qquad (41.12)$$

The quantity m_s is called the **spin quantum number.**

The z-component of the spin angular momentum vector is determined by the electron's orientation. The $m_s = +\frac{1}{2}$ state, with $S_z = +\frac{1}{2}\hbar$, is called the **spin-up** state and the $m_s = -\frac{1}{2}$ state is called the **spin-down** state. It is convenient to picture a little angular momentum vector that can be drawn $\uparrow$ for an $m_s = +\frac{1}{2}$ state and $\downarrow$ for an $m_s = -\frac{1}{2}$ state. We will use this notation in the next section. Because the electron must be either spin-up or spin-down, a hydrogen atom in the Stern-Gerlach experiment will be deflected either up or down. This causes the two groups of atoms seen in Figure 41.13. No atoms have $S_z = 0$, so there are no undeflected atoms in the center.

NOTE ▶ The atom has spin angular momentum *in addition* to any orbital angular momentum that the electrons may have. Only in s states, for which $L = 0$, can we see the effects of "pure spin." ◀

The equation for the spin angular momentum S is analogous to Equation 41.3 for L:

$$S = \sqrt{s(s+1)}\hbar = \frac{\sqrt{3}}{2}\hbar \qquad (41.13)$$

where s is a quantum number with the single value $s = \frac{1}{2}$. S is the *inherent* angular momentum of the electron. Because of the single value of s, physicists usually say that the electron has "spin one-half." Figure 41.14, which should be compared to Figure 41.3, shows that the terms "spin up" and "spin down" refer to S_z not the full spin angular momentum. As was the case with $\vec{L}$, it's not possible for $\vec{S}$ to point along the z-axis.

NOTE ▶ The term "spin" must be used with caution. Although a classical charged particle could generate a magnetic moment by spinning, the electron most assuredly is *not* a classical particle. It is not spinning in any literal sense. It simply has an inherent magnetic moment, just as it has an inherent mass and charge, and that magnetic moment makes it look *as if* the electron is spinning. It is a convenient figure of speech, not a factual statement. **The electron has a spin, but it is *not* a spinning electron!** ◀

The electron's spin has significant implications for atomic structure. The solutions to the Schrödinger equation could be described by the three quantum numbers n, l, and m, but the Stern-Gerlach experiment implies that this is not a complete description of an atom. Knowing that a ground-state atom has quantum numbers $n = 1$, $l = 0$, and $m = 0$ is not sufficient to predict whether the atom will be deflected up or down in a nonuniform magnetic field. We need to add the spin quantum number m_s to make our description complete. (Strictly speaking, we also need to add the quantum number s, but it provides no additional information because its value never changes.) So we really need *four* quantum numbers (n, l, m, m_s) to characterize the stationary states of the atom. The spin orientation does not affect the atom's energy, so a ground-state electron in hydrogen could be in either the $(1, 0, 0, +\frac{1}{2})$ spin-up state or the $(1, 0, 0, -\frac{1}{2})$ spin-down state.

The fact that s has the single value $s = \frac{1}{2}$ has other interesting implications. The correspondence principle tells us that a quantum particle begins to "act classical" in the limit of large quantum numbers. But s cannot become large! **The electron's spin is an intrinsic quantum property of the electron that has *no* classical counterpart.**

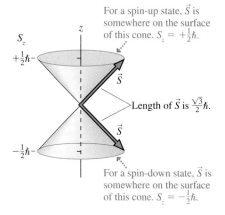

For a spin-up state, $\vec{S}$ is somewhere on the surface of this cone. $S_z = +\frac{1}{2}\hbar$.

Length of $\vec{S}$ is $\frac{\sqrt{3}}{2}\hbar$.

For a spin-down state, $\vec{S}$ is somewhere on the surface of this cone. $S_z = -\frac{1}{2}\hbar$.

FIGURE 41.14 The spin angular momentum has two possible orientations.

41.4 Multielectron Atoms

The Schrödinger-equation solution for the hydrogen atom matches the experimental evidence, but so did the Bohr hydrogen atom. The real test of Schrödinger's theory is how well it works for multielectron atoms.

A neutral multielectron atom consists of Z electrons surrounding a nucleus with Z protons and charge $+Ze$. Z, the *atomic number,* is the order in which elements are listed in the periodic table. Hydrogen is $Z = 1$, helium $Z = 2$, lithium $Z = 3$, and so on.

The potential-energy function of a multielectron atom consists of Z electrons interacting with the nucleus *and* of the Z electrons interacting *with each other.* The electron-electron interaction makes the atomic-structure problem more difficult than the solar-system problem, and it proved to be the downfall of the simple Bohr model. The planets in the solar system do exert attractive gravitational forces on each other, but their masses are so much less than that of the sun that these planet-planet forces are insignificant for all but the most precise calculations. Not so in an atom. The electron charge is the same as the proton charge, so the electron-electron repulsion is just as important to atomic structure as is the electron-nucleus attraction.

The potential energy due to electron-electron interactions fluctuates rapidly in value as the electrons move and the distances between them change. Rather than treat this interaction in detail, we can reasonably consider each electron to be moving in an *average* potential due to all the other electrons. That is, electron i has potential energy

$$U(r_i) = -\frac{Ze^2}{4\pi\epsilon_0 r_i} + U_{\text{elec}}(r_i) \qquad (41.14)$$

where the first term is the electron's interaction with the Z protons in the nucleus and U_{elec} is the average potential energy due to all the other electrons. Because each electron is treated independently of the other electrons, this approach is called the **independent particle approximation,** or IPA. This approximation allows the Schrödinger equation for the atom to be broken into Z separate equations, one for each electron.

A major consequence of the IPA is that **each electron can be described by a wave function having the same four quantum numbers n, l, m, and m_s used to describe the single electron of hydrogen.** Because m and m_s do not affect the energy, we can still refer to electrons by their n and l quantum numbers, using the same labeling scheme that we used for hydrogen.

A major difference, however, is that the energy of an electron in a multielectron atom depends on both n and l. Whereas the $2s$ and $2p$ states in hydrogen had the same energy, their energies are different in a multielectron atom. The difference arises from the electron-electron interactions that do not exist in a single-electron hydrogen atom.

Figure 41.15 shows an energy-level diagram for the electrons in a multielectron atom. For comparison, the hydrogen-atom energies are shown on the right edge of the figure. The comparison is quite interesting. States in a multielectron atom that have small values of l are significantly lower in energy than the corresponding state in hydrogen. For each n, the energy increases as l increases until the maximum-l state has an energy very nearly that of the same n in hydrogen. Can we understand this pattern?

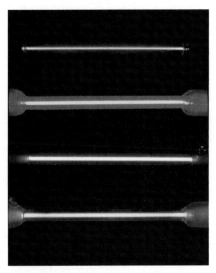

The distinctive color of these gas discharge tubes is due to the unique energy-level structure of each element in the periodic table. The discharges seen here are in hydrogen, neon, helium, and mercury.

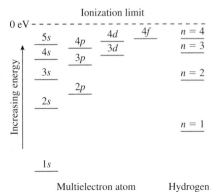

FIGURE 41.15 An energy-level diagram for electrons in a multielectron atom.

Indeed we can. Recall that states of lower l correspond to elliptical classical orbits and the highest-l state corresponds to a circular orbit. Except for the smallest values of n, an electron in a circular orbit spends most of its time *outside* the electron cloud of the remaining electrons. This is illustrated in Figure 41.16. The outer electron is orbiting a ball of charge consisting of Z protons and $(Z − 1)$ electrons. This ball of charge has *net* charge $q_{net} = +e$, so the outer electron "thinks" it is orbiting a proton. An electron in a maximum-l state is nearly indistinguishable from an electron in the hydrogen atom, thus its energy is very nearly that of hydrogen.

The low-l states correspond to elliptical orbits. A low-l electron penetrates in very close to the nucleus, which is no longer shielded by the other electrons. Its interaction with the Z protons in the nucleus is much stronger than the interaction it would have with the single proton in a hydrogen nucleus. This strong interaction *lowers* its energy in comparison to the same state in hydrogen.

As we noted earlier, a quantum electron does not really orbit. Even so, the probability density of a $3s$ electron has in-close peaks that are missing in the probability density of a $3d$ electron, as you should confirm by looking back at Figure 41.8. Thus a low-l electron really does have a likelihood of being at small r, where its interaction with the Z protons is strong, whereas a high-l electron remains farther from the nucleus.

The Pauli Exclusion Principle

By definition, the ground state of a quantum system is the state of lowest energy. What is the ground state of an atom having Z electrons and Z protons? Because the $1s$ state is the lowest energy state in the independent particle approximation, it seems that the ground state should be one in which all Z electrons are in the $1s$ state. However, this hypothesis is not consistent with the experimental evidence.

In 1925, the young Austrian physicist Wolfgang Pauli hypothesized that no two electrons in a quantum system can be in the same quantum state. That is, **no two electrons can have exactly the same set of quantum numbers (n, l, n, m_s).** If one electron is present in a state, it *excludes* all others. This statement is called the **Pauli exclusion principle.**

Wolfgang Pauli was a prodigy who burst onto the physics scene in 1921 when, at the age of 21, he wrote a masterful article on Einstein's theory of relativity. He made many contributions to theoretical physics in the 1920s and the 1930s, but he is most well known for his hypothesis that no two electrons can share the same quantum state. This turns out to be an extremely profound statement about the nature of matter.

The exclusion principle is not applicable to hydrogen, where there is a only a single electron. But in helium, with $Z = 2$ electrons, we must make sure that the two electrons are in different quantum states. This is not difficult. For a $1s$ state, with $l = 0$, the only possible value of the magnetic quantum number is $m = 0$. But there are *two* possible values of m_s, namely $+\frac{1}{2}$ and $-\frac{1}{2}$. If a first electron is in the spin-up $1s$ state $(1, 0, 0, +\frac{1}{2})$, a second $1s$ electron can still be added as long as it is in the spin-down state $(1, 0, 0, -\frac{1}{2})$. This is shown schematically in Figure 41.17a, where the dots represent electrons on the rungs of the "energy ladder" and the arrows represent spin up or spin down.

The Pauli exclusion principle does not prevent both electrons of helium from being in the $1s$ state as long as they have opposite values of m_s, so we predict this to be the ground state. A list of an atom's occupied energy levels is called its **electron configuration.** The electron configuration of the helium ground state is written $1s^2$, where the superscript 2 indicates two electrons in the $1s$ energy level. An excited state of the helium atom might be the electron configuration $1s2s$. This state is shown in Figure 41.17b. Here, because the two electrons have different values of n, there is no restriction on their values of m_s.

A high-l electron corresponds to a circular orbit. It stays outside the core of inner electrons and sees a net charge of $+e$, so it behaves like an electron in a hydrogen atom.

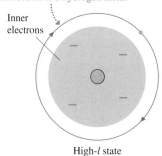

High-l state

A low-l electron corresponds to an elliptical orbit. It penetrates into the core and interacts strongly with the nucleus. The electron-nucleus force is attractive, so this interaction lowers the electron's energy.

Low-l state

FIGURE 41.16 High-l and low-l orbitals in a multielectron atom.

(a) He ground state

The horizontal lines are the allowed energies.

Each circle represents an electron in that energy level.

(b) He excited state

The arrow indicates whether the electron's spin is up ($m_s = +\frac{1}{2}$) or down ($m_s = -\frac{1}{2}$).

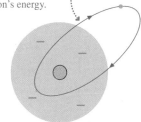

FIGURE 41.17 The ground state and first excited state of helium.

(a) Li ground state

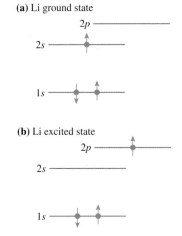

FIGURE 41.18 The ground state and first excited state of lithium.

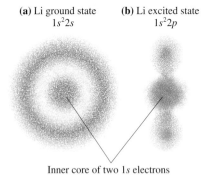

(a) Li ground state
$1s^2 2s$

(b) Li excited state
$1s^2 2p$

Inner core of two $1s$ electrons

FIGURE 41.19 Electron clouds for the lithium electron configurations $1s^2 2s$ and $1s^2 2p$.

The states $(1, 0, 0, +\frac{1}{2})$ and $(1, 0, 0, -\frac{1}{2})$ are the only two states with $n = 1$. The ground state of helium has one electron in each of these states, so all the possible $n = 1$ states are filled. Consequently, the electron configuration $1s^2$ is called a **closed shell.** Because the two electron magnetic moments point in opposite directions, we can predict that helium has *no* net magnetic moment and will be undeflected in a Stern-Gerlach apparatus. This prediction is confirmed by experiment.

The next element, lithium, has $Z = 3$ electrons. The first two electrons can go into $1s$ states, with opposite values of m_s, but what about the third electron? The $1s^2$ shell is closed, and there are no additional quantum states having $n = 1$. The only option for the third electron is the next energy state, $n = 2$. The $2s$ and $2p$ states had equal energies in the hydrogen atom, but they do *not* in a multielectron atom. As Figure 41.15 showed, a lower-l state has lower energy than a higher-l state with the same n. The $2s$ state of lithium is lower in energy than $2p$, so lithium's third ground-state electron will be $2s$. This requires $l = 0$ and $m = 0$ for the third electron, but the value of m_s is not relevant because there is only a single electron in $2s$. Figure 41.18a shows the electron configuration with the $2s$ electron being spin-up, but it could equally well be spin-down. The electron configuration for the lithium ground state is written $1s^2 2s$. This indicates two $1s$ electrons and a single $2s$ electron.

Figure 41.19a shows the probability density of electrons in the $1s^2 2s$ ground state of lithium. You can see the $2s$ electron shell surrounding the inner $1s^2$ core. For comparison, Figure 41.19b shows the *first excited state* of lithium, in which the $2s$ electron has been excited to the $2p$ energy level. This forms the $1s^2 2p$ configuration, also shown in Figure 41.18b.

The Schrödinger equation accurately predicts the energies of the $1s^2 2s$ and the $1s^2 2p$ configurations of lithium, but the Schrödinger equation does not tell us which states the electrons actually occupy. The electron spin and the Pauli exclusion principle were the final pieces of the puzzle. Once these were added to Schrödinger's theory, the initial phase of quantum mechanics was complete. Physicists finally had a successful theory for understanding the structure of atoms.

41.5 The Periodic Table of the Elements

The 19th century was a time when chemists were discovering new elements and studying their chemical properties. The century opened with the atomic model still not completely validated, with no clear distinction between atoms and molecules, and with no one having any idea how many elements there might be. But chemistry developed quickly, and by mid-century it was clear that there were dozens of elements, but not hundreds.

Several chemists in the 1860s began to point out the regular recurrence of chemical properties. For example, there are obvious similarities among the alkali metals lithium, sodium, potassium, and cesium. But attempts at organization were hampered by the fact that many elements had yet to be discovered.

The Russian chemist Dmitri Mendeléev was the first to propose, in 1867, a *periodic* arrangement of the elements. He did so by explicitly pointing out "gaps" where, according to his hypothesis, undiscovered elements should exist. He could then predict the expected properties of the missing elements. The subsequent discovery of these elements verified Mendeléev's organizational scheme, which came to be known as the *periodic table of the elements.*

Figure 41.20 shows a modern periodic table. A larger version is printed in Appendix B. The significance of the periodic table to a physicist is the implication that there is a basic regularity or periodicity to the *structure* of atoms. Any successful theory of the atom needs to explain *why* the periodic table looks the way it does.

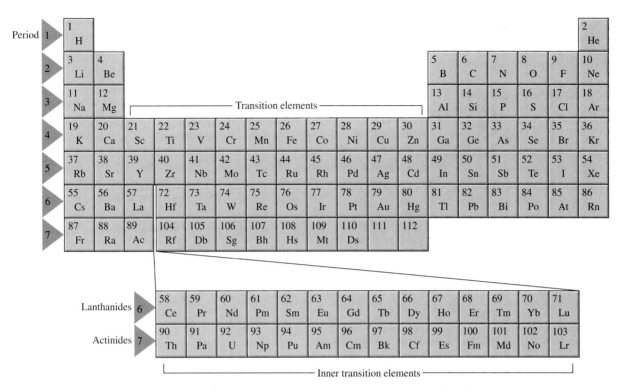

FIGURE 41.20 The modern periodic table of the elements, showing the atomic number *Z* of each.

The First Two Rows

Quantum mechanics successfully explains the structure of the periodic table. We need three basic ideas to see how this works:

1. The energy levels of an atom are found by solving the Schrödinger equation for multielectron atoms. Figure 41.15, a very important figure for understanding the periodic table, showed that the energy depends on quantum numbers *n* and *l*.
2. For each value *l* of the orbital quantum number, there are $2l + 1$ possible values of the magnetic quantum number *m* and, for each of these, two possible values of the spin quantum number m_s. Consequently, each energy *level* in Figure 41.15 is actually $2(2l + 1)$ different *states*. Each of these states has the same energy.
3. The ground state of the atom is the lowest-energy electron configuration that is consistent with the Pauli exclusion principle.

We used these ideas in the last section to look at the elements helium ($Z = 2$) and lithium ($Z = 3$). Four-electron beryllium ($Z = 4$) comes next. The first two electrons go into 1*s* states, forming a closed shell, and the third goes into 2*s*. There is room in the 2*s* level for a second electron as long as its spin is opposite that of the first 2*s* electron. Thus the third and fourth electrons occupy states $(2, 0, 0, +\frac{1}{2})$ and $(2, 0, 0, -\frac{1}{2})$. These are the only two possible 2*s* states. All of the states with the same values and *n* and *l* are called a **subshell,** so the fourth electron closes the 2*s* subshell. (The outer two electrons are called a subshell, rather than a shell, because they complete only the 2*s* possibilities. There are still spaces for 2*p* electrons.) The ground state of beryllium, shown in Figure 41.21, is $1s^2 2s^2$.

These principles can continue to be applied as we work our way through the elements. There are $2l + 1$ values of *m* associated with each value of *l*, and each of these can have $m_s = \pm\frac{1}{2}$. This gives, altogether, $2(2l + 1)$ distinct quantum states in each *nl* subshell. Table 41.2 lists the number of states in each subshell.

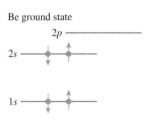

FIGURE 41.21 The ground state of beryllium ($Z = 4$).

TABLE 41.2 Number of states in each subshell of an atom

Subshell	*l*	Number of states
s	0	2
p	1	6
d	2	10
f	3	14

Boron ($1s^22s^22p$) opens the $2p$ subshell. The remaining possible $2p$ states are filled as we continue across the second row of the periodic table. These elements are shown in Figure 41.22. With neon ($1s^22s^22p^6$), which has six $2p$ electrons, the $n = 2$ shell is complete, and we have another closed shell. The second row of the periodic table is eight elements wide because of the two $2s$ electrons *plus* the six $2p$ electrons needed to fill the $n = 2$ shell.

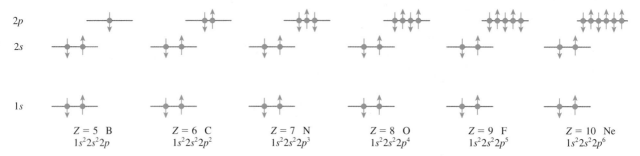

FIGURE 41.22 Filling the $2p$ subshell in the elements boron ($Z = 5$) through neon ($Z = 10$).

Elements with $Z > 10$

The third row of the periodic table is similar to the second. The two $3s$ states are filled in sodium and magnesium. The two columns on the left of the periodic table (groups I and II) represent the two electrons that can go into an s subshell. Then the six $3p$ states are filled, one by one, in aluminum through argon. The six columns on the right (groups III–VIII) represent the six electrons of the p subshell. Argon ($Z = 18$, $1s^22s^22p^63s^23p^6$) is another inert gas, although this seems perhaps surprising in that the $3d$ subshell is still open.

The fourth row is where the periodic table begins to get complicated. You might expect the closure of the $3p$ subshell in argon to be followed, starting with potassium ($Z = 19$), by filling the $3d$ subshell. But if you look back at Figure 41.15, where the energies of the different nl states are shown, you will see that the $3d$ state is slightly *higher* in energy than the $4s$ state. Because the ground state is the *lowest energy state* consistent with the Pauli exclusion principle, potassium finds it more favorable to fill a $4s$ state than to fill a $3d$ state. Thus the ground state configuration of potassium is $1s^22s^22p^63s^23p^64s$ rather than the expected $1s^22s^22p^63s^23p^63d$.

At this point, we begin to see a competition between increasing n and decreasing l. The highly elliptical characteristic of the $4s$ state brings part of its orbit in so close to the nucleus that its energy is less than that of the more circular $3d$ state. The $4p$ state, though, reverts to the "expected" pattern. We find that

$$E_{4s} < E_{3d} < E_{4p}$$

so the states across the fourth row are filled in the order $4s$, then $3d$, then finally $4p$.

Because there had been no previous d states, the $3d$ subshell "splits open" the periodic table to form the 10-element-wide group of *transition elements*. Most commonly occurring metals are transition elements, and their metallic properties are determined by their partially filled d subshell. The $3d$ subshell closes with zinc, at $Z = 30$, then the next six elements fill the $4p$ subshell up to krypton, at $Z = 36$.

Things get even more complex starting in the sixth row, but the ideas are familiar. The $l = 3$ subshell (f electrons) becomes a possibility with $n = 4$, but it turns out that the 5s, 5p, and 6s states are all lower in energy than 4f. Not until barium ($Z = 56$) fills the 6s subshell (and lanthanum ($Z = 57$) adds a 5d electron) is it energetically favorable to add a 4f electron. Immediately after lanthanum you have to switch down to the *lanthanides* at the bottom of the table. The lanthanides fill in the 4f states.

The 4f subshell is complete with $Z = 71$ lutetium. Then $Z = 72$ hafnium through $Z = 80$ mercury complete the transition-element 5d subshell, followed by the 6p subshell in the six elements thallium through radon at the end of the sixth row. Radon, the last inert gas, has $Z = 86$ electrons and the ground-state configuration

$$\text{radon } (Z = 86): 1s^2 2s^2 2p^6 3s^2 3p^6 4s^2 3d^{10} 4p^6 5s^2 4d^{10} 5p^6 6s^2 4f^{14} 5d^{10} 6p^6$$

This is frightening to behold, but we can now understand it!

EXAMPLE 41.4 The ground state of arsenic

Predict the ground-state electron configuration of arsenic.

SOLVE The periodic table shows that arsenic (As) has $Z = 33$, so we must identify the states of 33 electrons. Arsenic is in the fourth row, following the first group of transition elements. Argon ($Z = 18$) filled the 3p subshell, then calcium ($Z = 20$) filled the 4s subshell. The next 10 elements, through zinc ($Z = 30$) filled the 3d subshell. The 4p subshell starts filling with gallium ($Z = 31$), and arsenic is the third element in this group, so it will have three 4p electrons. Thus the ground-state configuration of arsenic is

$$1s^2 2s^2 2p^6 3s^2 3p^6 4s^2 3d^{10} 4p^3$$

The entire periodic table is well explained by quantum mechanics. Figure 41.23 summarizes the results, showing the subshells as they are filled. It is especially important to note the significance of the electron's spin. Although the introduction of the electron's spin and magnetic moment may have seemed obscure and unnecessary, we now find that the spin quantum number m_s is absolutely essential for understanding the periodic table.

FIGURE 41.23 Summary of the order in which subshells are filled in the periodic table.

Ionization Energies

Ionization energy is the energy needed to remove a ground-state electron from an atom and leave a positive ion behind. The ionization energy of hydrogen is 13.60 eV, because the ground-state energy is $E_1 = -13.60$ eV. Figure 41.24 shows the ionization energies of the first 60 elements in the periodic table.

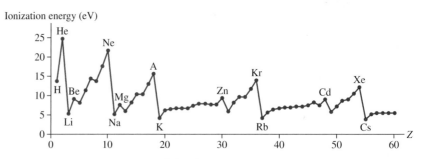

FIGURE 41.24 Ionization energies of the elements up to $Z = 60$.

The ionization energy is different for each element, but there's a clear pattern to the values. Ionization energies are ≈ 5 eV for the alkali metals, on the left edge of the periodic table, then increase steadily to ≥ 15 eV for the inert gases before plunging back to ≈ 5 eV. Can the quantum theory of atoms explain this recurring pattern in the ionization energies?

Indeed it can. The inert-gas elements (helium, neon, argon, . . .) in the right column of the periodic table have *closed shells*. A closed shell is a very stable structure, and that is why these elements are chemically nonreactive (i.e., inert). It takes a large amount of energy to pull an electron out of a stable closed shell, thus the inert gases have the largest ionization energies.

The alkali metals, in the left column of the periodic table, have a single *s*-electron outside a closed shell. This electron is easily disrupted, which is why these elements are highly reactive and have the lowest ionization energies. Between the edges of the periodic table are elements such as beryllium ($1s^2 2s^2$) with a closed 2s subshell. You can see in Figure 41.24 that the closed subshell gives beryllium a larger ionization energy than its neighbors lithium ($1s^2 2s$) or boron ($1s^2 2s^2 2p$). However, a closed subshell is not nearly as tightly bound as a closed shell, so the ionization energy of beryllium is much less than that of helium or neon.

All in all, you can see that the basic idea of shells and subshells, which follows from the Schrödinger-equation energy levels and the Pauli principle, provides a good understanding of the recurring features in the ionization energies.

STOP TO THINK 41.4   Is the electron configuration $1s^2 2s^2 2p^4 3s$ a ground-state configuration or an excited-state configuration?

  a. Ground-state
  b. Excited-state
  c. It's not possible to tell without knowing which element it is.

41.6 Excited States and Spectra

18.2   Activ Physics ONLINE

The periodic table organizes information about the *ground states* of the elements. These states are chemically most important because most atoms spend most of the time in their ground states. All the chemical ideas of valence, bonding, reactivity, and so on are consequences of these ground-state atomic structures. But the periodic table does not tell us anything about the excited states of atoms. It is the excited states that hold the key to understanding atomic spectra, and that is the topic to which we turn next.

Sodium ($Z = 11$) is a multielectron atom that we will use as a prototypical atom. The ground-state electron configuration of sodium is $1s^2 2s^2 2p^6 3s$. The first 10 electrons completely fill the $n = 1$ and $n = 2$ shells, creating a *neon core,* while the $3s$ electron is a valence electron. It is customary to represent this configuration as $[Ne]3s$ or, more simply, as just $3s$.

The excited states of sodium are produced by raising the valence electron to a higher energy level. The electrons in the neon core are unchanged. Thus the excited states can be labeled $[Ne]nl$ or, more simply, as just nl. Figure 41.25 is an energy-level diagram showing the ground state and some of the excited states of sodium. Notice that the $1s$, $2s$, and $2p$ states of the neon core are not shown on the diagram. These states are filled and unchanging, so only the states available to the valence electron are shown.

Figure 41.25 has a new feature: The zero of energy has been shifted to the ground state. As we have discovered many times, the zero of energy can be located where it is most convenient. When we solved the Schrödinger equation, it was most convenient to let zero energy represent the energy of an electron infinitely far away. But for analyzing spectra it is more convenient to let the ground state have $E = 0$. With this choice, the excited-state energies tell us how far each state is above the ground state. The ionization limit now occurs at the value of the atom's ionization energy, which is 5.14 eV for sodium.

The first energy level above $3s$ is $3p$, so the *first excited state* of sodium is $1s^2 2s^2 2p^6 3p$, written as $[Ne]3p$ or, more simply, $3p$. The valence electron is excited while the core electrons are unchanged. This state is followed, in order of increasing energy, by $[Ne]4s$, $[Ne]3d$, and $[Ne]4p$. Notice that the order of excited states is exactly the same order ($3p$-$4s$-$3d$-$4p$) that explained the fourth row of the periodic table.

Other atoms with a single valence electron have energy-level diagrams similar to that of sodium. Things get more complicated when there is more than one valence electron, so we'll defer those details to more advanced courses. The point to remember is that quantum mechanics provides the correct framework for classifying and understanding the many interactions that take place within an atom. You can *utilize* the information shown on an energy-level diagram without having to understand precisely *why* each level is where it is.

Excitation by Absorption

Left to itself, an atom will be in its lowest-energy ground state. How does an atom get into an excited state? The process of getting it there is called **excitation,** and there are two basic mechanisms: absorption and collision. We'll begin by looking at excitation by absorption.

One of the postulates of the basic Bohr model is that an atom can jump from one stationary state, of energy E_1, to a higher-energy state E_2 by absorbing a photon of frequency

$$f = \frac{\Delta E_{\text{atom}}}{h} = \frac{E_2 - E_1}{h} \tag{41.15}$$

Because we are interested in spectra, it is more useful to write Equation 41.15 in terms of the wavelength:

$$\lambda = \frac{c}{f} = \frac{hc}{\Delta E_{\text{atom}}} = \frac{1240 \text{ eV nm}}{\Delta E(\text{in eV})} \tag{41.16}$$

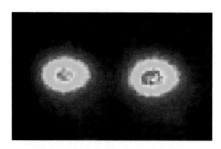

The dots of light are being emitted by two beryllium ions held in a device called an ion trap. Each ion, which is excited by an invisible ultraviolet laser, emits about 10^6 visible-light photons per second.

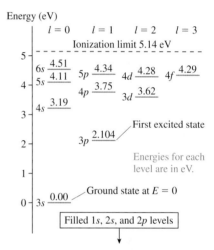

FIGURE 41.25 The $[Ne]3s$ ground state of the sodium atom and some of the excited states.

The final expression, which uses the value $hc = 1240\text{ eV nm}$, gives the wavelength in nanometers *if* ΔE_{atom} is in electron volts.

Bohr's idea of quantum jumps remains an integral part of our interpretation of the results of quantum mechanics. By absorbing a photon, an atom jumps from its ground state to one of its excited states. However, a careful analysis of how the electrons in an atom interact with a light wave shows that not every conceivable transition can occur. The **allowed transitions** must satisfy one or more **selection rules.**

The only selection rule that will concern us says that a transition (either absorption or emission) from a state in which the valence electron has orbital quantum number l_1 to another with orbital quantum number l_2 is allowed only if

$$\Delta l = |l_2 - l_1| = 1 \quad \text{(selection rule for emission and absorption)} \tag{41.17}$$

For example, this selection rule says that an atom in an *s* state ($l = 0$) can absorb a photon and be excited to a *p* state ($l = 1$) but *not* to another *s*-state or to a *d* state. An atom in a *p* state ($l = 1$) can emit a photon by dropping to a lower-energy *s* state *or* to a lower-energy *d* state but not to another *p* state.

EXAMPLE 41.5 Absorption in hydrogen
What is the longest wavelength in the absorption spectrum of hydrogen? What is the transition?

SOLVE The longest wavelength corresponds to the smallest energy change ΔE_{atom}. Because the atom starts from the $1s$ ground state, the smallest energy change occurs for absorption to the first $n = 2$ excited state. The energy change is

$$\Delta E_{\text{atom}} = E_2 - E_1 = \frac{-13.6\text{ eV}}{2^2} - \frac{-13.6\text{ eV}}{1^2} = 10.2\text{ eV}$$

The wavelength of this transition is

$$\lambda = \frac{1240\text{ eV nm}}{10.2\text{ eV}} = 122\text{ nm}$$

This is an ultraviolet wavelength. Because of the selection rule, the transition is $1s \rightarrow 2p$, not $1s \rightarrow 2s$.

EXAMPLE 41.6 Absorption in sodium
What is the longest wavelength in the absorption spectrum of sodium? What is the transition?

SOLVE The sodium ground state is $[\text{Ne}]3s$. The lowest excited state is the $3p$ state. $3s \rightarrow 3p$ is an allowed transition ($\Delta l = 1$), so this will be the longest wavelength. You can see from the data in Figure 41.25 that $\Delta E_{\text{atom}} = 2.104\text{ eV}$ for this transition.

The corresponding wavelength is

$$\lambda = \frac{1240\text{ eV nm}}{2.104\text{ eV}} = 589\text{ nm}$$

ASSESS This wavelength (yellow color) is a prominent feature in the spectrum of sodium. Because the ground state has $l = 0$, absorption *must* be to a *p* state. The *s* states and *d* states of sodium cannot be excited by absorption.

Collisional Excitation

An electron traveling with a speed of 1.0×10^6 m/s has a kinetic energy of 2.85 eV. If this electron collides with a ground-state sodium atom, a portion of its energy can be used to excite the atom to its $3p$ state. This process is called **collisional excitation** of the atom.

Collisional excitation differs from excitation by absorption in one very fundamental way. In absorption, the photon disappears. Consequently, *all* of the photon's energy must be transferred to the atom. Conservation of energy requires $E_{\text{photon}} = \Delta E_{\text{atom}}$. In contrast, the electron is still present after collisional excitation and can carry away some kinetic energy. That is, the electron does *not* have to transfer its entire energy to the atom. If the electron has an incident kinetic energy of 2.85 eV, it could transfer 2.10 eV to the sodium atom, thereby exciting it to the $3p$ state, and still depart the collision with a speed of 5.1×10^5 m/s and an energy of 0.75 eV.

To excite the atom, the incident energy of the electron (or any other matter particle) merely has to *exceed* ΔE_{atom}. That is $E_{particle} \geq \Delta E_{atom}$. There's a threshold energy for exciting the atom, but no upper limit. It is all a matter of energy conservation. Figure 41.26 shows the idea graphically.

Collisional excitation by electrons is the predominant method of excitation in electrical discharges such as fluorescent lights, street lights, and neon signs. A gas is placed in a tube at reduced pressure (≈ 1 mm of Hg), then a fairly high voltage (≈ 1000 V) between electrodes at the ends of the tube causes the gas to ionize, creating a current in which both ions and electrons are charge carriers. The mean free path of electrons between collisions is large enough for the electrons to gain several eV of kinetic energy as they accelerate in the electric field. This energy is then transferred to the gas atoms upon collision. The process does not work at atmospheric pressure because the mean free path between collisions is too short for the electrons to gain enough kinetic energy to excite the atoms.

NOTE ▶ In contrast to photon absorption, there are no selection rules for collisional excitation. Any state can be excited if the colliding particle has sufficient energy. ◀

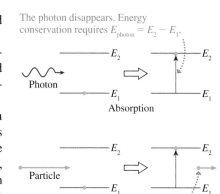

The photon disappears. Energy conservation requires $E_{photon} = E_2 - E_1$.

Absorption

Collisional excitation

The particle carries away energy. Energy conservation requires $E_{particle} \geq E_2 - E_1$.

FIGURE 41.26 Excitation by photon absorption and electron collision.

EXAMPLE 41.7 Excitation of hydrogen

Can an electron traveling at 2.0×10^6 m/s cause a hydrogen atom to emit the prominent red spectral line ($\lambda = 656$ nm) in the Balmer series?

MODEL The electron must have sufficient energy to excite the upper state of the transition.

SOLVE The electron's energy is $E_{elec} = \frac{1}{2}mv^2 = 11.4$ eV. This is significantly larger than the 1.89 eV energy of a photon with wavelength 656 nm, but don't confuse the energy of the photon with the energy of the excitation. The red spectral line in the

Balmer series is emitted by an $n = 3$ to $n = 2$ quantum jump with $\Delta E_{atom} = 1.89$ eV. But to cause this emission, the electron must excite an atom from its *ground state*, with $n = 1$, up to the $n = 3$ level. The necessary excitation energy is

$$\Delta E_{atom} = E_3 - E_1 = (-1.51 \text{ eV}) - (-13.60 \text{ eV})$$
$$= 12.09 \text{ eV}$$

The electron does *not* have sufficient energy to excite the atom to the state from which the emission would occur.

Emission Spectra

The absorption of light is an important process, but it is the emission of light that really gets our attention. The overwhelming bulk of sensory information that we perceive comes to us in the form of light. The recognition and appreciation of light and color has formed the basis of aesthetics and art since the days of prehistory. With the small exception of cosmic rays, all of our knowledge about the cosmos comes to us in the form of light and other electromagnetic waves emitted in various processes.

The discovery of discrete emission spectra helped bring down classical physics, and it was the understanding of discrete emission spectra that provided the first major triumph of quantum mechanics. Emission spectra are more than just scientific curiosities. Many of today's artificial light sources, from fluorescent lights to lasers, are applications of emission spectra.

Understanding emission hinges upon the three ideas shown in Figure 41.27. Once we have determined the energy levels of an atom, by solving the Schrödinger equation, we can immediately predict its emission spectrum. Conversely, we can use the measured emission spectrum to determine an atom's energy levels.

As an example, Figure 41.28a on the next page shows some of the transitions and wavelengths observed in the emission spectrum of sodium. This diagram makes the point that each wavelength represents a quantum jump between two well-defined energy levels. Notice that the selection rule $\Delta l = 1$ is being obeyed

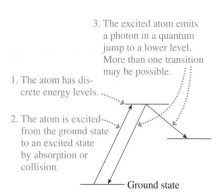

3. The excited atom emits a photon in a quantum jump to a lower level. More than one transition may be possible.

1. The atom has discrete energy levels.

2. The atom is excited from the ground state to an excited state by absorption or collision.

Ground state

FIGURE 41.27 Generation of an emission spectrum.

(a)

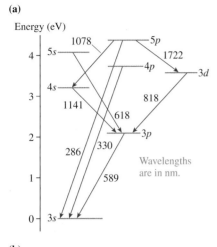

(b)

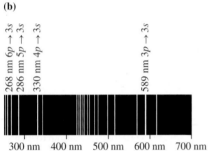

FIGURE 41.28 The emission spectrum of sodium.

The colors in a stained-glass window are due to the selective absorption of light.

in the sodium spectrum. The 5*p* levels can undergo quantum jumps to 3*s*, 4*s*, or 3*d* but *not* to 3*p* or 4*p*.

Figure 41.28b shows the emission spectrum of sodium as it would be recorded in a spectrometer. (Many of the lines seen in this spectrum start from higher excited states that are not seen in the rather limited energy-level diagram of Figure 41.28a.) By comparing the spectrum to the energy-level diagram, you can recognize the spectral lines at 589 nm, 330 nm, 286 nm, and 268 nm form a *series* of lines due to all the possible *np* → 3*s* transitions. They are the dominant features in the sodium spectrum.

The most obvious visual feature of sodium emission is its bright yellow color, produced by the emission wavelength of 589 nm. This is the basis of the *flame test* used in chemistry to test for sodium: A sample is held in a Bunsen burner, and a bright yellow glow indicates the presence of sodium. The 589 nm emission is also prominent in the pinkish-yellow glow of the common sodium-vapor street lights. These operate by creating an electrical discharge in sodium vapor. Most sodium-vapor lights use high-pressure lamps to increase their light output. The high pressure, however, causes the formation of Na_2 molecules, and these molecules emit the pinkish portion of the light.

Some cities close to astronomical observatories use low-pressure sodium lights, and these emit the distinctively yellow 589 nm light of sodium. The glow of city lights is a severe problem for astronomers, but the very specific 589 nm emission from sodium is easily removed with a *sodium filter*. The light from the telescope is passed through a container of sodium vapor, and the sodium atoms *absorb* just the unwanted 589 nm photons without disturbing any other wavelengths! However, this cute trick does not work for the other wavelengths emitted by high-pressure sodium lamps or light from other sources.

Color in Solids

It is worth concluding this section with a few remarks about color in solids. Whether it be the intense multihued colors of a stained glass window, the bright colors of flowers or paint, or the deep luminescent red of a ruby, most of the colors we perceive in our lives come from solids rather than free atoms. The basic principles are the same, but the details are different for solids.

An excited atom in a gas has little choice but to give up its energy by emitting a photon. Its only other option, which is rare for gas atoms, is to collide with another atom and transfer its energy into the kinetic energy of recoil. But the atoms in a solid are in intimate contact with each other at all times. Although an excited atom in a solid has the option of emitting a photon, it is often more likely that the energy will be converted, via interactions with neighboring atoms, to the thermal energy of the solid. A process in which an atom is de-excited without radiating is called a **nonradiative transition.**

This is what happens in pigments, such as those in paints, plants, and dyes. Pigments are molecules that absorb certain wavelengths of light but not other wavelengths. The energy-level structure of a molecule is complex, so the absorption consists of "bands" of wavelengths rather than discrete spectral lines. But instead of re-radiating the energy by photon emission, as a free atom would, the pigment molecules undergo nonradiative transitions and convert the energy into increased thermal energy. That is why darker objects get hotter in the sun than lighter objects.

When light falls on an object, it can be either absorbed or reflected. If *all* wavelengths are reflected, the object is perceived as white. Any wavelengths absorbed by the pigments are removed from the reflected light. A pigment with blue-absorbing properties converts the energy of blue-wavelength photons into thermal energy, but photons of other wavelengths are reflected without change. A blue-absorbing pigment reflects the red and yellow wavelengths, causing the object to be perceived as the color orange!

The mechanism for creating colors in colored glass and plastic is the selective absorption of some wavelengths, followed by nonradiative transitions. Blue glass contains molecules that absorb all wavelengths *except* blue. Nonradiative transitions convert the absorbed energy to the thermal energy of the glass. Only blue wavelengths pass through and are seen. The glass acts as a *filter* that passes some wavelengths but filters out others.

Some solids, though, are a little different. The color of many minerals and crystals is due to so-called *impurity atoms* embedded in them. For example, the gemstone ruby is a very simple and common crystal of aluminum oxide, called corundum, that happens to have chromium atoms present at the concentration of about one part in a thousand. Pure corundum is transparent, so all of a ruby's color comes from these chromium impurity atoms.

Figure 41.29 shows what happens when ruby is illuminated by white light. The chromium atoms have a group of excited states that absorb all wavelengths shorter than about 600 nm—that is, everything except orange and red. Unlike the pigments in red glass, which convert all the absorbed energy into thermal energy, the chromium atoms dissipate only a small amount of heat as they undergo a non-radiative transition to another excited state. From there they emit a photon with $\lambda = hc/(E_2 - E_1) \approx 690$ nm as they jump back to the ground state.

The net effect is that short-wavelength photons, rather than being completely absorbed, are *re-radiated* as longer-wavelength photons. This is why rubies sparkle and have such intense color, whereas red glass is a dull red color. The color of other minerals and gems is due to different impurity atoms, but the principle is the same.

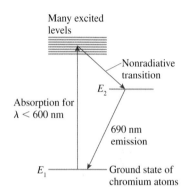

FIGURE 41.29 Absorption and emission in a crystal of ruby.

STOP TO THINK 41.5 In this hypothetical atom, what is the photon energy E_{photon} of the longest-wavelength photons emitted by atoms in the $5p$ state?

a. 1.0 eV
b. 2.0 eV
c. 3.0 eV
d. 4.0 eV

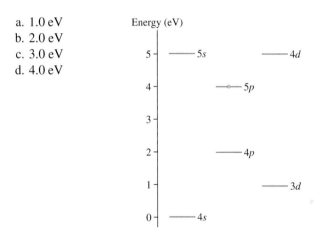

41.7 Lifetimes of Excited States

Excitation of an atom, by either absorption or collision, leaves it in an excited state. From there it jumps back to a lower energy level by emitting a photon. How long does this process take? There are actually two questions here. First, how long does an atom remain in an excited state before undergoing a quantum jump to a lower state? Second, how long does the transition last as the quantum jump is occurring?

Our best understanding of the quantum physics of atoms is that quantum jumps are instantaneous. The absorption or emission of a photon is an all-or-nothing event, so there is not a moment when a photon is "half emitted." The prediction that quantum jumps are instantaneous has troubled many physicists, but careful experimental tests have never revealed any evidence that the jump itself takes a measurable amount of time.

The time spent in the excited state, waiting to make a quantum jump, is another story. Figure 41.30 shows experimental data for the length of time that doubly charged xenon ions Xe^{++} spend in a certain excited state. In this experiment, a pulse of electrons was used to excite the atoms to the excited state. The number of excited-state atoms was then monitored by detecting the photons emitted—one-by-one!—as the excited atoms jumped back to the ground state. The number of photons emitted at time t is directly proportional to the number of excited-state atoms present at time t. As the figure shows, the number of atoms in the excited state decreases *exponentially* with time, and virtually all have decayed within 25 ms of their creation.

Figure 41.30 has two important implications. First, atoms spend time in the excited state before undergoing a quantum jump back to a lower state. Second, the length of time spent in the excited state is not a constant value but varies from atom to atom. If every excited xenon ion lived for 5 ms in the excited state, then we would detect *no* photons for 5 ms, a big burst right at 5 ms as they all decay, then no photons after that. Instead, the data tell us that there is a *range* of times spent in the excited state. Some undergo a quantum jump and emit a photon after 1 ms, others after 5 ms or 10 ms, and a few wait as long as 20 or 25 ms.

Consider an experiment in which N_0 excited atoms are created at time $t = 0$. As the solid curve in Figure 41.30 shows, the number of excited atoms remaining at time t is very well described by the exponential function

$$N_{\text{exc}} = N_0 e^{-t/\tau} \tag{41.18}$$

where τ is the point in time at which $e^{-1} = 0.368 = 36.8\%$ of the original atoms are left. Thus 63.2% of the atoms, nearly two-thirds, have emitted a photon and jumped to the lower state by time $t = \tau$. The interval of time τ is called the **lifetime** of the excited state. From Figure 41.30 we can deduce that the lifetime of this state in Xe^{++} is ≈ 4 ms because that is the point in time at which the curve has decayed to 36.8% of its initial value.

This lifetime in Xe^{++} is abnormally long, which is why the state was being studied. More typical excited-state lifetimes are a few nanoseconds. Table 41.3 gives some measured values of excited-state lifetimes. Whatever the value of τ, the number of excited-state atoms decreases exponentially. Can we understand why this is?

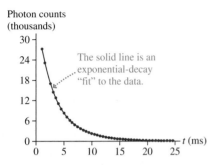

Photon counts (thousands)

The solid line is an exponential-decay "fit" to the data.

FIGURE 41.30 Experimental data for the photon emission rate from an excited state in Xe^{++}.

TABLE 41.3 Some excited-state lifetimes

Atom	State	Lifetime (ns)
Hydrogen	$2p$	1.6
Sodium	$3p$	17
Neon	$3p$	20
Potassium	$4p$	26

The Decay Equation

Quantum mechanics is about probabilities. We cannot say exactly where the electron is located, but we can use quantum mechanics to calculate the *probability* that the electron is located in a small interval Δx at position x. Similarly, we cannot say exactly when an excited electron will undergo a quantum jump and emit a photon. However, we can use quantum mechanics to find the *probability* that the electron will undergo a quantum jump during a small time interval Δt at time t.

Let us assume that the probability of an excited atom emitting a photon during time interval Δt is *independent* of how long the atom has been waiting in the excited state. For example, a newly excited atom may have a 10% probability of emitting a photon within the 1 ns interval from 0 ns to 1 ns. If it survives until $t = 7$ ns, our assumption is that it still has a 10% probability for emitting a photon during the 1 ns interval from 7 ns to 8 ns.

This assumption, which can be justified with a detailed analysis, is similar to flipping coins. The probability of a head on your first flip is 50%. If you flip seven heads in a row, the probability of a head on your eighth flip is still 50%. It is *unlikely* that you will flip seven heads in a row, but doing so does not influence the eighth flip. Likewise, it may be *unlikely* for an excited atom to live for 7 ns, but doing so does not affect its probability of emitting a photon during the next 1 ns.

If Δt is small, the probability of photon emission during time interval Δt is directly proportional to Δt. That is, if the emission probability in 1 ns is 1%, it will be 2% in 2 ns and 0.5% in 0.5 ns. (This logic fails if Δt gets too big. If the probability is 70% in 20 ns, we can *not* say that the probability would be 140% in 40 ns because a probability >1 is meaningless.) We will be interested in the limit $\Delta t \rightarrow dt$, so the concept is valid and we can write

$$\text{Prob(emission in } \Delta t \text{ at time } t) = r\Delta t \qquad (41.19)$$

where r is called the **decay rate** because the number of excited atoms decays with time. It is a probability *per second,* with units of s^{-1}, and thus is a rate. For example, if an atom has a 5% probability of emitting a photon during a 2 ns interval, its decay rate is

$$r = \frac{P}{\Delta t} = \frac{0.05}{2 \text{ ns}} = 0.025 \text{ ns}^{-1} = 2.5 \times 10^7 \text{ s}^{-1}$$

NOTE ▶ Equation 41.19 is directly analogous to Prob(found in Δx at x) = $P\Delta x$, where P, which had units of m^{-1}, was the probability density. ◀

Figure 41.31 shows N_{exc} atoms in an excited state. During a small time interval Δt, the number of these atoms that we expect to undergo a quantum jump and emit a photon is N_{exc} multiplied by the probability of decay. That is,

$$\text{number of photons in } \Delta t \text{ at time } t = N_{\text{exc}} \times \text{Prob(emission in } \Delta t \text{ at } t)$$
$$= rN_{\text{exc}}\Delta t \qquad (41.20)$$

Now the *change* in N_{exc} is the *negative* of Equation 41.20. For example, suppose 1000 excited atoms are present at time t and that each has a 5% probability of emitting a photon in the next 1 ns. On average, the number of photons emitted during the next 1 ns will be $1000 \times 0.05 = 50$. Consequently, the number of excited atoms changes by $\Delta N_{\text{exc}} = -50$, with the minus sign indicating a decrease.

Thus the *change* in the number of atoms in the excited state is

$$\Delta N_{\text{exc}}(\text{in } \Delta t \text{ at } t) = -N_{\text{exc}} \times \text{Prob(decay in } \Delta t \text{ at } t) = -rN_{\text{exc}}\Delta t \qquad (41.21)$$

Now let $\Delta t \rightarrow dt$. Then $\Delta N_{\text{exc}} \rightarrow dN_{\text{exc}}$ and Equation 41.21 becomes

$$\frac{dN_{\text{exc}}}{dt} = -rN_{\text{exc}} \qquad (41.22)$$

Equation 41.22 is a *rate equation* because it describes the *rate* at which the excited-state population changes. If r is large, the population will decay at a rapid rate and will have a short lifetime. Conversely, a small value of r implies that the population decays slowly and will live a long time.

The rate equation is a differential equation, but we've solved similar equations before. First, rewrite Equation 41.22 as

$$\frac{dN_{\text{exc}}}{N_{\text{exc}}} = -r \, dt$$

Then integrate both sides from $t = 0$, when the initial excited-state population is N_0, to an arbitrary time t when the population is N_{exc}. That is,

$$\int_{N_0}^{N_{\text{exc}}} \frac{dN_{\text{exc}}}{N_{\text{exc}}} = -r \int_0^t dt \qquad (41.23)$$

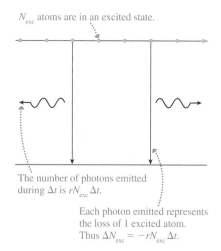

N_{exc} atoms are in an excited state.

The number of photons emitted during Δt is $rN_{\text{exc}}\Delta t$.

Each photon emitted represents the loss of 1 excited atom. Thus $\Delta N_{\text{exc}} = -rN_{\text{exc}}\Delta t$.

FIGURE 41.31 The number of atoms that emit photons during Δt is directly proportional to the number of excited atoms.

Both are well-known integrals, giving

$$\ln N_{\text{exc}} \Big|_{N_0}^{N_{\text{exc}}} = \ln N_{\text{exc}} - \ln N_0 = \ln\left(\frac{N_{\text{exc}}}{N_0}\right) = -rt$$

We can solve for the number of excited atoms at time t by taking the exponential of both sides, then multiplying by N_0. Doing so gives

$$N_{\text{exc}} = N_0 e^{-rt} \tag{41.24}$$

Notice that $N_{\text{exc}} = N_0$ at $t = 0$, as expected. Equation 41.24, the *decay equation,* shows that the excited-state population decays exponentially with time, as we saw in the experimental data of Figure 41.30.

It will be more convenient to write Equation 41.24 as

$$N_{\text{exc}} = N_0 e^{-t/\tau} \tag{41.25}$$

where

$$\tau = \frac{1}{r} = \text{the } \textit{lifetime} \text{ of the excited state} \tag{41.26}$$

This is the definition of the lifetime we used in Equation 41.18 to describe the experimental results. Now we see explicitly that the lifetime is the inverse of the decay rate r.

EXAMPLE 41.8 The lifetime of an excited state in mercury

The mercury atom has two valence electrons. One is always in the $6s$ state, the other is in a state with quantum numbers n and l. One of the excited states in mercury is the state designated $6s6p$. The decay rate of this state is $7.7 \times 10^8 \text{ s}^{-1}$.

a. What is the lifetime of this state?
b. If 1.0×10^{10} mercury atoms are created in the $6s6p$ state at $t = 0$, how many photons will be emitted during the first 1.0 ns?

SOLVE

a. The lifetime is

$$\tau = \frac{1}{r} = \frac{1}{7.7 \times 10^8 \text{ s}^{-1}} = 1.3 \times 10^{-9} \text{ s} = 1.3 \text{ ns}$$

b. If there are $N_0 = 10^{10}$ excited atoms at $t = 0$, the number still remaining at $t = 1.0$ ns is

$$N_{\text{exc}} = N_0 e^{-t/\tau} = (1.0 \times 10^{10}) e^{-(1.0 \text{ ns})/(1.3 \text{ ns})} = 4.63 \times 10^9$$

This result implies that 5.37×10^9 atoms undergo quantum jumps during the first 1.0 ns. Each of these atoms emits one photon, so the number of photons emitted during the first 1.0 ns is 5.37×10^9.

The decay rates R for excited states can be calculated in quantum mechanics and compared to experimentally measured lifetimes of excited states. The agreement is very good, thus providing another validation of the quantum-mechanical description of atoms.

STOP TO THINK 41.6 An equal number of excited A atoms and excited B atoms are created at $t = 0$. The decay rate of B atoms is twice that of A atoms: $r_B = 2r_A$. At $t = \tau_A$ (i.e., after one lifetime of A atoms has elapsed), the ratio N_B/N_A of the number of excited B atoms to the number of excited A atoms is

a. >2.   b. 2.   c. 1.   d. $\frac{1}{2}$.   e. $<\frac{1}{2}$.

41.8 Stimulated Emission and Lasers

We have seen that an atom can jump from a lower-energy level E_1 to a higher-energy level E_2 by absorbing a photon. Figure 41.32a illustrates the basic absorption process, with a photon of frequency $f = \Delta E_{atom}/h$ disappearing as the atom jumps from level 1 to level 2. Once in level 2, as shown in Figure 41.32b, the atom can emit a photon of the same frequency as it jumps back to level 1. Because this transition occurs spontaneously, without the introduction of outside energy, it is called **spontaneous emission.**

In 1917, four years after Bohr's proposal of stationary states in atoms but still prior to de Broglie and Schrödinger, Einstein was puzzled by how quantized atoms reach thermodynamic equilibrium in the presence of electromagnetic radiation. Einstein found that the process of absorption and spontaneous emission were not sufficient to allow a collection of atoms to reach thermodynamic equilibrium. To resolve this difficulty, Einstein proposed a third mechanism for the interaction of atoms with light.

The left half of Figure 41.32c shows a photon with frequency $f = \Delta E_{atom}/h$ approaching an *excited* atom. If a photon can induce the $1 \rightarrow 2$ transition of absorption, then Einstein proposed that it should also be able to induce a $2 \rightarrow 1$ transition. In a sense, this transition is a *reverse absorption*. But to undergo a reverse absorption, the atom must *emit* a photon of frequency $f = \Delta E_{atom}/h$. The end result, as seen in the right half of Figure 41.32c, is an atom in level 1 plus *two* photons! Because the first photon induced the atom to emit the second photon, this process is called **stimulated emission.**

Stimulated emission occurs only if the first photon's frequency exactly matches the $E_2 - E_1$ energy difference of the atom. This is precisely the same condition that absorption has to satisfy. More interesting, the emitted photon is *identical* to the incident photon. This means that as the two photons leave the atom they have exactly the same frequency and wavelength, are traveling in exactly the same direction, and are exactly in phase with each other. In other words, **stimulated emission produces a second photon that is an exact clone of the first.**

Stimulated emission is of no importance in most practical situations. Atoms typically spend only a few nanoseconds in an excited state before undergoing spontaneous emission, so the atom would need to be in an extremely intense light wave for stimulated emission to occur prior to spontaneous emission. Ordinary light sources are not nearly intense enough for stimulated emission to be more than a minor effect, hence it was many years before Einstein's prediction was confirmed. No one had doubted Einstein, because he had clearly demonstrated that stimulated emission was necessary to make the energy equations balance, but it seemed no more important than would pennies to a millionaire balancing her checkbook. At least, that is, until 1960, when a revolutionary invention appeared that made explicit use of stimulated emission: the laser.

Lasers

The word **laser** is an acronym for the phrase **l**ight **a**mplification by the **s**timulated **e**mission of **r**adiation. Lasers were an outgrowth of research with microwaves during the 1950s, which culminated in the invention of the *maser*, short for **m**icrowave **a**mplification by the **s**timulated **e**mission of **r**adiation. The first laser, a ruby laser, was demonstrated in 1960, and several other kinds of lasers appeared within a few months. The driving force behind much of the research was the American physicist Charles Townes. Townes was awarded the Nobel prize in 1964 for the invention of the maser and his theoretical work leading to the laser.

Today, lasers do everything from being the light source in fiber-optic communications to measuring the distance to the moon and from playing your CD to performing delicate eye surgery. But what is a laser? Basically it is a device that

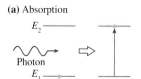
(a) Absorption

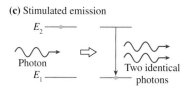
(b) Spontaneous emission

(c) Stimulated emission

FIGURE 41.32 Three types of radiative transitions.

Activ Physics 18.3

Charles Townes.

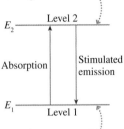

N_2 atoms in level 2. Photons of energy $E_{\text{photon}} = E_2 - E_1$ can cause these atoms to undergo stimulated emission.

N_1 atoms in level 1. These atoms can absorb photons of energy $E_{\text{photon}} = E_2 - E_1$.

FIGURE 41.33 Energy levels 1 and 2, with populations N_1 and N_2.

produces a beam of highly *coherent* and essentially monochromatic (single-color) light as a result of stimulated emission. **Coherent** light is light in which all the electromagnetic waves have the same phase, direction, and amplitude. It is the coherence of a laser beam that allows it to be very tightly focused or to be rapidly modulated for communications.

Let's take a brief look at how a laser works. Figure 41.33 represents a system of atoms that have a lower energy level E_1 and a higher energy level E_2. Suppose that there are N_1 atoms in level 1 and N_2 atoms in level 2. Left to themselves, all the atoms would soon end up in level 1 because of the spontaneous emission $2 \rightarrow 1$. To prevent this, we can imagine that some type of excitation mechanism, perhaps an electrical discharge, is continuing to produce new excited atoms in level 2.

Let a photon of frequency $f = (E_2 - E_1)/h$ be incident on this group of atoms. Because it has the correct frequency, it could be absorbed by one of the atoms in level 1. Another possibility is that it could cause stimulated emission from one of the level 2 atoms. Ordinarily $N_2 \ll N_1$, so absorption events far outnumber stimulated emission events. Even if a few photons were generated by stimulated emission, they would quickly be absorbed by the vastly larger group of atoms in level 1.

But what if we could somehow arrange to place *every* atom in level 2, making $N_1 = 0$. Then the incident photon, upon encountering its first atom, will cause stimulated emission. Where there was initially one photon of frequency f, now there are two. These will strike two additional excited-state atoms, again causing stimulated emission. Then there will be four photons. As Figure 41.34 shows, there will be a *chain reaction* of stimulated emission until all N_2 atoms emit a photon of frequency f.

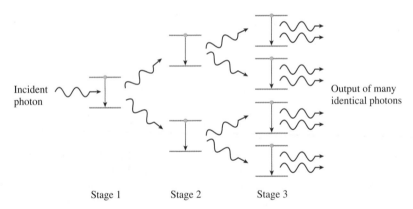

Incident photon

Output of many identical photons

Stage 1   Stage 2   Stage 3

FIGURE 41.34 Stimulated emission creates a chain reaction of photon production in a population of excited atoms.

In stimulated emission, each emitted photon is *identical* to the incident photon. The chain reaction of Figure 41.34 will lead not just to N_2 photons of frequency f, but to N_2 *identical* photons, all traveling together in the same direction with the same phase. If N_2 is a large number, as would be the case in any practical device, the one initial photon will have been *amplified* into a gigantic, coherent pulse of light! A collection of excited-state atoms is referred to as an *optical amplifier*.

The stimulated emission is sustained by placing the *lasing medium*—the sample of atoms that emits the light—in an **optical cavity** consisting of two facing mirrors. As Figure 41.35 shows, the photons interact repeatedly with the atoms in the medium as they bounce back and forth. This repeated interaction is necessary for the light intensity to build up to a high level. One of the mirrors will be partially transmitting so that some of the light emerges as the *laser beam*.

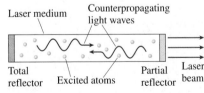

Laser medium   Counterpropagating light waves

Total reflector   Excited atoms   Partial reflector   Laser beam

FIGURE 41.35 Lasing takes place in an optical cavity.

Although the chain reaction of Figure 41.34 illustrates the idea most clearly, it is not necessary for every atom to be in level 2 for amplification to occur. All that is needed is to have $N_2 > N_1$ so that stimulated emission exceeds absorption. Such a situation is called a **population inversion.** The process of obtaining a population inversion is called **pumping,** and we will look at two specific examples. Pumping is the technically difficult part of designing and building a laser because normal excitation mechanisms do not create population inversions. In fact, lasers would likely have been discovered accidentally long before 1960 if population inversions were easy to create.

The Ruby Laser

The first laser to be developed was a ruby laser. Figure 41.36a shows the energy-level structure of the chromium atoms that gives ruby its optical properties. Normally, the number of atoms in the ground-state level E_1 far exceeds the number of excited-state atoms with energy E_2. That is, $N_2 \ll N_1$. Under these circumstances 690 nm light is absorbed rather than amplified. But suppose that we could *rapidly* excite more than half the chromium atoms to level E_2. Then we would have a population inversion ($N_2 > N_1$) between levels E_1 and E_2.

This can be accomplished by *optically pumping* the ruby with a very intense pulse of white light from a *flashlamp*. A flashlamp is like a camera flash, only vastly more intense. In the basic arrangement of Figure 41.36b, a helical flashlamp is coiled around a ruby rod that has mirrors bonded to its end faces. The lamp is fired by discharging a high-voltage capacitor through it, creating a very intense light pulse that lasts just a few microseconds. This intense light excites nearly all the chromium atoms from the ground state to the upper energy levels. From there, they quickly ($\approx 10^{-8}$ s) decay nonradiatively to level 2. With $N_2 > N_1$, a population inversion has been created.

Once a photon initiates the laser pulse, the light intensity builds quickly into a brief but incredibly intense burst of light. A typical output pulse lasts 10 ns and has an energy of 1 J. This gives a *peak power* of

$$P = \frac{\Delta E}{\Delta t} = \frac{1 \text{ J}}{10^{-8} \text{ s}} = 10^8 \text{ W} = 100 \text{ MW}$$

One hundred megawatts of light power! That is more than the electrical power consumed by a small city. The difference, of course, is that a city consumes that power continuously but the laser pulse lasts a mere 10 ns. The laser cannot fire again until the capacitor is recharged and the laser rod cooled. A typical firing rate is a few pulses per second, so the laser is "on" only a few billionths of a second out of each second.

Ruby lasers have been replaced by other pulsed lasers that, for various practical reasons, are easier to operate. However, they all operate with the same basic idea of rapid optical pumping to upper states, rapid nonradiative decay to level 2 where the population inversion is formed, then rapid build-up of an intense optical pulse.

The Helium-Neon Laser

The familiar red laser used in lecture demonstrations, laboratories, and supermarket checkout scanners is the helium-neon laser, often called a HeNe laser. Its output is a *continuous,* rather than pulsed, wavelength of 632.8 nm. The medium of a HeNe laser is a mixture of $\approx 90\%$ helium and $\approx 10\%$ neon gases. As Figure 41.37a on the next page shows, the gases are sealed in a glass tube, then an electrical discharge is established along the bore of the tube. Two mirrors are bonded to the ends of the discharge tube, one a total reflector and the other having $\approx 2\%$ transmission so that the laser beam can be extracted.

(a)

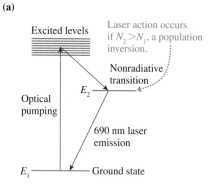

(b)

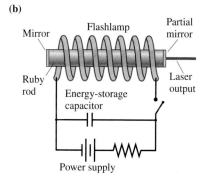

FIGURE 41.36 A flashlamp-pumped ruby laser.

(a) He/Ne gas mixture   Laser beam

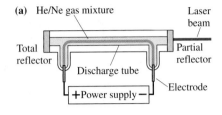

Total reflector
Partial reflector
Discharge tube
+Power supply–
Electrode

(b)   Excitation transfer

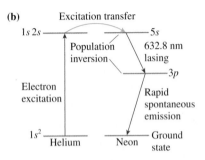

$1s\,2s$ ——————— $5s$
Population inversion
632.8 nm lasing
$3p$
Electron excitation
Rapid spontaneous emission
$1s^2$ ——————— Ground state
Helium   Neon

FIGURE 41.37 A HeNe laser.

The atoms that lase in a HeNe are the neon atoms, but the pumping method involves the helium atoms. The electrons in the discharge easily excite the $1s2s$ excited state of helium. This state has a very small spontaneous decay rate (i.e., a very long lifetime), so it is possible to build up a fairly large population (but not an inversion) of excited helium atoms in this state. The energy of the $1s2s$ state is 20.6 eV.

Interestingly, an excited state of neon, the $5s$ state, also has an energy of 20.6 eV. If a $1s2s$ excited helium atom collides with a ground state neon atom, as frequently happens, the excitation energy can be transferred from one atom to the other! Written as a chemical reaction, the process is

$$He^* + Ne \rightarrow He + Ne^*$$

where the asterisk indicates the atom is in an excited state. This process, called **excitation transfer,** is very efficient for the $5s$ state because the process is *resonant*—a perfect energy match. Thus the two-step process of collisional excitation of helium, followed by excitation transfer between helium and neon, pumps the neon atoms into the excited $5s$ state. This is shown in Figure 41.37b.

The $5s$ energy level in neon is ≈ 1.95 eV above the $3p$ state. The $3p$ state is very nearly empty of population, both because it is not efficiently populated in the discharge and because it undergoes very rapid spontaneous emission to the $3s$ states. Thus the large number of atoms pumped into the $5s$ state creates a population inversion with respect to the lower $3p$ state. These are the necessary conditions for laser action.

Because the lower level of the laser transition is normally empty of population, or very nearly so, placing only a small fraction of the neon atoms in the $5s$ state creates a population inversion. Thus a fairly modest pumping action is sufficient to create the inversion and start the laser. Furthermore, a HeNe laser can maintain a *continuous* inversion and thus sustain continuous lasing. The electrical discharge continuously creates $5s$ excited atoms in the upper level, and the rapid spontaneous decay of the $3p$ atoms from the lower level keeps its population low enough to sustain the inversion.

A typical helium-neon laser has a power output of 1 mW = 10^{-3} J/s at 632.8 nm in a 1-mm-diameter laser beam. As you can show in a homework problem, this output corresponds to the emission of 3.2×10^{15} photons per second. Other continuous lasers operate by similar principles, but can produce much more power. The argon laser, which is widely used in scientific research, can produce up to 20 W of power at green and blue wavelengths. The carbon dioxide laser produces output power in excess of 1000 W at the infrared wavelength of 10.6 μm. It is used in industrial applications for cutting and welding.

EXAMPLE 41.9 An ultraviolet laser

An ultraviolet laser generates a 10 MW, 5.0-ns-long light pulse at a wavelength of 355 nm. How many photons are in each pulse?

SOLVE The energy of each light pulse is the power multiplied by the duration:

$$E_{\text{pulse}} = P\Delta t = (1.0 \times 10^7 \text{ W})(5.0 \times 10^{-9} \text{ s}) = 0.050 \text{ J}$$

Each photon in the pulse has energy

$$E_{\text{photon}} = hf = \frac{hc}{\lambda} = 3.50 \text{ eV} = 5.60 \times 10^{-19} \text{ J}$$

Because $E_{\text{pulse}} = NE_{\text{photon}}$, the number of photons is

$$N = \frac{E_{\text{pulse}}}{E_{\text{photon}}} = 8.9 \times 10^{16} \text{ photons}$$

SUMMARY

The goal of Chapter 41 has been to understand the structure and properties of atoms.

IMPORTANT CONCEPTS

Hydrogen Atom

The three-dimensional Schrödinger equation has stationary-state solutions for the hydrogen atom potential energy only if three conditions are satisfied:

- Energy $E_n = -13.60 \text{ eV}/n^2$ $n = 1, 2, 3, \ldots$
- Angular momentum $L = \sqrt{l(l+1)}\,\hbar$ $l = 0, 1, 2, 3, \ldots, n-1$
- z-component of angular momentum
  $L_z = m\hbar$ $m = -l, -l+1, \ldots, 0, \ldots, l-1, l$

Each state is characterized by **quantum numbers** (n, l, m), but the energy depends only on n.

The probability of finding the electron within a small distance interval δr at distance r is

$$\text{Prob(in } \delta r \text{ at r)} = P_r(r)\,\delta r$$

where $P_r(r) = 4\pi r^2 |R_{nl}(r)|^2$ is the **radial probability density.**

Graphs of $P_r(r)$ suggest that the electrons are arranged in shells.

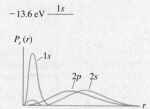

Multielectron Atoms

The potential energy is electron-nucleus plus electron-electron. In the **independent particle approximation,** each electron is described by the same quantum numbers (n, l, m, m_s) used for the hydrogen atom. The energy of a state depends on n and l. For each n, energy increases as l increases.

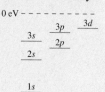

- High-l states correspond to circular orbits. These stay outside the core.
- Low-l states correspond to elliptical orbits. These penetrate the core to interact more strongly with the nucleus. This interaction lowers their energy.

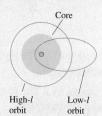

Electron spin

The electron has an inherent angular momentum $\vec{S}$ and magnetic moment $\vec{\mu}$ *as if* it were spinning. The spin angular momentum has a fixed magnitude $S = \sqrt{s(s+1)}\,\hbar$, where $s = \frac{1}{2}$. The z-component is $S_z = m_s\hbar$, where $m_s = \pm\frac{1}{2}$. These two states are called **spin-up** and **spin-down.** Each atomic state is fully characterized by the four quantum numbers (n, l, m, m_s).

The Pauli exclusion principle says that no more than one electron can occupy each quantum state. The periodic table of the elements is based on the fact that the ground state is the lowest-energy electron configuration compatible with the Pauli principle.

APPLICATIONS

Atomic spectra are generated by excitation followed by a photon-emitting quantum jump.

- Excitation by absorption or collision.
- Quantum-jump selection rule $\Delta l = \pm 1$.

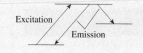

Lifetimes of excited states

The excited-state population decreases exponentially as

$$N_{\text{exc}} = N_0 e^{-t/\tau}$$

where $\tau = 1/r$ is the **lifetime** and r is the **decay rate.** It's not possible to predict when a particular atom will decay, but the *probability* is

$$\text{Prob(in } \delta t \text{ at t)} = r\,\delta t.$$

Stimulated emission of an excited state can be caused by a photon with $E_{\text{photon}} = E_2 - E_1$. Laser action can occur if $N_2 > N_1$, a condition called a **population inversion.**

TERMS AND NOTATION

principal quantum number, n	spin-down	nonradiative transition
orbital quantum number, l	independent particle	lifetime, τ
magnetic quantum number, m	approximation (IPA)	decay rate, r
ionization energy	Pauli exclusion principle	spontaneous emission
electron cloud	electron configuration	stimulated emission
radial wave function, $R_{nl}(r)$	closed shell	laser
radial probability density, $P_r(r)$	subshell	coherent
shell model	excitation	optical cavity
spin	allowed transition	population inversion
spin quantum number, m_s	selection rule	pumping
spin-up	collisional excitation	excitation transfer

EXERCISES AND PROBLEMS

Exercises

Section 41.1 The Hydrogen Atom: Angular Momentum and Energy

Section 41.2 The Hydrogen Atom: Wave Functions and Probabilities

1. What is the angular momentum of a hydrogen atom in (a) a $4p$ state and (b) a $5f$ state? Give your answers as a multiple of $\hbar$.

2. List the quantum numbers, excluding spin, of (a) all possible $3p$ states and (b) all possible $3d$ states.

3. A hydrogen atom has orbital angular momentum 3.65×10^{-34} J s.
   a. What letter (s, p, d, or f) describes the electron?
   b. What is the atom's minimum possible energy? Explain.

4. What is the maximum possible angular momentum L (as a multiple of $\hbar$) of a hydrogen atom with energy -0.544 eV?

5. What are E and L (as a multiple of $\hbar$) of a hydrogen atom in the $6f$ state?

6. Explain the difference between l and L.

Section 41.3 The Electron's Spin

7. How many lines of atoms would you expect to see on the collector plate of a Stern-Gerlach apparatus if the experiment is done with (a) lithium and (b) beryllium? Explain.

8. When all quantum numbers are considered, how many different quantum states are there for a hydrogen atom with $n = 1$? With $n = 2$? With $n = 3$? List the quantum numbers of each state.

9. Explain the difference between s and S.

Section 41.4 Multielectron Atoms

Section 41.5 The Periodic Table of the Elements

10. Predict the ground-state electron configurations of Mg, Sr, and Ba.

11. Predict the ground-state electron configurations of Si, Ge, and Pb.

12. Identify the element for each of these electron configurations. Then determine whether this configuration is the ground state or an excited state.
    a. $1s^2 2s^2 2p^5$
    b. $1s^2 2s^2 2p^6 3s^2 3p^6 3d^{10} 4s^2 4p$

13. Identify the element for each of these electron configurations. Then determine whether this configuration is the ground state or an excited state.
    a. $1s^2 2s^2 2p^4 3d$
    b. $1s^2 2s^2 2p^6 3s^2 3p^6 3d^8 4s^2$

14. Figure 41.24 shows that the ionization energy of cadmium ($Z = 48$) is larger than that of its neighbors. Why is this?

Section 41.6 Excited States and Spectra

15. What is the electron configuration of the second excited state of lithium?

16. Show that $hc = 1240$ eV nm.

17. A neon discharge emits a bright reddish-orange spectrum. But a glass tube filled with neon is completely transparent. Why doesn't the neon in the tube absorb orange and red wavelengths?

18. The two spectra shown in Figure Ex41.18 belong to the same element, a fictional Element X. Explain why they are different.

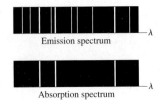

FIGURE EX41.18

Emission spectrum

Absorption spectrum

19. a. Is a $4p \rightarrow 4s$ transition allowed in sodium? If so, what is its wavelength? If not, why not?
    b. Is a $3d \rightarrow 4s$ transition allowed in sodium? If so, what is its wavelength? If not, why not?

Section 41.7 Lifetimes of Excited States

20. An atom in an excited state has a 1.0% chance of emitting a photon in 0.10 ns. What is the lifetime of the excited state?
21. An excited state of an atom has a 25 ns lifetime. What is the probability that an excited atom will emit a photon during a 0.50 ns interval of time?
22. 1.0×10^6 sodium atoms are excited to the $3p$ state at $t = 0$ s. How many of these atoms remain in the $3p$ state at (a) $t = 10$ ns, (b) $t = 30$ ns, and (c) $t = 100$ ns?
23. 1.0×10^6 atoms are excited to an upper energy level at $t = 0$ s. At the end of 20 ns, 90% of these atoms have undergone a quantum jump to the ground state.
    a. How many photons have been emitted?
    b. What is the lifetime of the excited state?

Section 41.8 Stimulated Emission and Lasers

24. A 1.0 mW helium neon laser emits a visible laser beam with a wavelength of 633 nm. How many photons are emitted per second?
25. A 1000 W carbon dioxide laser emits an infrared laser beam with a wavelength of 10.6 μm. How many photons are emitted per second?

Problems

26. a. Draw a diagram similar to Figure 41.3 to show all the possible orientations of the angular momentum vector $\vec{L}$ for the case $l = 3$. Label each $\vec{L}$ with the appropriate value of m.
    b. What is the minimum angle between $\vec{L}$ and the z-axis?
27. There exist subatomic particles whose spin is characterized by $s = 1$, rather than the $s = \frac{1}{2}$ of electrons. These particles are said to have a spin of one.
    a. What is the magnitude of the spin angular momentum S for a particle with a spin of one?
    b. What are the possible values of the spin quantum number?
    c. Draw a vector diagram similar to Figure 41.14 to show the possible orientations of $\vec{S}$.
28. For a hydrogen atom with $l = 2$, what are the (a) minimum and (b) maximum values of the quantity $(L_x^2 + L_y^2)^{1/2}$?
29. A hydrogen atom in its fourth excited state emits a photon with a wavelength of 1282 nm. What is the atom's maximum possible orbital angular momentum after the emission?
30. Calculate (a) the radial wave function and (b) the radial probability density at $r = \frac{1}{2}a_B$ for an electron in the $1s$ state of hydrogen. Give your answers in terms of a_B.
31. For an electron in the $1s$ state of hydrogen, what is the probability of being in a spherical shell of thickness $0.01a_B$ at distance (a) $\frac{1}{2}a_B$, (b) a_B, and (c) $2a_B$ from the proton?
32. The hydrogen atom $1s$ wave function is a maximum at $r = 0$. But the $1s$ radial probability density, shown in Figure 41.8, peaks at $r = a_B$ and is zero at $r = 0$. Explain this paradox.
33. Prove that the normalization constant of the $1s$ radial wave function of the hydrogen atom is $(\pi a_B^3)^{-1/2}$, as given in Equation 41.7.
    Hint: A useful definite integral is
    $$\int_0^\infty x^n e^{-\alpha x}\,dx = \frac{n!}{\alpha^{n+1}}$$

34. Prove that the normalization constant of the $2p$ radial wave function of the hydrogen atom is $(24\pi a_B^3)^{-1/2}$, as shown in Equation 41.7.
    Hint: See the hint in Problem 33.
35. Prove that the radial probability density peaks at $r = a_B$ for the $1s$ state of hydrogen.
36. a. Calculate and graph the hydrogen radial wave function $R_{2p}(r)$ over the interval $0 \leq r \leq 8a_B$.
    b. Determine the value of r (in terms of a_B) for which $R_{2p}(r)$ is a maximum.
    c. Example 41.3 and Figure 41.8 showed that the radial probability density for the $2p$ state is a maximum at $r = 4a_B$. Explain why this differs from your answer to part b.
37. In general, an atom can have both orbital angular momentum and spin angular momentum. The *total* angular momentum is defined to be $\vec{J} = \vec{L} + \vec{S}$. The total angular momentum is quantized in the same way as $\vec{L}$ and $\vec{S}$. That is, $J = \sqrt{j(j+1)}\hbar$, where j is the total angular momentum quantum number. The z-component of $\vec{J}$ is $J_z = L_z + S_z = m_j\hbar$, where m_j goes in integer steps from $-j$ to $+j$. Consider a hydrogen atom in a p state, with $l = 1$.
    a. L_z has three possible values and S_z has two. List all possible combinations of L_z and S_z. For each, compute J_z and determine the quantum number m_j. Put your results in a table.
    b. The number of values of J_z that you found in part a is too many to go with a single value of j. But you should be able to divide the values of J_z into two groups that correspond to two values of j. What are the allowed values of j? Explain. In a classical atom, there would be no restrictions on how the two angular momenta $\vec{L}$ and $\vec{S}$ can combine. Quantum mechanics is different. You've now shown that there are only two allowed ways to add these two angular momenta.
38. In a multielectron atom, the lowest-l state for each n ($1s$, $2s$, $3s$, etc.) is significantly lower in energy than the hydrogen state having the same n. But the highest-l state for each n ($2p$, $3d$, $4f$, etc.) is very nearly equal in energy to the hydrogen state with the same n. Explain.
39. Draw a series of pictures, similar to Figure 41.22, for the ground states of K, Ti, Fe, Ge, and Br.
40. Draw a series of pictures, similar to Figure 41.22, for the ground states of Ca, Sc, Co, Zn, and Kr.
41. a. What downward transitions are possible for a sodium atom in the $6s$ state? (See Figure 41.25.)
    b. What are the wavelengths of the photons emitted in each of these transitions?
42. The ionization energy of an atom is known to be 5.5 eV. The emission spectrum of this atom contains only the four wavelengths 310.0 nm, 354.3 nm, 826.7 nm, and 1240.0 nm. Draw an energy-level diagram with the fewest possible energy levels that agrees with these experimental data. Label each level with an appropriate l quantum number.
    Hint: Don't forget about the Δl selection rule.
43. Suppose you put five electrons into a 0.50-nm-wide one-dimensional rigid box (i.e., an infinite potential well).
    a. Use an energy-level diagram to show the electron configuration of the ground state.
    b. What is the ground-state energy?
44. The $5d \rightarrow 3p$ transition in the emission spectrum of sodium has a wavelength of 499 nm. What is the energy of the $5d$ state?

45. A sodium atom emits a photon with wavelength 818 nm shortly after being struck by an electron. What minimum speed did the electron have before the collision?

46. Figure P41.46 shows a few energy levels of the mercury atom.
    a. Make a table showing all the allowed transitions in the emission spectrum. For each transition, indicate the photon wavelength, in nm.
    b. What minimum speed must an electron have to excite the 492-nm-wavelength blue emission line in the Hg spectrum?

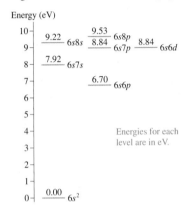

FIGURE P41.46

47. Figure P41.47 shows the first few energy levels of the lithium atom. Make a table showing all the allowed transitions in the emission spectrum. For each transition, indicate
    a. The wavelength, in nm.
    b. Whether the transition is in the infrared, the visible, or the ultraviolet spectral region.
    c. Whether or not the transition would be observed in the lithium absorption spectrum.

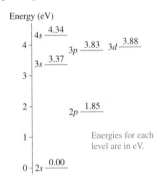

FIGURE P41.47

48. Three electrons are in a one-dimensional rigid box (i.e., an infinite potential well) of length 0.5 nm. Two are in the $n = 1$ state and one is in the $n = 6$ state. The selection rule for the rigid box allows only those transitions for which Δn is odd.
    a. Draw an energy-level diagram. On it, show the filled levels and show all transitions that could emit a photon.
    b. What are all the possible wavelengths that could be emitted by this system?

49. a. What is the decay rate for the $2p$ state of hydrogen?
    b. During what interval of time will 10% of a sample of $2p$ hydrogen atoms decay?

50. A hydrogen atom is in the $2p$ state. How much time must elapse for there to be a 1% chance that this atom will undergo a quantum jump to the ground state?

51. An atom in an excited state has a 1.0% chance of emitting a photon in 0.20 ns. How long will it take for 25% of a sample of excited atoms to decay?

52. a. Find an expression in terms of τ for the *half-life* $t_{1/2}$ of a sample of excited atoms. The half-life is the time at which half of the excited atoms have undergone a quantum jump and emitted a photon.
    b. What is the half-life of the $3p$ state of sodium?

53. An electrical discharge in a neon-filled tube maintains a *steady* population of 1.0×10^9 atoms in an excited state with $\tau = 20$ ns. How many photons are emitted per second from atoms in this state?

54. A ruby laser emits a 100 MW, 10-ns-long pulse of light with a wavelength of 690 nm. How many chromium atoms undergo stimulated emission to generate this pulse?

55. A laser emits 1.0×10^{19} photons per second from an excited state with energy $E_2 = 1.17$ eV. The lower energy level is $E_1 = 0$ eV.
    a. What is the wavelength of this laser?
    b. What is the power output of this laser?

Challenge Problems

56. Two excited energy levels are separated by the very small energy difference ΔE. As atoms in these levels undergo quantum jumps to the ground state, the photons they emit have nearly identical wavelengths λ.
    a. Show that the wavelengths differ by
    $$\Delta\lambda = \frac{\lambda^2}{hc}\Delta E$$
    b. In the Lyman series of hydrogen, what is the wavelength difference between photons emitted in the $n = 20$ to $n = 1$ transition and photons emitted in the $n = 21$ to $n = 1$ transition?

57. What is the probability of finding a $1s$ hydrogen electron at distance $r > a_B$ from the proton?

58. What is the probability of finding a $1s$ hydrogen electron at distance $r < \frac{1}{2}a_B$ from the proton?

59. Prove that the most probable distance from the proton of an electron in the $2s$ state of hydrogen is $5.236a_B$.

60. Find the distance, in terms of a_B, between the two peaks in the radial probability density of the $2s$ state of hydrogen.

61. Suppose you have a machine that gives you pieces of candy when you push a button. Eighty percent of the time, pushing the button gets you two pieces of candy. Twenty percent of the time, pushing the button yields 10 pieces. The *average* number of pieces per push is $N_{avg} = 2 \times 0.80 + 10 \times 0.20 = 3.6$. That is, 10 pushes should get you, on average, 36 pieces. Mathematically, the average value when the probabilities differ is $N_{avg} = \Sigma(N_i \times \text{Probability of } i)$. We can do the same thing in quantum mechanics, with the difference that the sum becomes an integral. If you measured the distance of the electron from the proton in many hydrogen atoms, you would get many values, as indicated by the radial probability density. But the *average* value of r would be
    $$r_{avg} = \int_0^\infty rP_r(r)\,dr$$
    Calculate the average value of r in terms of a_B for the electron in the $1s$ and the $2p$ states of hydrogen.

62. The 1997 Nobel prize in physics went to Steven Chu, Claude Cohen-Tannoudji, and William Phillips for their development of techniques to slow, stop, and "trap" atoms with laser light. To see how this works, consider a beam of rubidium atoms (mass 1.4×10^{-25} kg) traveling at 500 m/s after being evaporated out of an oven. A laser beam with a wavelength of 780 nm is directed against the atoms. This is the wavelength of the $5s \rightarrow 5p$ transition in rubidium, with $5s$ being the ground state, so the photons in the laser beam are easily absorbed by the atoms. After an average time of 15 ns, an excited atom spontaneously emits a 780-nm-wavelength photon and returns to the ground state.

a. The energy-momentum-mass relationship of Einstein's theory of relativity is $E^2 = p^2c^2 + m^2c^4$. A photon is massless, so the momentum of a photon is $p = E_{photon}/c$. Assume that the atoms are traveling in the positive x-direction and the laser beam in the negative x-direction. What is the initial momentum of an atom leaving the oven? What is the momentum of a photon of light?

b. The total momentum of the atom and the photon must be conserved in the absorption processes. As a consequence, how many photons must be absorbed to bring the atom to a halt?

NOTE ▶ Momentum is also conserved in the emission processes. However, spontaneously emitted photons are emitted in random directions. Averaged over many absorption/emission cycles, the net recoil of the atom due to emission is zero and can be ignored. ◀

c. Assume that the laser beam is so intense that a ground-state atom absorbs a photon instantly. How much time is required to stop the atoms?

d. Use Newton's second law in the form $F = \Delta p/\Delta t$ to calculate the force exerted on the atoms by the photons. From this, calculate the atoms' acceleration as they slow.

e. Over what distance is the beam of atoms brought to a halt?

STOP TO THINK ANSWERS

Stop to Think 41.1: $n = 3, l = 1$, or a $3p$ state.

Stop to Think 41.2: 4. You can see in Figure 41.8 that the ns state has n maxima.

Stop to Think 41.3: No. $m_s = \pm\frac{1}{2}$, so the z-component S_z cannot be zero.

Stop to Think 41.4: b. The atom would have less energy if the $3s$ electron were in a $2p$ state.

Stop to Think 41.5: c. Emission is a quantum jump to a lower-energy state. The $5p \rightarrow 4p$ transition is not allowed because $\Delta l = 0$ violates the selection rule. The lowest-energy allowed transition is $5p \rightarrow 3d$, with $E_{photon} = \Delta E_{atom} = 3.0$ eV.

Stop to Think 41.6: e. Because $r_B = 2r_A$, the ratio is $e^{-2}/e^{-1} = e^{-1} < \frac{1}{2}$.

42 Nuclear Physics

The technique known as *carbon dating* uses the radioactive decay of the naturally occurring carbon isotope ^{14}C to determine the age of fossils and archeological artifacts.

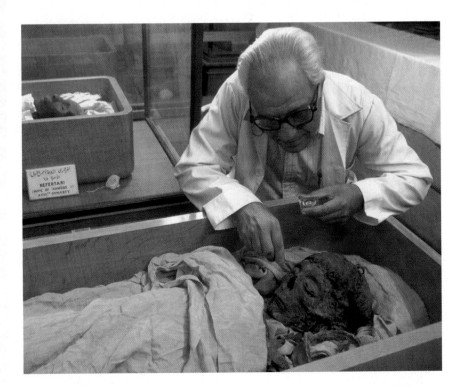

▶ **Looking Ahead**
The goal of Chapter 42 is to understand the physics of the nucleus and some of the applications of nuclear physics. In this chapter you will learn to:

- Interpret the basic structure of the nucleus.
- Understand how the strong force holds the nucleus together.
- Understand why some nuclei are unstable and undergo radioactive decay.
- Calculate the half-lives of radioactive decay.
- Apply nuclear physics to biology and medicine.

◀ **Looking Back**
The material in this chapter depends on basic atomic structure and on the quantized energy levels in potential-energy wells. Please review:

- Sections 37.6–37.7 Rutherford's model of the nucleus.
- Section 40.6 Finite potential-energy wells.

The nucleus of the atom is extremely remote from our everyday experience. Thus it comes as something of a surprise to notice the extent to which nuclear physics has become part of our modern technology and contemporary vocabulary: nuclear power and nuclear weapons, nuclear medicine and nuclear waste, nuclear fission and nuclear fusion.

Rutherford's discovery of the atomic nucleus marked the beginning of nuclear physics. Other physicists were soon designing experiments to probe within the nucleus and learn the properties of nuclear matter. In this final chapter, we'll explore the physics of the nucleus and look at some of the applications of nuclear physics.

One interesting application, shown in the photo, is the use of naturally occurring radioactivity to date archeological samples and geological formations. We'll also examine how exposure to nuclear radiation is measured and look at some of the uses of nuclear physics in medicine.

42.1 Nuclear Structure

The 1890s was a decade of mysterious rays. Cathode rays were being studied in several laboratories, and, in 1895, Röntgen discovered x rays. In 1896, after hearing of Röntgen's discovery, the French scientist A. H. Becquerel wondered if mineral crystals that fluoresce after exposure to sunlight were emitting x rays. He put a piece of film in an opaque envelope, then placed a crystal on top and left it in the sun. To his delight, the film in the envelope was exposed.

Becquerel thought he had discovered x rays coming from crystals, but his joy was short lived. He soon found that the film could be exposed equally well simply by being stored in a closed drawer with the crystals. Further investigation showed that the crystal, which happened to be a mineral containing uranium, was spontaneously emitting some new kind of ray. Rather than finding x rays, as he had hoped, Becquerel had discovered what became known as *radioactivity*.

Ernest Rutherford soon took up the investigation and found not one but three distinct kinds of rays emitted from crystals containing uranium. Not knowing what they were, he named them for their ability to penetrate matter and ionize air. The first, which caused the most ionization and penetrated the least, he called *alpha rays*. The second, with intermediate penetration and ionization, were *beta rays*, and the third, with the least ionization but the largest penetration, became *gamma rays*.

Within a few years, Rutherford was able to show that alpha rays are helium nuclei emitted from the crystal at very high velocities. These became the projectiles that he used in 1909 to probe the structure of the atom. The outcome of that experiment, as you learned in Chapter 37, was Rutherford's discovery that atoms have a very small, dense nucleus at the center.

Rutherford's discovery of the nucleus may have settled the question of atomic structure, but it raised many new issues for scientific research. Foremost among them were

- What is nuclear matter? What are its properties?
- What holds the nucleus together? Why doesn't the repulsive electrostatic force blow it apart?
- What is the connection between the nucleus and radioactivity?

These were the beginnings of **nuclear physics,** the study of the properties of the atomic nucleus.

Nucleons

The nucleus is a tiny speck in the center of a vastly larger atom. As Figure 42.1 shows, the nuclear diameter of roughly 10^{-14} m is only about 1/10,000 the diameter of the atom. What we call *matter* is overwhelming empty space!

You learned in Chapter 37 that the nucleus is composed of two types of particles: *protons* and *neutrons.* Together, these are referred to as **nucleons.** The role of the neutrons, which have nothing to do with keeping electrons in orbit, is an important issue that we'll address in this chapter. Table 42.1 summarizes the basic properties of protons and neutrons.

As you can see, protons and neutrons are virtually identical other than the fact that the proton has one unit of the fundamental charge e whereas the neutron is electrically neutral. The neutron is slightly more massive than the proton, but the difference is very small, only about 0.1%. Notice that the proton and neutron, like the electron, have an *inherent angular momentum* and magnetic moment with spin quantum number $s = \frac{1}{2}$. As a consequence, protons and neutrons obey the Pauli exclusion principle.

The number of protons Z is the element's **atomic number.** In fact, an element is identified by the number of protons in the nucleus, not by the number of orbiting electrons. Electrons are easily added and removed, forming negative and positive ions, but doing so doesn't change the element. The **mass number** A is defined to be $A = Z + N$, where N is the **neutron number.** The mass number is the total number of nucleons in a nucleus.

NOTE ▶ The mass number, which is dimensionless, is *not* the same thing as the atomic mass m. We'll look at actual atomic masses below. ◀

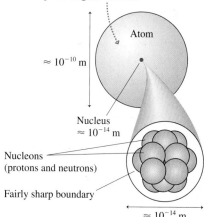

This picture of an atom would need to be 10 m in diameter if it were drawn to the same scale as the dot representing the nucleus.

Atom

$\approx 10^{-10}$ m

Nucleus $\approx 10^{-14}$ m

Nucleons (protons and neutrons)

Fairly sharp boundary

$\approx 10^{-14}$ m

FIGURE 42.1 The nucleus is a tiny speck within an atom.

TABLE 42.1 Protons and neutrons

	Proton	**Neutron**
Number	Z	N
Charge q	$+e$	0
Spin s	$\frac{1}{2}$	$\frac{1}{2}$
Mass, in u	1.00728	1.00866

Isotopes and Isobars

It was discovered early in the 20th century that not all atoms of the same element (same Z) have the same mass. There are a *range* of neutron numbers that happily form a nucleus with Z protons, creating a series of nuclei having the same Z-value (i.e., they are all the same chemical element) but different A-values. Each A-value in a series of nuclei with the same Z-value is called an **isotope.**

Chemical behavior is determined by the orbiting electrons. All isotopes of one element have the same number of orbiting electrons (if the atoms are electrically neutral) and thus have the same chemical properties, but different isotopes of the same element can have quite different nuclear properties. In addition, macroscopic behavior that depends on mass, such as the diffusion of a gas, can slightly favor one isotope over another.

The notation used to label isotopes is $^A Z$, where the mass number A is given as a *leading* superscript. The proton number Z is not specified by an actual number but, equivalently, by the chemical symbol for that element. Hence ordinary carbon, which has six protons and six neutrons in the nucleus, is written ^{12}C and pronounced "carbon twelve." The radioactive form of carbon used in carbon dating is ^{14}C. It has six protons, making it carbon, and eight neutrons.

More than 3000 isotopes are known. The majority of these are **radioactive,** meaning that the nucleus is not stable but, after some period of time, will either fragment or emit some kind of subatomic particle in an effort to reach a more stable state. Many of these radioactive isotopes are created by nuclear reactions in the laboratory and have only a fleeting existence. Only 266 isotopes are **stable** (i.e., nonradioactive) and occur in nature. We'll begin to look at the issue of nuclear stability in the next section.

The *naturally occurring* nuclei include the 266 stable isotopes and a handful of radioactive isotopes with such long half-lives, measured in billions of years, that they also occur naturally. The most well-known example of a naturally occurring radioactive isotope is the uranium isotope ^{238}U. For each element, the fraction of naturally occurring nuclei represented by one particular isotope is called the **natural abundance** of that isotope.

Although there are many radioactive isotopes of the element iodine, iodine occurs *naturally* only as ^{127}I. Consequently, we say that the natural abundance of ^{127}I is 100%. Most elements have multiple naturally occurring isotopes. The natural abundance of ^{14}N is 99.6%, meaning that 996 out of every 1000 naturally occurring nitrogen atoms are the isotope ^{14}N. The remaining 0.4% of naturally occurring nitrogen is the isotope ^{15}N, with one extra neutron.

A series of nuclei having the same A-value (the same mass number) but different values of Z and N are called **isobars.** For example, the three nuclei ^{14}C, ^{14}N, and ^{14}O are isobars with $A = 14$. Only ^{14}N is stable; the other two are radioactive.

Atomic Mass

You learned in Chapter 16 that atomic masses are specified in terms of the *atomic mass unit* u, defined such that the atomic mass of the isotope ^{12}C is exactly 12 u. The conversion to SI units is

$$1\,\text{u} = 1.6605 \times 10^{-27}\,\text{kg}$$

Alternatively, we can use Einstein's $E_0 = mc^2$ to express masses in terms of their energy equivalent. The energy equivalent of 1 u of mass is

$$E_0 = (1.6605 \times 10^{-27}\,\text{kg})(2.9979 \times 10^8\,\text{m/s})^2$$
$$= 1.4924 \times 10^{-10}\,\text{J} = 931.49\,\text{MeV}$$

(42.1)

Thus the atomic mass unit can be written

$$1\,\text{u} = 931.49\,\text{MeV}/c^2$$

It may seem unusual, but the units MeV/c^2 are units of mass.

NOTE ► We're using more significant figures than usual. Many nuclear calculations look for the small difference between two masses that are almost the same. Those two masses must be calculated or specified to four or five significant figures if their difference is to be meaningful. ◄

Table 42.2 shows the atomic masses of the electron, the nucleons, and three important light elements. Appendix C contains a more complete list. Notice that the mass of a hydrogen atom is the sum of the masses of a proton and an electron. But a quick calculation shows that the mass of a helium atom (2 protons, 2 neutrons, and 2 electrons) is 0.03038 u *less* than the sum of the masses of its constituents. The difference is due to the *binding energy* of the nucleus, a topic we'll look at in Section 42.2.

The isotope ^{2}H is a hydrogen atom in which the nucleus is not simply a proton but a proton and a neutron. Although the isotope is a form of hydrogen, it is called **deuterium.** The natural abundance of deuterium is 0.015%, or about 1 out of every 6700 hydrogen atoms. Water made with deuterium (sometimes written D_2O rather than H_2O) is called *heavy water.*

NOTE ► Don't let the name *deuterium* cause you to think this is a different element. Deuterium is an isotope of hydrogen. Chemically, it behaves just like ordinary hydrogen. ◄

The *chemical* atomic mass shown on the periodic table of the elements is the *weighted average* of the atomic masses of all naturally occurring isotopes. For example, chlorine has two stable isotopes: ^{35}Cl, with $m = 34.97$ u, is 75.8% abundant and ^{37}Cl, at 36.97 u, is 24.2% abundant. The average, weighted by abundance, is $0.758 \times 34.97 + 0.242 \times 36.97 = 35.45$. This is the value shown on the periodic table and is the correct value for most chemical calculations, but it is not the mass of any particular isotope of chlorine.

NOTE ► The atomic masses of the proton and the neutron are both ≈ 1 u. Consequently, the value of the mass number A is *approximately* the atomic mass in u. The approximation $m \approx A$ u is sufficient in many contexts, such as when we're calculating the masses of atoms in the kinetic theory of gases, but in nuclear physics calculations, we almost always need the more accurate mass values that you find in Table 42.2 or Appendix C. ◄

TABLE 42.2 Some atomic masses

Particle	Symbol	Mass (u)	Mass (MeV/c^2)
Electron	e	0.00055	0.51
Proton	p	1.00728	938.28
Neutron	n	1.00866	939.57
Hydrogen	^{1}H	1.00783	938.79
Deuterium	^{2}H	2.01410	1876.12
Helium	^{4}He	4.00260	3728.40

Nuclear Size and Density

Unlike the atom's electron cloud, which is quite diffuse, the nucleus has a fairly sharp boundary. Experimentally, the radius of a nucleus with mass number A is found to be

$$r = r_0 A^{1/3} \tag{42.2}$$

where $r_0 = 1.2$ fm. Recall that 1 fm = 1 femtometer = 10^{-15} m.

As Figure 42.2 shows, the radius is proportional to $A^{1/3}$, but the volume of the nucleus (proportional to r^3) is directly proportional to A, the number of nucleons. A nucleus with twice as many nucleons will occupy twice as much volume. This finding has several implications:

- Nucleons are incompressible. Adding more nucleons doesn't squeeze the inner nucleons into a smaller volume.
- The nucleons are tightly packed, looking much like the drawing in Figure 42.1.
- Nuclear matter has a constant density.

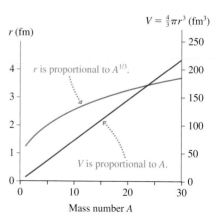

FIGURE 42.2 The nuclear radius and volume as a function of A.

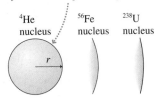

Imagine the nucleus is a drop of liquid. Its density is the same up to the edge of the drop.

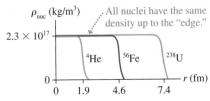

FIGURE 42.3 Density profiles of three nuclei.

In fact, we can use Equation 42.2 to estimate the density of nuclear matter. Consider a nucleus with mass number A. Its mass, within 1%, is A atomic mass units. Thus

$$\rho_{nuc} \approx \frac{A\,\text{u}}{\frac{4}{3}\pi r^3} = \frac{A\,\text{u}}{\frac{4}{3}\pi r_0^3 A} = \frac{1\,\text{u}}{\frac{4}{3}\pi r_0^3} = \frac{1.66 \times 10^{-27}\,\text{kg}}{\frac{4}{3}\pi(1.2 \times 10^{-15}\,\text{m})^3} \tag{42.3}$$

$$= 2.3 \times 10^{17}\,\text{kg/m}^3$$

The fact that A cancels means that **all nuclei have this density.** It is a staggeringly large density, roughly 10^{14} times larger than the density of familiar liquids and solids. One early objection to Rutherford's model of a nuclear atom was that matter simply can't have a density this high. Although we have no direct experience with such matter, nuclear matter really is this dense.

Figure 42.3 shows the density profiles of three nuclei. The constant density right to the edge is analogous to that of a drop of incompressible liquid, and, indeed, one successful model of many nuclear properties is called the **liquid-drop model.** Notice that the range of nuclear radii, from small helium to large uranium, is not quite a factor of 4. The fact that ^{56}Fe is a fairly typical atom in the middle of the periodic table is the basis for our earlier assertion that the nuclear diameter is roughly 10^{-14} m, or 10 fm.

STOP TO THINK 42.1 Three electrons orbit a neutral ^{6}Li atom. How many electrons orbit a neutral ^{7}Li atom?

42.2 Nuclear Stability

We've already noted that less than 10% of the known nuclei are stable (i.e., not radioactive). Because nuclei are characterized by two independent numbers, N and Z, it is useful to show the known nuclei on a plot of neutron number N versus proton number Z. Figure 42.4 shows such a plot. Stable nuclei are represented by blue diamonds and unstable, radioactive nuclei by red dots.

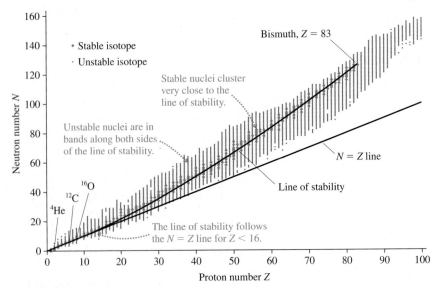

FIGURE 42.4 Stable and unstable nuclei shown on a plot of neutron number N versus proton number Z.

We can make several observations from this graph:

- The stable nuclei cluster very close to the curve called the **line of stability.**
- There are no stable nuclei with $Z > 83$ (bismuth).
- Unstable nuclei are in bands along both sides of the line of stability.
- The lightest elements, with $Z < 16$, are stable when $N \approx Z$. The familiar elements ^{4}He, ^{12}C, and ^{16}O all have equal numbers of protons and neutrons.
- As Z increases, the number of neutrons needed for stability grows increasingly larger than the number of protons. The N/Z ratio is ≈ 1.2 at $Z = 40$ but has grown to ≈ 1.5 at $Z = 80$.

These observations—especially $N \approx Z$ for small Z but $N > Z$ for large Z—cry out for an explanation. The quantum-mechanical model of the nucleus that we'll develop in Section 42.4 will provide the explanation we seek.

STOP TO THINK 42.2   The isobars corresponding to one specific value of A are found on the plot of Figure 42.4 along

a. A vertical line.
b. A horizontal line.
c. A diagonal line that goes up and to the right.
d. A diagonal line that goes up and to the left.

Binding Energy

A nucleus is a *bound system.* That is, you would need to supply energy to disperse the nucleons by breaking the nuclear bonds between them. Figure 42.5 shows this idea schematically.

You learned a similar idea in atomic physics. The energy levels of the hydrogen atom are negative numbers because the bound system has less energy than a free proton and electron. The energy you must supply to an atom to remove an electron is called the *ionization energy.*

In much the same way, the energy you would need to supply to a nucleus to disassemble it into individual protons and neutrons is called the **binding energy.** Whereas ionization energies of atoms are only a few eV, the binding energies of nuclei are tens or hundreds of MeV, energies large enough that their mass equivalent is not negligible.

Consider a nucleus with mass m_{nuc}. It is found experimentally that m_{nuc} is *less* than the total mass $Zm_p + Nm_n$ of the Z protons and N neutrons that form the nucleus, where m_p and m_n are the masses of the proton and neutron. That is, the energy equivalent $m_{nuc}c^2$ of the nucleus is less than the energy equivalent $(Zm_p + Nm_n)c^2$ of the individual nucleons. The binding energy B of the nucleus (not the entire atom) is defined as

$$B = (Zm_p + Nm_n - m_{nuc})c^2 \tag{42.4}$$

This is the energy you would need to supply to disassemble the nucleus into its pieces.

The practical difficulty is that laboratory scientists use mass spectroscopy to measure *atomic* masses, not nuclear masses. The atomic mass m_{atom} is m_{nuc} plus the mass Zm_e of Z orbiting electrons. (Strictly speaking, we should allow for the binding energy of the electrons, but these binding energies are roughly 10^6 smaller than the nuclear binding energies and can be neglected in all but the most precise measurements and calculations.)

Fortunately, we can switch from the nuclear mass to the atomic mass by the simple trick of both adding and subtracting Z electron masses. Begin by writing Equation 42.4 in the equivalent form

$$B = (Zm_p + Zm_e + Nm_n - m_{nuc} - Zm_e)c^2 \tag{42.5}$$

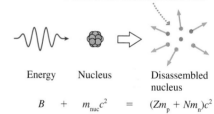

The binding energy is the energy that would be needed to disassemble a nucleus into individual nucleons.

Energy   Nucleus   Disassembled nucleus

$B \quad + \quad m_{nuc}c^2 \quad = \quad (Zm_p + Nm_n)c^2$

FIGURE 42.5 The nuclear binding energy.

Actⁱv Physⁱcs ONLINE   19.2

Now $m_{nuc} + Zm_e = m_{atom}$, the atomic mass, and $Zm_p + Zm_e = Z(m_p + m_e) = Zm_H$, where m_H is the mass of a hydrogen *atom*. Finally, use the conversion factor $1 \text{ u} = 931.49 \text{ MeV}/c^2$ to write $c^2 = 931.49 \text{ MeV/u}$. The binding energy is then

$$B = (Zm_H + Nm_n - m_{atom}) \times (931.49 \text{ MeV/u}) \quad \text{(42.6)}$$
(binding energy)

where all the three masses are in atomic mass units.

EXAMPLE 42.1 The binding energy of iron

What is the binding energy of the ^{56}Fe nucleus?

SOLVE The isotope ^{56}Fe has $Z = 26$ and $N = 30$. The atomic mass of ^{56}Fe, found in Appendix C, is 55.9349 u. Thus the mass difference between the ^{56}Fe nucleus and its constituents is

$B = 26(1.0078 \text{ u}) + 30(1.0087 \text{ u}) - 55.9349 \text{ u} = 0.528 \text{ u}$

where, from Table 42.2, 1.0078 u is the mass of the hydrogen atom. Thus the binding energy of ^{56}Fe is

$B = (0.538 \text{ u}) \times (931.49 \text{ MeV/u}) = 492 \text{ MeV}$

ASSESS The binding energy is extremely large, the energy equivalent of more than half the mass of a proton or a neutron.

The nuclear binding energy increases as A increases simply because there are more nuclear bonds. A more useful measure for comparing one nucleus to another is the quantity B/A, called the *binding energy per nucleon*. Iron, with $B = 492$ MeV and $A = 56$, has 8.79 MeV per nucleon. This is the amount of energy, on average, you would need to supply in order to remove *one* nucleon from the nucleus. Nuclei with larger values of B/A are more tightly held together than nuclei with smaller values of B/A.

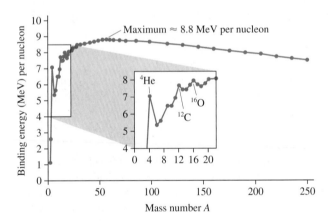

FIGURE 42.6 The curve of binding energy.

Figure 42.6 is a graph of the binding energy per nucleon versus mass number A. The dotted line connecting the points is often called the **curve of binding energy**. This curve has several important features:

- There are peaks in the binding energy curve at $A = 4$, 12, and 16. The one at $A = 4$, corresponding to ^{4}He, is especially pronounced. As you'll see, these peaks, which represent nuclei more tightly bound than their neighbors, are due to *closed shells* in much the same way that the graph of atomic ionization energies (see Figure 41.24) peaked for closed electron shells.
- The binding energy per nucleon is *roughly* constant at ≈ 8 MeV per nucleon for $A > 20$. This suggests that, as a nucleus grows, there comes a point where the nuclear bonds are *saturated*. Each nucleon interacts only with its nearest neighbors, the ones it's actually touching. This, in turn, implies that the nuclear force is a *short-range* force.

■ The curve has a broad maximum at $A \approx 60$. This will be important for our understanding of radioactivity. In principle, heavier nuclei could become *more* stable (more binding energy per nucleon) by breaking into smaller pieces. Lighter nuclei could become *more* stable by fusing together into larger nuclei. There may not always be a mechanism for such nuclear transformations to take place, but *if* there is a mechanism, it is energetically favorable for it to occur.

42.3 The Strong Force

Rutherford's discovery of the atomic nucleus was not immediately accepted by all scientists. Their primary objection was that the protons would blow themselves apart at tremendously high speeds due to the extremely large electrostatic forces between them at a separation of a few femtometers. No known force could hold the nucleus together.

It soon became clear that a previously unknown force of nature operates within the nucleus to hold the nucleons together. This new force had to be stronger than the repulsive electrostatic force, hence it was named the **strong force.** It is also called the *nuclear force.*

The strong force has several important properties:

1. It is an *attractive* force between any two nucleons.
2. It does not act on electrons.
3. It is a *short-range* force, acting only over nuclear distances.
4. Over the range where it acts, it is *stronger* than the electrostatic force that tries to push two protons apart.

The fact that the strong force is short-range, in contrast to the long-range $1/r^2$ electric, magnetic, and gravitational forces, is apparent from the fact that we see no evidence for nuclear forces outside the nucleus.

Figure 42.7 summarizes the three interactions that take place within the nucleus. Whether the strong force between two protons is the same strength as the force between two neutrons or between a proton and a neutron is an important question that can be answered experimentally. The primary means of investigating the strong force is to accelerate a proton to very high velocity, using a cyclotron or some other particle accelerator, then to study how the proton is scattered by various target materials.

The conclusion of many decades of research is that the strong force between two nucleons is independent of whether they are protons or neutrons. Charge is the basis for electromagnetic interactions, but it is of no relevance to the strong force. Protons and neutrons are identical as far as nuclear forces are concerned.

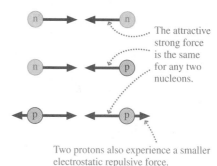

The attractive strong force is the same for any two nucleons.

Two protons also experience a smaller electrostatic repulsive force.

FIGURE 42.7 The strong force is the same between any two nucleons.

Potential Energy

Unfortunately, there's no simple formula to calculate the strong force or the potential energy of two nucleons interacting via the strong force. Figure 42.8 is an experimentally determined potential-energy diagram for two interacting nucleons, with r the distance between their centers. The potential-energy minimum at $r \approx 1$ fm is a point of stable equilibrium.

Recall that the force is the negative of the slope of a potential-energy diagram. The steeply rising potential for $r < 1$ fm represents a strongly repulsive force. That is, the nucleon "cores" strongly repel each other if they get too close together. The force is attractive for $r > 1$ fm, where the slope is positive, and it is strongest where the slope is steepest, at $r \approx 1.5$ fm. Notice that the strong force quickly decreases for $r > 1.5$ fm and is zero for $r > 3$ fm. That is, the strong force represented by this potential energy is effective only over a very short range of distances.

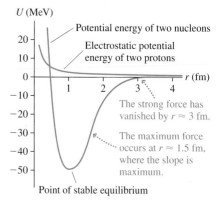

FIGURE 42.8 The potential-energy diagram for two nucleons interacting via the strong force.

Figure 42.8 also shows the electrostatic potential energy of two protons for comparison. Notice how small the electrostatic energy is in comparison to the potential energy of the strong force. At $r \approx 1.5$ fm, where the strong force is maximum, the aptly named attractive strong force is ≈ 100 times larger than the repulsive electrostatic force.

A question asked earlier was why the nucleus has neutrons at all. The answer is related to the short range of the strong force. Protons throughout the nucleus exert repulsive electrostatic forces on each other, but, because of the short range of the strong force, a proton feels an attractive force only from the very few other protons with which it is in close contact. Even though the strong force at its maximum is much larger than the electrostatic force, there wouldn't be enough attractive nuclear bonds for an all-proton nucleus to be stable. Because neutrons participate in the strong force but exert no repulsive forces, **the neutrons provide the extra "glue" that holds the nucleus together.** In small nuclei, where most nucleons are in contact, one neutron per proton is sufficient for stability. Hence small nuclei have $N \approx Z$. But as the nucleus grows, the repulsive force increases faster than the binding energy. More neutrons are needed for stability, causing heavy nuclei to have $N > Z$.

42.4 The Shell Model

Figure 42.8 is the potential energy of *two* interacting nucleons. To solve Schrödinger's equation for the nucleus, we would need to know the total potential energy of *all* interacting nucleon pairs within the nucleus, including both the strong force and the electrostatic force. This is far too complex to be a tractable problem.

We faced a similar situation with multielectron atoms. Calculating an atom's exact potential energy, including electron-electron repulsion, is exceedingly complicated. To simplify the problem, we made a *model* of the atom in which each electron moves independently with an *average* potential energy due to the nucleus and all other electrons. That model, although not perfect, led to the correct prediction of electron shells and explained the periodic table of the elements.

The **shell model** of the nucleus, using multielectron atoms as an analogy, was proposed in 1949 by Marie Goeppert-Mayer, one of the first prominent women in physics. The shell model considers each nucleon to move independently with an *average* potential energy due to the strong force of all the other nucleons. For the protons, we also have to include the electrostatic potential energy due to the other protons.

Figure 42.9 shows the average potential energy of a neutron and a proton. You can see that, to a good approximation, a nucleon appears to be a particle in a finite potential well, a quantum-mechanics problem you learned how to deal with in Chapter 40.

Marie Goeppert-Mayer shows her Nobel Prize.

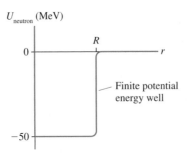

The average neutron potential energy is due to the strong force.

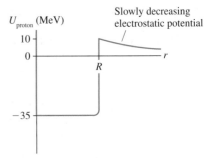

The average proton potential energy is due to the strong force and the electric force. This potential-well depth is for $Z \approx 30$.

FIGURE 42.9 The average potential energy of a neutron and a proton.

Several observations are worthwhile:

1. The depth of the neutron's potential-energy well is ≈ 50 MeV for all nuclei. The radius of the potential-energy well is the nuclear radius $R = r_0 A^{1/3}$.
2. For protons, the positive electrostatic potential energy "lifts" the potential-energy well. The lift varies from essentially none for very light elements to a significant fraction of the well depth for very heavy elements. The potential-energy well shown in the figure would be appropriate for a nucleus with $Z \approx 30$.
3. Outside the nucleus, where the strong force has vanished, a proton's potential energy is $U = (Z - 1)e^2/4\pi\epsilon_0 r$ due to its electrostatic interaction with the $(Z - 1)$ other protons within the nucleus. This positive potential energy decreases slowly with increasing distance.

The task of quantum mechanics is to solve for the energy levels and wave functions of the nucleons in these potential-energy wells. Once the energy levels are found, we build up the nuclear state, just as we did with atoms, by placing all the nucleons in the lowest energy levels consistent with the Pauli principle. The Pauli principle affects nucleons, just as it did electrons, because they are spin-$\frac{1}{2}$ particles. Each energy level can hold only a certain number of spin-up particles and spin-down particles, depending on the quantum numbers. Additional nucleons have to go into higher energy levels.

Low-Z Nuclei

As an example, we'll consider the energy levels of low-Z nuclei ($Z < 8$). Because these nuclei have so few protons, we can use a reasonable approximation that neglects the electrostatic potential energy due to proton-proton repulsion and considers only the much larger nuclear potential energy. In that case, the proton and neutron potential-energy wells and energy levels are the same.

Figure 42.10 shows the three lowest energy levels and the maximum number of nucleons that the Pauli principle allows in each. Energy values vary from nucleus to nucleus, but the spacing between these levels is several MeV. It's customary to draw the proton and neutron potential-energy diagrams and energy levels back to back. Notice that the radial axis for the proton potential-energy well points to the right, while the radial axis for the neutron potential-energy well points to the left.

Let's apply this model to the $A = 12$ isobar. Recall that an isobar is a series of nuclei with the same number of neutrons and protons. Figure 42.11 shows the energy-level diagrams of ^{12}B, ^{12}C, and ^{12}N. Look first at ^{12}C, a nucleus with six protons and six neutrons. You can see that exactly six protons are allowed in the $n = 1$ and $n = 2$ energy levels. Likewise for the six neutrons. Thus ^{12}C has a closed $n = 2$ proton shell and a closed $n = 2$ neutron shell.

NOTE ▶ Protons and neutrons are different particles, so the Pauli principle is not violated if a proton and a neutron have the same quantum numbers. ◀

The neutron radial distance is measured to the left. The proton potential energy is nearly identical to the neutron potential energy when Z is small.

These are the first three allowed energy levels. They are spaced several MeV apart.

These are the maximum number of nucleons allowed by the Pauli principle.

FIGURE 42.10 The three lowest energy levels of a low-Z nucleus. The neutron energy levels are on the left, the proton energy levels on the right.

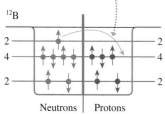

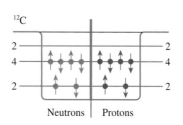

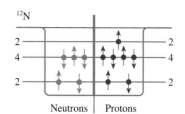

FIGURE 42.11 The $A = 12$ isobar has to place 12 nucleons in the lowest available energy levels.

^{12}N has seven protons and five neutrons. The sixth proton fills the $n = 2$ proton shell, so the seventh proton has to go into the $n = 3$ energy level. The $n = 2$ neutron shell has one vacancy because there are only five neutrons. ^{12}B is just the opposite, with the seventh neutron in the $n = 3$ energy level. You can see from the diagrams that the ^{12}B and ^{12}N nuclei have significantly more energy—by several MeV—than ^{12}C.

In atoms, electrons in higher energy levels decay to lower energy levels by emitting a photon as the electron undergoes a quantum jump. That can't happen here because the higher-energy nucleon in ^{12}B is a neutron whereas the vacant lower energy level is that of a proton. But an analogous process could occur *if* a neutron could somehow turn into a proton. And that's exactly what happens! We'll explore the details in Section 42.6, but both ^{12}B and ^{12}N decay into ^{12}C in the process known as *beta decay*.

^{12}C is just one of three low-Z nuclei in which both the proton and neutron shells are full. The other two are ^{4}He (filling both $n = 1$ shells with $Z = 2$, $N = 2$) and ^{16}O (filling both $n = 3$ shells with $Z = 8$, $N = 8$). If the analogy with closed electron shells is valid, these nuclei should be more tightly bound than nuclei with neighboring values of A. And indeed, we've already noted that the curve of binding energy (Figure 42.6) has peaks at $A = 4$, 12, and 16. The shell model of the nucleus satisfactorily explains these peaks. Unfortunately, the shell model quickly becomes much more complex as we go beyond $n = 3$. Heavier nuclei do have closed shells, but there's no evidence for them in the curve of binding energy.

High-Z Nuclei

We can use the shell model to give a qualitative explanation for one more observation, although the details are beyond the scope of this text. Figure 42.12 shows the neutron and proton potential-energy wells of a high-Z nucleus. In a nucleus with many protons, the electrostatic potential energy lifts the proton potential-energy well higher than the neutron potential-energy well. Protons and neutrons now have a different set of energy levels.

As a nucleus is "built," by the adding of protons and neutrons, the proton energy well and the neutron energy well must fill to just about the same height. If there were neutrons in energy levels above vacant proton levels, the nucleus would lower its energy by using beta decay to change the neutron into a proton. Similarly, beta decay would change a proton into a neutron if there were a vacant neutron energy level beneath a filled proton level. **The net result of beta decay is to keep the filled levels on both sides at just about the same height.**

Because the neutron potential-energy well starts at a lower energy, *more neutron states* are available than proton states. Consequently, a high-Z nucleus will have more neutrons than protons. This conclusion is consistent with our observation in Figure 42.2 that $N > Z$ for heavy nuclei.

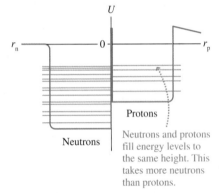

FIGURE 42.12 The proton energy levels are displaced upward in a high-Z nucleus.

Neutrons and protons fill energy levels to the same height. This takes more neutrons than protons.

42.5 Radiation and Radioactivity

Becquerel's 1896 discovery of "rays" from crystals of uranium prompted a burst of activity. Becquerel was soon joined in France by Marie Curie and Pierre Curie. They focused on isolating the element or elements responsible for the radiation, and, in the process, discovered the element radium.

In England, J. J. Thompson and, especially, his student and protégé Ernest Rutherford worked to identify the unknown rays. Using combinations of electric and magnetic fields, much as Thompson had done in his investigations of cathode rays, they found three distinct types of radiation. Figure 42.13 shows the basic experimental procedure, and Table 42.3 summarizes the results.

Marie Curie.

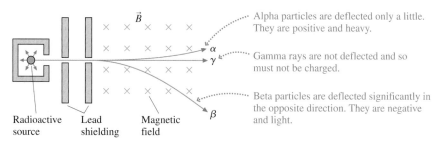

FIGURE 42.13 Identifying radiation by its deflection in a magnetic field.

TABLE 42.3 Three types of radiation

Radiation	Identification	Charge	Stopped by
Alpha, α	^{4}He nucleus	$+2e$	Sheet of paper
Beta, β	Electron	$-e$	Few mm of aluminum
Gamma, γ	High-energy photon	0	Many cm of lead

Within a few years, as Rutherford and others deduced the basic structure of the atom, it became clear that these emissions of radiation were coming from the atomic nucleus. We now define *radioactivity* or *radioactive decay* to be the spontaneous emission of particles or high-energy photons from unstable nuclei as they decay from higher-energy to lower-energy states. Radioactivity has nothing to do with the orbiting valence electrons.

NOTE ▶ The term "radiation" merely means something that is *radiated outward,* similar to the word "radial." Electromagnetic waves are often called *electromagnetic radiation.* Infrared waves from a hot object are referred to as "thermal radiation." Thus it was no surprise that these new "rays" were also called radiation. Unfortunately, the general public has come to associate the word "radiation" with *nuclear radiation,* something to be feared. It is important, when you use the term, to be sure you're not conveying a wrong impression to a listener or a reader. ◀

Ionizing Radiation

Electromagnetic waves, from microwaves through ultraviolet radiation, are *absorbed by* matter. The absorbed energy increases an object's thermal energy and its temperature, which is why objects sitting in the sun get warm.

In contrast to visible-light photon energies of a few eV, the energies of the alpha and beta particles and the gamma-ray photons of nuclear decay are typically in the range 0.1–10 MeV, a factor of roughly 10^6 larger. These energies are much larger than the ionization energies of atoms and molecules. Rather than simply being absorbed and increasing an object's thermal energy, nuclear radiation *ionizes* matter and *breaks* molecular bonds. Nuclear radiation and x rays, which behave much the same in matter, are called **ionizing radiation.**

An alpha or beta particle traveling through matter creates a trail of ionization, as shown in Figure 42.14a. Because the ionization energy of an atom is $\approx$10 eV, a particle with 1 MeV of kinetic energy can ionize $\approx$100,000 atoms or molecules before finally stopping. The low-mass electrons are kicked sideways, but the much more massive positive ions barely move and form the trail. This behavior is the basis for the *cloud chamber* and the *hydrogen bubble chamber,* where microscopic water droplets or hydrogen gas bubbles coalesce around the positive ions to make the trail visible. Figure 42.14b is a picture of the ionization trails of high-energy particles in a bubble chamber. The curvature of the trajectories is due to a magnetic field.

(a)

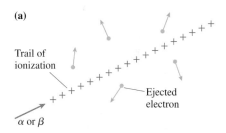

(b)

FIGURE 42.14 Alpha and beta particles create a trail of ionization as they pass through matter. This is the basis for the hydrogen bubble chamber.

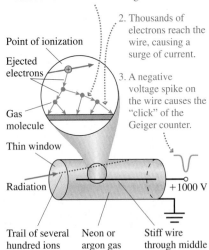

1. Ejected electrons cause a chain reaction of ionization of the gas.

2. Thousands of electrons reach the wire, causing a surge of current.

3. A negative voltage spike on the wire causes the "click" of the Geiger counter.

Point of ionization

Ejected electrons

Gas molecule

Thin window

Radiation

+1000 V

Trail of several hundred ions

Neon or argon gas

Stiff wire through middle

FIGURE 42.15 A Geiger counter.

Ionization is also the basis for the **Geiger counter,** one of the most well-known detectors of nuclear radiation. Figure 42.15 shows how a Geiger counter works. The important thing to remember is that a Geiger counter detects only *ionizing radiation.*

Ionizing radiation damages materials. Ions drive chemical reactions that wouldn't otherwise occur. Broken molecular bonds alter the workings of molecular machinery, especially in large biological molecules. It is through these mechanisms—ionization and bond breaking—that nuclear radiation can cause mutations or tumors. We'll look at the biological issues in Section 42.7.

NOTE ▶ Ionizing radiation causes structural damage to materials, but **irradiated objects do not become radioactive.** Ionization drives chemical processes involving the electrons. An object could become radioactive only if its nuclei were somehow changed, and that does not happen. ◀

STOP TO THINK 42.3   A very bright spotlight shines on a Geiger counter. Does it click?

Nuclear Decay and Half-Lives

Rutherford was the first to find that the number of radioactive atoms in a sample decreases exponentially with time. This is the expected time dependence if the decay is a *random process.* But to say that a process is random doesn't mean there are no patterns. Tossing a coin is a random process because you can't predict what one coin will do. Even so, if you tossed 1000 coins into the air, you'd certainly find very nearly 500 heads and 500 tails. Nuclear decay is similar.

Let r be the probability that one particular nucleus will decay in the next 1 s by emitting an alpha or beta particle or a gamma-ray photon. For example, $r = 0.010 \text{ s}^{-1}$ means that a nucleus has a 1% chance of decay in the next second. Notice that r, which is called the **decay rate,** has units s^{-1}, making it a *rate*.

The probability that a nucleus decays during the small interval of time Δt is

$$\text{Prob(in } \Delta t) = r \Delta t \qquad (42.7)$$

For example, a nucleus with $r = 0.010 \text{ s}^{-1}$ has a 0.1% chance of decay (Prob = 0.001) during a 0.1 s interval. If there are N independent nuclei, the number of nuclei expected to decay during Δt is

$$\text{number of decays} = N \times \text{Probability of decay} = rN\Delta t \qquad (42.8)$$

This is like saying you expect 500 heads when tossing 1000 coins, each coin with a 50% probability of landing heads up.

Each decay *decreases* the number of radioactive nuclei in the sample, hence the change in the number of radioactive nuclei during Δt is

$$\Delta N = -rN\Delta t \qquad (42.9)$$

The negative sign shows that N, the number of nuclei, decreases due to the decays. Finally, if we let $\Delta t \rightarrow dt$, Equation 42.9 becomes

$$\frac{dN}{dt} = -rN \qquad (42.10)$$

The *rate of change* in the number of radioactive nuclei depends both on the decay rate (a larger probability of decay per second means more decays per second) and on the number of radioactive nuclei present (more nuclei means that more are available to decay). And dN/dt is negative because N is decreasing.

Equation 42.10 is the same equation we solved in Chapter 31, with different symbols, for the voltage decay in an *RC* circuit. First, separate the variables onto opposite sides of the equation:

$$\frac{dN}{N} = -r\,dt \tag{42.11}$$

We need to integrate this equation, starting from $N = N_0$ nuclei at $t = 0$. Thus

$$\int_{N_0}^{N} \frac{dN}{N} = -r \int_{0}^{t} dt \tag{42.12}$$

By carrying out the integrations we find

$$\ln N - \ln N_0 = \ln\left(\frac{N}{N_0}\right) = -rt \tag{42.13}$$

We can now solve for *N* by taking the exponential of both sides and multiplying by N_0. The result is

$$N = N_0 e^{-rt} \tag{42.14}$$

Equation 42.13 predicts that the number of radioactive nuclei will decrease exponentially, a prediction that has been borne out in countless experiments during the last hundred years.

It is useful to define the **time constant** τ as

$$\tau = \frac{1}{r}$$

With this definition, Equation 42.13 becomes

$$N = N_0 e^{-t/\tau} \tag{42.15}$$

Figure 42.16 shows the decrease of *N* with time. The number of radioactive nuclei decreases from N_0 at $t = 0$ to $e^{-1}N_0 = 0.368N_0$ at time $t = \tau$. In practical terms, the number decreases by roughly two-thirds during one time constant.

NOTE ▶ An important aspect of exponential decay is that you can choose any instant you wish to be $t = 0$. The number of radioactive nuclei present at that instant is N_0. If at one instant you have 10,000 radioactive nuclei whose time constant is $\tau = 10$ min, you'll have roughly 3680 nuclei 10 min later. The fact that you may have had more than 10,000 nuclei earlier isn't relevant. ◀

Equation 42.14 is useful in the theoretical sense that we can relate τ directly to the probability of decay. But in practice, it's much easier to measure the time at which half of a sample has decayed than the time at which 36.8% has decayed. Let's define the **half-life** $t_{1/2}$ as the time interval in which half of a sample of radioactive atoms decays. The half-life is shown in Figure 42.16.

The half-life is easily related to the time constant τ because we know, by definition, that $N = \frac{1}{2}N_0$ at $t = t_{1/2}$. Thus, according to Equation 42.15

$$\frac{N_0}{2} = N_0 e^{-t_{1/2}/\tau} \tag{42.16}$$

The N_0 cancels, and we can then take the natural logarithm of both sides to find

$$\ln\left(\frac{1}{2}\right) = -\ln 2 = -\frac{t_{1/2}}{\tau} \tag{42.17}$$

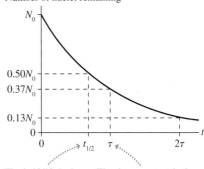

Number of nuclei remaining

The half-life is the time in which half the nuclei decay.

The time constant is the time at which the number of nuclei is e^{-1}, or 37%, of the initial number.

FIGURE 42.16 The number of radioactive atoms decreases exponentially with time.

With one final rearrangement we have

$$t_{1/2} = \tau \ln 2 = 0.693\tau \tag{42.18}$$

We'll leave it as a homework problem for you to show that Equation 42.15 can be written in terms of the half-life as

$$N = N_0\left(\frac{1}{2}\right)^{t/t_{1/2}} \tag{42.19}$$

Thus $N = N_0/2$ at $t = t_{1/2}$, $N = N_0/4$ at $t = 2t_{1/2}$, $N = N_0/8$ at $t = 3t_{1/2}$, and so on. **No matter how many nuclei there are, the number decays by half during the next half-life.**

NOTE ▶ Half the nuclei decay during one half-life, but don't fall into the trap of thinking that all will have decayed after two half-lives. ◀

Figure 42.17 shows the half-life graphically. This figure also conveys two other important ideas:

1. Nuclei don't vanish when they decay. The decayed nuclei have merely become some other kind of nuclei.
2. The decay process is random. We can predict that half the nuclei will decay in one half-life, but we can't predict which ones.

Each radioactive isotope, such as ^{14}C, has its own half-life. That half-life doesn't change with time as a sample decays. If you've flipped a coin 10 times and, against all odds, seen 10 heads, you may feel that a tail is overdue. Nonetheless, the probability that the next flip will be a head is still 50%. After 10 half-lives have gone by, $(1/2)^{10} = 1/1024$ of a radioactive sample is still there. There was nothing special or distinctive about these nuclei, and, despite their longevity, each remaining nucleus has exactly a 50% chance of decay during the next half-life.

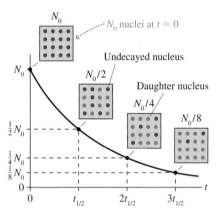

FIGURE 42.17 Half the nuclei decay during each half-life.

EXAMPLE 42.2 The decay of iodine
The iodine isotope ^{131}I, which has an eight-day half-life, is used in nuclear medicine. A sample of ^{131}I containing 2.00×10^{12} atoms is created in a nuclear reactor.

a. How many ^{131}I atoms remain 36 hours later when the sample is delivered to a hospital?
b. Although the sample is constantly getting weaker, it remains usable as long as there are at least 5.0×10^{11} ^{131}I atoms. What is the maximum delay before the sample is no longer usable?

MODEL The number of ^{131}I atoms decays exponentially.

SOLVE

a. The half-life is $t_{1/2} = 8$ days $= 192$ hr. After 36 hr have elapsed,

$$N = (2.00 \times 10^{12})\left(\frac{1}{2}\right)^{36/192} = 1.76 \times 10^{12} \text{ nuclei}$$

b. The time after creation at which 5.0×10^{11} ^{131}I atoms remain is given by

$$5.0 \times 10^{11} = 0.50 \times 10^{12} = (2.0 \times 10^{12})\left(\frac{1}{2}\right)^{t/8 \text{ days}}$$

To solve for t, first write this as

$$\frac{0.50}{2.00} = 0.25 = \left(\frac{1}{2}\right)^{t/8 \text{ days}}$$

Now take the logarithm of both sides. Either natural logarithms or base-10 logarithms can be used, but we'll use natural logarithms:

$$\ln(0.25) = -1.39 = \frac{t}{t_{1/2}}\ln(0.5) = -0.693\frac{t}{t_{1/2}}$$

Solving for t gives

$$t = 2.00t_{1/2} = 16 \text{ days}$$

ASSESS The weakest usable sample is one-quarter of the initial sample. You saw in Figure 42.17 that a radioactive sample decays to one-quarter of its initial number in 2 half-lives.

Activity

The **activity** R of a radioactive sample is the number of decays per second, or the decay rate. The decay rate is simply the absolute value of dN/dt, or

$$R = \left| \frac{dN}{dt} \right| = rN = rN_0 e^{-t/\tau} = R_0 e^{-t/\tau} = R_0 \left(\frac{1}{2} \right)^{t/t_{1/2}} \quad (42.20)$$

where $R_0 = rN_0$ is the activity at $t = 0$. The activity of a sample decreases exponentially along with the number of remaining nuclei.

The SI unit of activity is the **becquerel,** defined as

$$1 \text{ becquerel} = 1 \text{ Bq} \equiv 1 \text{ decay/s or } 1 \text{ s}^{-1}$$

An older unit of activity, but one that continues in widespread use, is the **curie.** The curie was originally defined as the activity of 1 g of radium. Today, the conversion factor is

$$1 \text{ curie} = 1 \text{ Ci} \equiv 3.7 \times 10^{10} \text{ Bq}$$

1 Ci is a substantial amount of radiation. The radioactive samples used in laboratory experiments are typically $\approx 1 \ \mu\text{Ci}$, or, equivalently, $\approx 40,000$ Bq. These samples can be handled with only minor precautions. Larger sources of activity require lead shielding and special precautions to prevent exposure to high levels of radiation.

EXAMPLE 42.3 A laboratory source
The isotope ^{137}Cs is a standard laboratory source of gamma rays. The half-life of ^{137}Cs is 30 years.

a. How many ^{137}Cs atoms are in a 5.0 μCi source?
b. What is the activity of the source 10.0 years later?

MODEL The number of ^{137}Cs atoms decays exponentially.

SOLVE

a. The number of atoms can be found from $N_0 = R_0/r$. The activity in SI units is

$$R = 5.0 \times 10^{-6} \text{ Ci} \times \frac{3.7 \times 10^{10} \text{ Bq}}{1 \text{ Ci}} = 1.85 \times 10^5 \text{ Bq}$$

To find the decay rate, first convert the half-life to seconds:

$$t_{1/2} = 30 \text{ years} \times \frac{3.15 \times 10^7 \text{ s}}{1 \text{ year}} = 9.45 \times 10^8 \text{ s}$$

Then

$$r = \frac{1}{\tau} = \frac{\ln 2}{t_{1/2}} = 7.33 \times 10^{-10} \text{ s}^{-1}$$

Thus the number of ^{137}Cs atoms is

$$N_0 = \frac{R_0}{r} = \frac{1.85 \times 10^5 \text{ Bq}}{7.33 \times 10^{-10} \text{ s}^{-1}} = 2.52 \times 10^{14} \text{ atoms}$$

b. The activity decreases exponentially, just like the number of nuclei. After 10 years,

$$R = R_0 \left(\frac{1}{2} \right)^{t/t_{1/2}} = (5.0 \ \mu\text{Ci}) \left(\frac{1}{2} \right)^{10/30} = 4.0 \ \mu\text{Ci}$$

ASSESS Although N_0 is a very large number, it is a very small fraction ($\approx 10^{-10}$) of a mole. The sample is about 60 ng (nanograms) of ^{137}Cs.

Radioactive Dating

Many geological and archeological samples can be dated by measuring the decays of naturally occurring radioactive isotopes. Because we have no way to know N_0, the initial number of radioactive nuclei, radioactive dating depends on the use of ratios.

The most well-known dating technique is carbon dating. The carbon isotope ^{14}C has a half-life of 5730 years, so any ^{14}C present when the earth formed 4.5 billion years ago would long since have decayed away. Nonetheless, ^{14}C is present in atmospheric carbon dioxide because high-energy cosmic rays collide with gas molecules high in the atmosphere. These cosmic rays are energetic enough to create ^{14}C nuclei from nuclear reactions with nitrogen and oxygen nuclei. The creation and decay of ^{14}C have reached a steady state in which the $^{14}\text{C}/^{12}\text{C}$ ratio is

1.3×10^{-12}. That is, atmospheric carbon dioxide has ^{14}C at the concentration of 1.3 parts per trillion. As small as this is, it's easily measured by modern chemical techniques.

All living organisms constantly exchange carbon dioxide with the atmosphere, so the $^{14}C/^{12}C$ ratio in living organisms is also 1.3×10^{-12}. As soon as an organism dies, the ^{14}C in its tissue begins to decay and no new ^{14}C is added. Objects are dated by comparing the measured $^{14}C/^{12}C$ ratio to the 1.3×10^{-12} value of living material.

Carbon dating is used to date skeletons, wood, paper, fur, food material, and anything else made of organic matter. It is quite accurate for ages to about 15,000 years. Beyond that, the difficulty of measuring such a small ratio and some uncertainties about the cosmic ray flux in the past combine to decrease the accuracy. Even so, items are dated to about 50,000 years with a fair degree of reliability.

Other isotopes with longer half-lives are used to date geological samples. Potassium-argon dating, using ^{40}K with a half-life of 1.25 billion years, is especially useful for dating rocks of volcanic origin.

EXAMPLE 42.4 **Carbon dating**
Archeologists excavating an ancient hunters' camp have recovered a 5.0 g piece of charcoal from a fireplace. Measurements on the sample find that the ^{14}C activity is 0.35 Bq. What is the approximate age of the camp?

MODEL Charcoal, from burning wood, is almost pure carbon. The number of ^{14}C atoms in the wood has decayed exponentially since the branch fell off a tree. Because wood rots, it is reasonable to assume that there was no significant delay from when the branch fell off the tree and when the hunters burned it.

SOLVE The $^{14}C/^{12}C$ ratio was 1.3×10^{-12} when the branch fell from the tree. We first need to determine the present ratio, then use the known ^{14}C half-life $t_{1/2} = 5730$ years to calculate the time needed to reach the present ratio. The number of ordinary ^{12}C nuclei in the sample is

$$N(^{12}C) = \left(\frac{5.0 \text{ g}}{12 \text{ g/mol}}\right) 6.02 \times 10^{23} \text{ atoms/mol}$$

$$= 2.5 \times 10^{23} \text{ nuclei}$$

The number of ^{14}C nuclei can be found from the activity to be $N(^{14}C) = R/r$, but we need to determine the ^{14}C decay rate r. After converting the half-life to seconds, $t_{1/2} = 5730$ years $= 1.807 \times 10^{11}$ s, we can compute

$$r = \frac{1}{\tau} = \frac{1}{t_{1/2}/\ln 2} = 3.84 \times 10^{-12} \text{ s}^{-1}$$

Thus

$$N(^{14}C) = \frac{R}{r} = \frac{0.35 \text{ Bq}}{3.84 \times 10^{-12} \text{ s}^{-1}} = 9.1 \times 10^{10} \text{ nuclei}$$

and the present $^{14}C/^{12}C$ ratio is $N(^{14}C)/N(^{12}C) = 0.36 \times 10^{-12}$. Because this ratio has been decaying with a half-life of 5730 years, the time needed to reach the present ratio is found from

$$0.36 \times 10^{-12} = 1.3 \times 10^{-12} \left(\frac{1}{2}\right)^{t/t_{1/2}}$$

To solve for t, first write this as

$$\frac{0.36}{1.3} = 0.277 = \left(\frac{1}{2}\right)^{t/t_{1/2}}$$

Now take the logarithm of both sides:

$$\ln(0.277) = -1.28 = \frac{t}{t_{1/2}}\ln(0.5) = -0.693\frac{t}{t_{1/2}}$$

Thus the age of the hunters' camp is

$$t = 1.85 t_{1/2} = 10,600 \text{ years}$$

ASSESS This is a realistic example of how radioactive dating is done.

STOP TO THINK 42.4   A sample starts with 1000 radioactive atoms. How many half-lives have elapsed when 750 atoms have decayed?

a. 0.25
b. 1.5
c. 2.0
d. 2.5

42.6 Nuclear Decay Mechanisms

This section will look in more detail at the mechanisms of the three types of radioactive decay.

Activ
Physics ONLINE   19.4

Alpha Decay

An alpha particle, symbolized as α, is a ^{4}He nucleus, a strongly bound system of two protons and two neutrons. An unstable nucleus that ejects an alpha particle will lose two protons and two neutrons, so we can write the decay as

$$^A X_Z \rightarrow {}^{A-4} Y_{Z-2} + \alpha + \text{energy} \qquad (42.21)$$

Figure 42.18 shows the alpha-decay process. The original nucleus X is called the **parent nucleus** and the decay-product nucleus Y is the **daughter nucleus.** This reaction can occur only when the mass of the parent nucleus is greater than the mass of the daughter nucleus plus the mass of an alpha particle. This requirement is met for heavy, high-Z nuclei well above the maximum on the Figure 42.6 curve of binding energy. It is energetically favorable for these nuclei to eject an alpha particle because the daughter nucleus is more tightly bound than the parent nucleus.

Although the mass requirement is based on the nuclear masses, we can express it—as we did the binding energy equation—in terms of atomic masses. The energy released in an alpha decay, essentially all of which goes into the alpha particle's kinetic energy, is

$$\Delta E \approx K_\alpha = (m_X - m_Y - m_{He})c^2 \qquad (42.22)$$

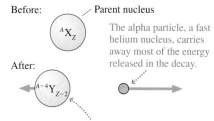

Before: — Parent nucleus

$^A X_Z$

The alpha particle, a fast helium nucleus, carries away most of the energy released in the decay.

After:

$^{A-4} Y_{Z-2}$

The daughter nucleus has two fewer protons and four fewer nucleons. It has a small recoil.

FIGURE 42.18 Alpha decay.

EXAMPLE 42.5 Alpha decay of uranium
The uranium isotope ^{238}U undergoes alpha decay to ^{234}Th. The atomic masses are 238.0508 u for ^{238}U and 234.0436 u for ^{234}Th. What is the kinetic energy, in MeV, of the alpha particle?

MODEL Essentially all of the energy release ΔE goes into the alpha particle's kinetic energy.

SOLVE The atomic mass of helium, from Table 42.2, is 4.0026 u. Thus

$$K_\alpha = (238.0508 \text{ u} - 234.0436 \text{ u} - 4.0026 \text{ u})c^2$$

$$= \left(0.0046 \text{ u} \times \frac{931.5 \text{ MeV}/c^2}{1 \text{ u}}\right)c^2 = 4.3 \text{ MeV}$$

ASSESS This is a typical alpha-particle energy. Notice how the c^2 canceled from the calculation so that we never had to evaluate c^2.

Alpha decay is a purely quantum-mechanical effect. Figure 42.19 shows the potential energy of an alpha particle, where the ^{4}He nucleus of an alpha particle is so tightly bound that we can think of it as existing "prepackaged" inside the parent nucleus. Both the depth of the energy well and the height of the Coulomb barrier are twice that of a proton because the charge of an α particle is $2e$.

Because of the high Coulomb barrier (alpha decay occurs only in high-Z nuclei), there may be one or more allowed energy levels with $E > 0$. Energy levels with $E < 0$ are completely bound, but an alpha particle in an energy level with $E > 0$ can *tunnel* through the Coulomb barrier and escape. That is exactly how alpha decay occurs.

Energy must be conserved, so the kinetic energy of the escaping α particle is the height of the energy level above $E = 0$. That is, potential energy is transformed into kinetic energy as the particle escapes. Notice that the width of the barrier decreases as E increases. The tunneling probability depends very sensitively on the barrier width, as you learned in conjunction with the scanning tunneling microscope. Thus an alpha particle in a higher energy level should have a

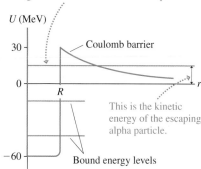

An alpha particle in this energy level can tunnel through the Coulomb barrier and escape.

This is the kinetic energy of the escaping alpha particle.

Bound energy levels

FIGURE 42.19 The potential-energy diagram of an alpha particle in the parent nucleus.

shorter half-life and escape with *more kinetic energy*. The full analysis is beyond the scope of this text, but this prediction is in excellent agreement with measured energies and half-lives.

Beta Decay

Beta decay was initially associated with the emission of an electron e^-, the beta particle. It was later discovered that some nuclei can undergo beta decay by emitting a positron e^+, the antiparticle of the electron, although this decay mode is not as common. A positron is identical to an electron except that it has a positive charge. To be precise, the emission of an electron is called *beta-minus decay* and the emission of a positron is *beta-plus decay*.

A typical example of beta decay occurs in the carbon isotope ^{14}C, which undergoes the beta-decay process ^{14}C → ^{14}N + e^-. Carbon has $Z = 6$ and nitrogen has $Z = 7$. Because Z increases by 1 but A doesn't change, it appears that a neutron within the nucleus has changed itself into a proton and electron. That is, the basic beta-minus decay process appears to be

$$n \rightarrow p^+ + e^- \tag{42.23}$$

The electron is ejected from the nucleus but the proton is not. Thus the decay process, shown in Figure 42.20a, is

$$^A X_Z \rightarrow {}^A Y_{Z+1} + e^- + \text{energy} \quad \text{(beta-minus decay)} \tag{42.24}$$

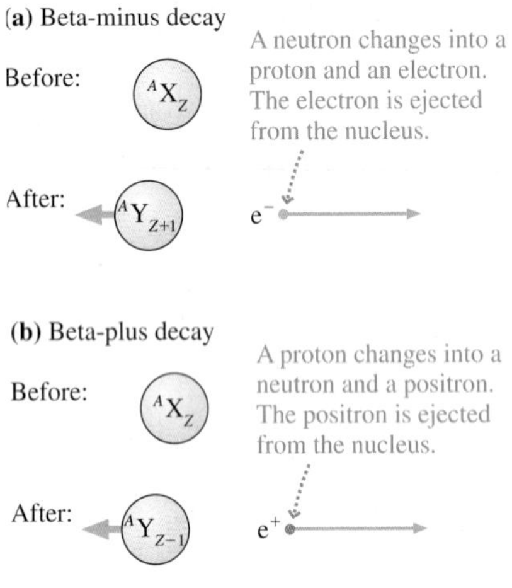

Indeed, a free neutron turns out to *not* be a stable particle. It decays with a half-life of approximately 10 min into a proton and an electron. This decay is energetically allowed because $m_n > m_p + m_e$. Furthermore, it conserves charge.

Whether a neutron *within* a nucleus can decay depends not only on the masses of the neutron and proton but also on the masses of the parent and daughter nuclei, because energy has to be conserved for the entire nuclear system. **Beta decay occurs only if $m_X > m_Y$.** ^{14}C can undergo beta decay to ^{14}N because $m(^{14}\text{C}) > m(^{14}\text{N})$. But $m(^{12}\text{C}) < m(^{12}\text{N})$, so ^{12}C is stable and its neutrons cannot decay.

Beta-plus decay is the conversion of a proton into a neutron and a positron:

$$p^+ \rightarrow n + e^+ \tag{42.25}$$

The full decay process, shown in Figure 42.20b, is

$$^A X_Z \rightarrow {}^A Y_{Z-1} + e^+ + \text{energy} \quad \text{(beta-plus decay)} \tag{42.26}$$

Beta-plus decay does *not* happen for a free proton because $m_p < m_n$. It *can* happen within a nucleus as long as energy is conserved for the entire nuclear system.

In our earlier discussion of Figure 42.11 we noted that the ^{12}B and ^{12}N nuclei could reach a lower energy state if a proton could change into a neutron and vice versa. Now we see that such a change can occur if the energy conditions are favorable. And, indeed, ^{12}B undergoes beta-minus decay to ^{12}C while ^{12}N undergoes beta-plus decay to ^{12}C.

In general, beta decay is a process used by nuclei with too many neutrons or too many protons in order to move closer to the line of stability in Figure 42.4.

NOTE ▶ The electron emitted in beta decay has nothing to do with the atom's valence electrons. The beta particle is created in the nucleus and ejected directly from the nucleus when a neutron is transformed into a proton and an electron. ◀

A third form of beta decay occurs in some nuclei that have too many protons but not enough mass to undergo beta-plus decay. In this case, a proton changes

(a) Beta-minus decay

Before:

A neutron changes into a proton and an electron. The electron is ejected from the nucleus.

After: e^-

(b) Beta-plus decay

Before:

A proton changes into a neutron and a positron. The positron is ejected from the nucleus.

After: e^+

FIGURE 42.20 Beta decay.

into a neutron by "capturing" an electron from the innermost shell of orbiting electrons (an $n = 1$ electron). The process is

$$\text{p}^+ + \text{orbital e}^- \rightarrow \text{n} \qquad (42.27)$$

This form of beta decay is called **electron capture,** abbreviated EC. The net result, $^A\text{X}_Z \rightarrow {}^A\text{Y}_{Z-1}$, is the same as beta-plus decay but without the emission of a positron. Electron capture is the only nuclear decay mechanism that involves the orbital electrons.

The Weak Interaction

We've presented beta decay as if it were perfectly normal for one kind of matter to change spontaneously into a completely different kind of matter. For example, it would be energetically favorable for a large truck to spontaneously turn into a Cadillac and a VW Beetle, ejecting the Beetle at high speed. But it doesn't happen.

Once you stop to think of it, the process $\text{n} \rightarrow \text{p}^+ + \text{e}^-$ seems ludicrous, not because it violates mass-energy conservation but because we have no idea *how* a neutron could turn into a proton. Alpha decay may be a strange process because tunneling in general goes against our commonsense notions, but it is a perfectly ordinary quantum-mechanical process. Now we're suggesting that one of the basic building blocks of matter can somehow morph into a different basic building block.

To make matters more confusing, measurements in the 1930s found that beta decay didn't seem to conserve either energy or momentum. Faced with these difficulties, the Italian physicist Enrico Fermi made two bold suggestions:

1. A previously unknown fundamental force of nature is responsible for beta decay. This force, which has come to be known as the **weak interaction,** has the ability to turn a neutron into a proton and vice versa.
2. The beta-decay process emits a particle that, at that time, had not been detected. This new particle has to be electrically neutral, in order to conserve charge, and it has to be much smaller than an electron. Fermi called it the **neutrino,** meaning "little neutral one." Energy and momentum really are conserved, but the neutrino carries away some of the energy and momentum of the decaying nucleus. Thus experiments that detect only the electron seem to violate conservation laws.

The neutrino is represented by the symbol ν, a lowercase Greek nu. The beta-decay processes that Fermi proposed are

$$\begin{aligned} \text{n} &\rightarrow \text{p}^+ + \text{e}^- + \bar{\nu} \\ \text{p}^+ &\rightarrow \text{n} + \text{e}^+ + \nu \end{aligned} \qquad (42.28)$$

The symbol $\bar{\nu}$ is an *antineutrino,* although the reason why one is a neutrino and the other an antineutrino need not concern us here. Figure 42.21 shows that the electron and antineutrino (or positron and neutrino) *share* the energy released in the decay.

The neutrino interacts with matter so weakly that a neutrino can pass straight through the earth with only a very slight chance of a collision. Thousands of neutrinos created by nuclear fusion reactions in the core of the sun are passing through your body every second. Neutrino interactions are so rare that the first laboratory detection did not occur until 1956, over 20 years after Fermi's proposal.

It was initially thought that the neutrino had not only zero charge but zero mass. However, experiments within the last few years have shown that the neutrino mass, although very tiny, is not zero. The best current evidence suggests a mass about one-millionth the mass of an electron. Experiments now underway will attempt to determine a more accurate value. The result will have far more importance than simply understanding beta decay. Neutrinos are the most numerous of all particles in the universe, so it may turn out that the neutrino mass has cosmological significance for the evolution of the universe.

The Super Kamiokande neutrino detector in Japan looks for the neutrinos emitted from nuclear fusion reactions in the core of the sun.

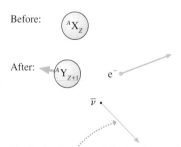

If only the electron and the daughter nucleus are measured, energy and momentum appear not to be conserved. The "missing" energy and momentum are carried away by the undetected antineutrino.

FIGURE 42.21 A more accurate picture of beta decay includes neutrinos.

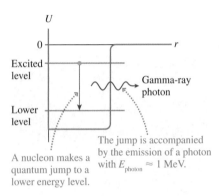

FIGURE 42.22 Gamma decay.

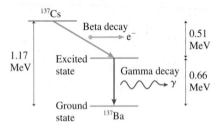

FIGURE 42.23 The decay of ^{137}Cs involves both beta and gamma decay.

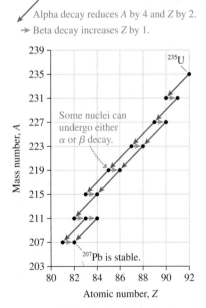

FIGURE 42.24 The nuclear decay series of ^{235}U.

EXAMPLE 42.6 **Beta decay of ^{14}C**
How much energy is released in the beta-minus decay of ^{14}C?

MODEL The decay is ^{14}C $\rightarrow$ ^{14}N + e$^-$ + $\bar{\nu}$.

SOLVE In Appendix C we find $m(^{14}$C$) = 14.003\,242$ u and $m(^{14}$N$) = 14.003\,074$ u. The mass difference is a mere $0.000\,168$ u, but this is the mass that is converted into the kinetic energy of the escaping particles. The energy released is

$$E = (\Delta m)c^2 = (0.000\,168 \text{ u}) \times (931.5 \text{ MeV/u})$$

$$= 0.156 \text{ MeV}$$

ASSESS This energy is shared between the electron and the antineutrino.

Gamma Decay

Gamma decay is the easiest form of nuclear decay to understand. You learned that an atomic system can emit a photon with $E_{photon} = \Delta E_{atom}$ when an electron undergoes a quantum jump from an excited energy level to a lower energy level. Nuclei are no different. A proton or a neutron in an excited nuclear state, such as the one shown in Figure 42.22, can undergo a quantum jump to a lower-energy state by emitting a high-energy photon. This is the gamma-decay process.

The spacing between atomic energy levels is only a few eV. Nuclear energy levels, by contrast, are typically 1 MeV apart. Hence gamma-ray photons have $E_{gamma} \approx 1$ MeV. Photons with this much energy have tremendous penetrating power and deposit an extremely large amount of energy at the point where they are finally absorbed.

Nuclei left to themselves are usually in their ground states and thus cannot emit gamma-ray photons. However, alpha and beta decay often leave the daughter nucleus in an excited nuclear state, so gamma emission is usually found to accompany alpha and beta emission.

The cesium isotope ^{137}Cs is a good example. We noted earlier that ^{137}Cs is used as a laboratory source of gamma rays. Actually, ^{137}Cs undergoes beta-minus decay to ^{137}Ba. Figure 42.23 shows the full process. A ^{137}Cs nucleus undergoes beta-minus decay by emitting an electron and an antineutrino, which share between them a total energy of 0.51 MeV. The half-life for this process is 30 years. This leaves the daughter ^{137}Ba nucleus in an excited state 0.66 MeV above the ground state. The excited Ba nucleus then decays within a few seconds to the ground state by emitting a 0.66 MeV gamma-ray photon. Thus a ^{137}Cs sample *is* a source of gamma-ray photons, but the photons are actually emitted by barium nuclei rather than cesium nuclei.

Decay Series

A radioactive nucleus decays into a daughter nucleus. In many cases, the daughter nucleus is also radioactive and decays to produce its own daughter nucleus. The process continues until reaching a daughter nucleus that is stable. The sequence of isotopes, starting with the original unstable isotope and ending with the stable isotope, is called a **decay series.**

Decay series are especially important for very heavy nuclei. As an example, Figure 42.24 shows the decay series of ^{235}U, an isotope of uranium with a 700-million-year half-life. This is a very long time, but it is only about 15% the age of the earth and most (but not all) of the ^{235}U nuclei present when the earth was formed have now decayed. There are many unstable nuclei along the way, but all ^{235}U nuclei eventually end as the ^{207}Pb isotope of lead, a stable nucleus.

Notice that some nuclei can decay by either alpha *or* beta decay. Thus there are a variety of paths that a decay can follow, but they all end at the same point.

STOP TO THINK 42.5   The cobalt isotope ^{60}Co ($Z = 27$) decays to the nickel isotope ^{60}Ni ($Z = 28$). The decay process is

a. Alpha decay.   b. Beta-minus decay.   c. Beta-plus decay.
d. Electron capture.   e. Gamma decay.

42.7 Biological Applications of Nuclear Physics

Nuclear physics has brought both peril and promise to society. Radioactivity can cause tumors. At the same time, radiation can be used to cure some cancers. This section is a brief survey of medical and biological applications of nuclear physics.

Radiation Dose

Nuclear radiation, which is ionizing radiation, disrupts a cell's machinery by altering and damaging the biological molecules. The consequences of this disruption vary from genetic mutations to uncontrolled cell multiplication (i.e., tumors) to cell death.

Beta and gamma radiation can penetrate the entire body and damage internal organs. Alpha radiation has less penetrating ability, but it deposits all its energy in a very small, localized volume. Internal organs are usually safe from alpha radiation, but the skin is very susceptible, as are the lungs if radioactive dust is inhaled.

Biological effects of radiation depend upon the **dose**, the amount of radiation received. Two factors enter into determining the dose. The first is the physical factor of energy absorbed by the body. The second is the biological factor of how tissue reacts to different forms of radiation.

The **rad,** an acronym for **r**adiation **a**bsorbed **d**ose, measures the energy deposited in an irradiated material. It is defined as

$$1 \text{ rad} \equiv 0.010 \text{ J/kg of absorbed energy}$$

The number of rads depends only on the energy absorbed, not at all on the type of radiation or on what the absorbing material is.

Biologists and biophysicists have found that a 1 rad dose of gamma rays and a 1 rad dose of alpha particles have different biological consequences. To account for such differences, the **relative biological effectiveness,** RBE, is defined as the biological effect of a given dose relative to the biological effect of an equal dose of x rays.

Table 42.4 shows the relative biological effectiveness of different forms of radiation. Larger values correspond to larger biological effects. Alpha and beta radiation have a range of values because the biological effect varies with the energy of the particle. Alpha radiation has the largest RBE because the energy is deposited in a smaller volume.

Combining these two measures, the **biologically equivalent dose** is the product of the energy dose in rads with the relative biological effectiveness. The biologically equivalent dose is measured in **rem,** an acronym for **r**öntgen **e**quivalent in **m**an. (This term is based on historical usage. The *röntgen* is a unit for measuring the ionization ability of x rays, but we need not be concerned with its definition.) To be precise,

$$\text{biologically effective dose in rem} \equiv \text{dose in rad} \times \text{RBE}$$

One rem of radiation produces the same biological damage regardless of the type of radiation.

TABLE 42.4 Relative biological effectiveness of radiation

Radiation type	RBE
X rays	1
Gamma rays	1
Beta particles	1–2
Alpha particles	10–20

EXAMPLE 42.7 Radiation exposure

A 75 kg laboratory technician working with the radioactive isotope ^{137}Cs receives an accidental 100 mrem (millirem) exposure. ^{137}Cs emits 0.66 MeV gamma-ray photons. How many gamma-ray photons are absorbed in the technician's body?

MODEL The radiation dose is a combination of deposited energy and biological effectiveness. The RBE for gamma rays is 1. Gamma rays are penetrating, so this is a whole-body exposure.

SOLVE The dose in rads is the dose in rems divided by the RBE. In this case, because the RBE = 1, the dose is 100 mrad = 0.10 rad = 0.0010 J/kg. This is a whole-body exposure, so the total energy deposited in the technician's body is 0.075 J. The energy of each absorbed photon is 0.66 MeV, but this value must be converted into joules. The number of photons in 0.075 J is

$$N = \frac{0.075 \text{ J}}{(6.6 \times 10^5 \text{ eV/photon})(1.60 \times 10^{-19} \text{ J/eV})}$$
$$= 7.1 \times 10^{11} \text{ photons}$$

ASSESS The energy deposited, 0.075 J, is very small. Radiation does its damage not by thermal effects, which would require substantially more energy, but by ionization.

TABLE 42.5 Radiation exposure

Radiation source	Typical exposure (mrem/year)
Natural background	300
Mammogram x ray	80
Chest x ray	30
Dental x ray	3

Table 42.5 shows some basic information about radiation exposure. We are all exposed to a continual *natural background* of radiation from cosmic rays and from naturally occurring radioactive atoms (uranium and atoms in the uranium decay series) in the ground, the atmosphere, and even the food we eat. This background averages about 300 mrem per year, although there are wide regional variations depending on the soil type and the elevation. (Higher elevations have a larger exposure to cosmic rays.)

Medical x rays vary significantly. The average person in the United States receives approximately 60 mrem per year from all medical sources. All other sources, such as fallout from atmospheric nuclear tests many decades ago, nuclear power plants, and industrial uses of radioactivity, amount to <10 mrem per year.

The question inevitably arises, "What is a safe dose?" This remains a controversial topic and the subject of ongoing research. The effects of large doses of radiation are easily observed. The effects of small doses are hard to distinguish from other natural and environmental causes. Thus there's no simple or clear definition of a safe dose. A prudent policy is to avoid unnecessary exposure to radiation but not to worry over exposures less than the natural background. It's worth noting that the μCi radioactive sources used in laboratory experiments provide exposures *much* less than the natural background, even if used on a regular basis.

Medical Uses of Radiation

Radiation can be put to good use killing cancer cells. This area of medicine is called *radiation therapy*. Gamma rays are the most common form of radiation, often from the isotope ^{60}Co. As Figure 42.25 shows, the gamma rays are directed along many different lines, all of which intersect the tumor. The goal is to provide a lethal dose to the cancer cells without overexposing nearby tissue. The patient and the radiation source are rotated around each other under careful computer control to deliver the proper dose.

Other tumors are treated by surgically implanting radioactive "seeds" within or next to the tumor. Alpha particles, which are very damaging locally but don't penetrate far, can be used in this fashion.

Radioactive isotopes are also used as *tracers* in diagnostic procedures. This technique is based on the fact that all isotopes of an element have identical chemical behavior. As an example, a radioactive isotope of iodine is used in the diagnosis of certain thyroid conditions. Iodine is an essential element in the body, and it concentrates in the thyroid gland. A doctor who suspects a malfunctioning thyroid gland gives the patient a small dose of sodium iodide in which some of the normal ^{127}I atoms have been replaced with ^{131}I. (Sodium iodide, which is harmless, dissolves in water and can simply be drunk.) The ^{131}I isotope, with a half-life of eight days, undergoes beta decay, and subsequently emits a gamma-ray photon.

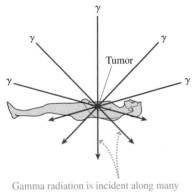

Gamma radiation is incident along many lines, all of which intersect the tumor.

FIGURE 42.25 Radiation therapy is designed to deliver a lethal dose to the tumor without damaging nearby tissue.

The radioactive iodine concentrates within the thyroid gland within a few hours. The doctor then monitors the gamma-ray photon emissions over the next few days to see how the iodine is being processed within the thyroid and how quickly it is eliminated from the body.

Other important radioactive tracers include the chromium isotope ^{51}Cr, which is taken up by red blood cells and can be used to monitor blood flow, and the xenon isotope ^{133}Xe, which is inhaled to reveal lung functioning. Radioactive tracers are *noninvasive*, meaning that the doctor can monitor the inside of the body without surgery.

Magnetic Resonance Imaging

The proton, like the electron, has an inherent angular momentum (spin) and an inherent magnetic moment. You can think of the proton as being like a little compass needle that can be in one of two positions, the positions we call spin up and spin down.

A compass needle aligns itself with an external magnetic field. This is the needle's lowest-energy position. Turning a compass needle by hand is like rolling a ball uphill; you're giving it energy, but, like the ball rolling downhill, it will realign itself with the lowest-energy position when you remove your finger. There is, however, an *unstable equilibrium* position in which the needle is anti-aligned with the field. The slightest jostle will cause it to flip around, but the needle will be steady in its upside-down configuration if you can balance it perfectly.

A proton in a magnetic field behaves similarly, but with a major difference: Because the proton's energy is quantized, the proton cannot assume an intermediate position. It's either aligned with the magnetic field (the spin-up orientation) or anti-aligned (spin-down). Figure 42.26a shows these two quantum states. Turning on a magnetic field lowers the energy of a spin-up proton and increases the energy of an anti-aligned, spin-down proton. In other words, the magnetic field creates an *energy difference* between these states.

Radiation therapy is a beneficial use of nuclear physics.

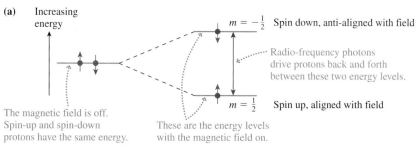

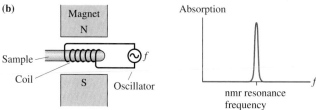

FIGURE 42.26 Nuclear magnetic resonance is possible because spin-up and spin-down protons have slightly different energies in a magnetic field.

The energy difference is very tiny, only about 10^{-7} eV. Nonetheless, photons whose energy matches the energy difference cause the protons to move back and forth between these two energy levels as the photons are absorbed and emitted. In effect, the photons are causing the proton's spin to flip back and forth rapidly. The

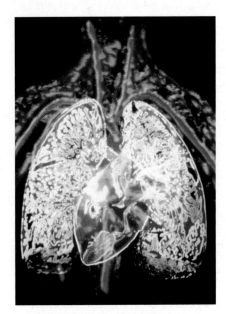

FIGURE 42.27 Magnetic resonance imaging shows internal organs in exquisite detail.

photon frequency, which depends on the magnetic field strength, is typically about 100 MHz, similar to FM radio frequencies.

Figure 42.26b shows how this behavior is put to use. A sample containing protons is placed in a magnetic field. A coil is wrapped around the sample, and a variable-frequency AC source drives a current through this coil. The protons absorb power from the coil when its frequency is just right to flip the spin back and forth; otherwise, no power is absorbed. A *resonance* is seen by scanning the coil through a small range of frequencies.

This technique of observing the spin flip of nuclei (the technique works for nuclei other than hydrogen) in a magnetic field is called **nuclear magnetic resonance,** or *nmr*. It has many applications in physics, chemistry, and materials science. Its medical use exploits the fact that tissue is mostly water, and two out of the three nuclei in a water molecule are protons. Thus the human body is basically a sample of protons, with the proton density varying as tissue density varies.

The medical procedure known as **magnetic resonance imaging,** or MRI, places the patient in a spatially varying magnetic field. The variations in the field cause the proton absorption frequency to vary from point to point. From the known shape of the field and measurements of the frequencies that are absorbed, and how strongly, sophisticated computer software can transform the raw data into detailed images such as the one shown in Figure 42.27.

As an interesting footnote, the technique was still being called *nuclear magnetic resonance* when it was first introduced into medicine. Unfortunately, doctors soon found that many patients were afraid of it because of the word "nuclear." Hence the alternative term "magnetic resonance imaging" was coined. It is true that the public perception of nuclear technology is not always positive, but equally true that nuclear physics has made many important and beneficial contributions to society.

SUMMARY

The goal of Chapter 42 has been to understand the physics of the nucleus and some of the applications of nuclear physics.

GENERAL PRINCIPLES

The Nucleus

The nucleus is a small, dense, positive core at the center of an atom.

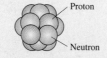

Z protons: charge $+e$, spin $\frac{1}{2}$

N neutrons: charge 0, spin $\frac{1}{2}$

The mass number is $A = Z + N$

The nuclear radius is $r = r_0 A^{1/3}$, where $r_0 = 1.2$ fm. Typical radii are a few fm.

Nuclear forces

Attractive strong force	Repulsive electric force
• Acts between any two nucleons	• Acts between two protons
• Is short range, <3 fm	• Is long range
• Is felt between nearest neighbors	• Is felt across the nucleus

Nuclear Stability

Most nuclei are not stable. Unstable nuclei undergo radioactive decay. Stable nuclei cluster along the **line of stability** in a plot of the isotopes.

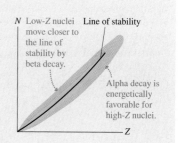

Three mechanisms by which unstable nuclei decay:

Decay	Particle	Mechanism	Energy	Penetration
α	^{4}He nucleus	tunneling	few MeV	low
β	e^-	$n \rightarrow p^+ + e^-$	≈ 1 MeV	medium
	e^+	$p^+ \rightarrow n + e^+$	≈ 1 MeV	medium
γ	photon	quantum jump	≈ 1 MeV	high

IMPORTANT CONCEPTS

Shell model

Each nucleon moves with an average potential energy due to all other nucleons.

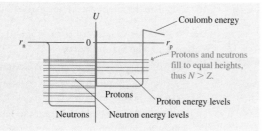

Curve of binding energy

The average binding energy per nucleon has a broad maximum at $A \approx 60$.

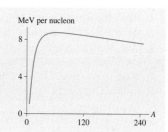

APPLICATIONS

Radioactive decay

The number of undecayed nuclei decreases exponentially with time t.

$$N = N_0 \exp(-t/\tau)$$
$$= N_0 (1/2)^{t/t_{1/2}}$$

The **time constant** τ is $1/r$, where r is the **decay rate**. The **half-life**

$$t_{1/2} = \tau \ln 2 = 0.693\tau$$

is the time in which half of any sample decays.

Measuring radiation

The **activity** $R = rN$ of a radioactive sample, measured in becquerels or curies, is the number of decays per second.

The radiation **dose** is measured in rad, where

$$1 \text{ rad} \equiv 0.010 \text{ J/kg of absorbed energy}$$

The **relative biological effectiveness** RBE is the biological effect of a dose relative to the biological effects of x rays. The **biologically equivalent dose** is measured in rem, where rem = rad $\times$ RBE. One rem of radiation produces the same biological effect regardless of the type of radiation.

TERMS AND NOTATION

nuclear physics	curve of binding energy	parent nucleus
nucleon	strong force	daughter nucleus
atomic number, Z	shell model	electron capture
mass number, A	alpha decay	weak interaction
neutron number, N	beta decay	neutrino
isotope	gamma decay	decay series
radioactive	ionizing radiation	dose
stable	Geiger counter	rad
natural abundance	decay rate, r	relative biological effectiveness (RBE)
isobar	time constant, τ	biologically equivalent dose
deuterium	half-life, $t_{1/2}$	rem
liquid-drop model	activity, R	nuclear magnetic resonance
line of stability	becquerel, Bq	magnetic resonance imaging (MRI)
binding energy, B	curie, Ci	

EXERCISES AND PROBLEMS

See Appendix C for data on atomic masses, isotopic abundance, radioactive decay modes, and half-lives.

Exercises

Section 42.1 Nuclear Structure

1. How many protons and how many neutrons are in (a) ^{3}H, (b) ^{40}Ar, (c) ^{40}Ca, and (d) ^{239}Pu?

2. How many protons and how many neutrons are in (a) ^{3}He, (b) ^{20}Ne, (c) ^{60}Co, and (d) ^{226}Ra?

3. Calculate the nuclear diameters of (a) ^{4}He, (b) ^{40}Ar, and (c) ^{220}Rn.

4. Which stable nuclei have a diameter of 8.84 fm?

5. Estimate the number of protons and the number of neutrons in 1 m^3 of air.

6. Estimate the number of protons and the number of neutrons in your body.

7. What would be the mass of a 1.0-cm-diameter marble if it had nuclear density?

Section 42.2 Nuclear Stability

8. Use data in Appendix C to make your own chart of stable and unstable nuclei, similar to Figure 42.4, for all nuclei with $Z \leq 8$. Use a blue or black dot to represent stable isotopes, a red dot to represent isotopes that undergo beta-minus decay, and a green dot to represent isotopes that undergo beta-plus decay or electron-capture decay.

9. a. What is the smallest value of A for which there are two stable nuclei? What are they?
   b. For which values of A less than this are there *no* stable nuclei?

10. Calculate (in MeV) the total binding energy and the binding energy per nucleon for ^{3}H and for ^{3}He.

11. Calculate (in MeV) the total binding energy and the binding energy per nucleon for ^{40}Ar and for ^{40}Ca.

12. Calculate (in MeV) the binding energy per nucleon for ^{3}He and ^{4}He. Which is more tightly bound?

13. Calculate (in MeV) the binding energy per nucleon for ^{12}C and ^{13}C. Which is more tightly bound?

14. Calculate (in MeV) the binding energy per nucleon for (a) ^{14}N, (b) ^{56}Fe, and (c) ^{207}Pb.

15. Calculate the chemical atomic mass of neon.

16. Calculate the chemical atomic mass of magnesium.

Section 42.3 The Strong Force

17. Use the potential-energy diagram in Figure 42.8 to estimate the strength of the strong force between two nucleons separated by 1.5 fm.

18. Use the potential-energy diagram in Figure 42.8 to sketch an approximate graph of the strong force between two nucleons versus the distance r between their centers.

19. What is the ratio of the gravitational potential energy to the nuclear potential energy for two neutrons separated by 1.0 fm?

Section 42.4 The Shell Model

20. a. Draw energy-level diagrams, similar to Figure 42.11, for all $A = 10$ nuclei listed in Appendix C. Show all the occupied neutron and proton levels.
    b. Which of these nuclei are stable? What is the decay mode of any that are radioactive?

21. a. Draw energy-level diagrams, similar to Figure 42.11, for all $A = 14$ nuclei listed in Appendix C. Show all the occupied neutron and proton levels.
    b. Which of these nuclei are stable? What is the decay mode of any that are radioactive?

Section 42.5 Radiation and Radioactivity

22. The barium isotope ^{133}Ba has a half-life of 10.5 years. A sample begins with 1.0×10^{10} ^{133}Ba atoms. How many are left after (a) 2 years, (b) 20 years, and (c) 200 years?

23. The cadmium isotope ^{109}Cd has a half-life of 462 days. A sample begins with 1.0×10^{12} ^{109}Cd atoms. How many are left after (a) 50 days, (b) 500 days, and (c) 5000 days?

24. The radioactive hydrogen isotope ^{3}H is called *tritium*.
    a. What are the decay mode and the daughter nucleus of tritium?
    b. What are the time constant and the decay rate of tritium?

25. How many half-lives must elapse until (a) 90% and (b) 99% of a radioactive sample of atoms has decayed?

26. What is the age in years of a bone in which the ^{14}C/^{12}C ratio is measured to be 1.65×10^{-13}?

27. What is the half-life in days of a radioactive sample with 5.0×10^{15} atoms and an activity of 5.0×10^{8} Bq?

Section 42.6 Nuclear Decay Mechanisms

28. Identify the unknown isotope X in the following decays.
    a. ^{234}U $\rightarrow$ X + α.
    b. ^{32}P $\rightarrow$ X + e^{-} + $\bar{\nu}$
    c. X $\rightarrow$ ^{30}Si + e^{+} + ν
    d. ^{24}Na $\rightarrow$ ^{24}Mg + e^{-} + $\bar{\nu}$ $\rightarrow$ X + γ

29. Identify the unknown isotope X in the following decays.
    a. X $\rightarrow$ ^{224}Ra + α
    b. X $\rightarrow$ ^{207}Pb + e^{-} + $\bar{\nu}$
    c. ^{7}Be + e^{-} $\rightarrow$ X + ν
    d. X $\rightarrow$ ^{60}Ni + γ

30. What is the energy (in MeV) released in the alpha decay of ^{239}Pu?

31. What is the energy (in MeV) released in the alpha decay of ^{228}Th?

32. What is the total energy (in MeV) released in the beta-minus decay of ^{3}H?
    Hint: The daughter $^{A}Y_{Z-1}$ is a positive ion. Tabulated masses are for neutral atom.

33. What is the total energy (in MeV) released in the beta-minus decay of ^{19}O? See the hint for Problem 32.

34. What is the total energy (in MeV) released in the beta decay of a neutron?

Section 42.7 Biological Applications of Nuclear Physics

35. A 50 kg laboratory worker is exposed to 20 mJ of beta radiation with RBE = 1.5. What is the dose in mrem?

36. How many rad of gamma-ray photons cause the same biological damage as 30 rad of alpha radiation with an RBE of 15?

37. 150 rad of gamma radiation are directed into a 150 g tumor during radiation therapy. How much energy does the tumor absorb?

38. The doctors planning a radiation therapy treatment have determined that a 100 g tumor needs to receive 0.20 J of gamma radiation. What is the dose in rads?

Problems

39. a. What initial speed must an alpha particle have to just touch the surface of a ^{197}Au gold nucleus before being turned back?

b. What is the initial energy (in MeV) of the alpha particle?
    Hint: The alpha particle is not a point particle.

40. Particle accelerators fire protons at target nuclei for investigators to study the nuclear reactions that occur. In one experiment, the proton needs to have 20 MeV of kinetic energy as it impacts a ^{207}Pb nucleus. With what initial kinetic energy (in MeV) must the proton be fired toward the lead target?
    Hint: The proton is not a point particle.

41. Stars are powered by nuclear reactions that fuse hydrogen into helium. The fate of many stars, once most of the hydrogen is used up, is to collapse, under gravitational pull, into a *neutron star*. The force of gravity becomes so large that protons and electrons are fused into neutrons in the reaction p^{+} + e^{-} $\rightarrow$ n + ν. The entire star is then a tightly packed ball of neutrons with the density of nuclear matter.
    a. Suppose the sun collapses into a neutron star. What will its radius be? Give your answer in km.
    b. The sun's rotation period is now 27 days. What will its rotation period be after it collapses?
    Rapidly rotating neutron stars emit pulses of radio waves at the rotation frequency and are known as *pulsars.*

42. The chemical atomic mass of hydrogen, with the two stable isotopes ^{1}H and ^{2}H (deuterium), is 1.00798. Use this value to determine the natural abundance of these two isotopes.

43. You learned in Chapter 41 that the binding energy of the electron in a hydrogen atom is 13.6 eV.
    a. By how much does the mass decrease when a hydrogen atom is formed from a proton and an electron? Give your answer both in atomic mass units and as a percentage of the mass of the hydrogen atom.
    b. By how much does the mass decrease when a helium nucleus is formed from two protons and two neutrons? Give your answer both in atomic mass units and as a percentage of the mass of the helium nucleus.
    c. Compare your answers to parts a and b. Why do you hear it said that mass is "lost" in nuclear reactions but not in chemical reactions?

44. Use the graph of binding energy to estimate the total energy released if a nucleus with mass number 240 fissions into two nuclei with mass number 120.

45. Use the graph of binding energy to estimate the total energy released if three ^{4}He nuclei fuse together to form a ^{12}C nucleus.

46. Could a ^{56}Fe nucleus fission into two ^{28}Al nuclei? Your answer, which should include some calculations, should be based on the curve of binding energy.

47. a. What are the isotopic symbols of all $A = 17$ isobars?
    b. Which of these are stable nuclei?
    c. For those that are not stable, identify both the decay mode and the daughter nucleus.

48. a. What are the isotopic symbols of all $A = 19$ isobars?
    b. Which of these are stable nuclei?
    c. For those that are not stable, identify both the decay mode and the daughter nucleus.

49. Derive Equation 42.19 from Equation 42.15.

50. What energy (in MeV) alpha particle has a de Broglie wavelength equal to the diameter of a ^{238}U nucleus?

51. What is the activity in Bq and in Ci of a 2.0 mg sample of ^{3}H?

52. The activity of an ^{39}Ar sample is 1.25×10^{9} Bq. What is the mass of the sample?

53. The activity of a sample of the cesium isotope ^{137}Cs is 2.0×10^8 Bq. Many years later, after the sample has fully decayed, how many beta particles will have been emitted?

54. A 115 mCi radioactive tracer is made in a nuclear reactor. When it is delivered to a hospital 16 hours later its activity is 95 mCi. The lowest usable level of activity is 10 mCi.
    a. What is the tracer's half-life?
    b. For how long after delivery is the sample usable?

55. The radium isotope ^{223}Ra, an alpha emitter, has a half-life of 11.43 days. You happen to have a 1.0 g cube of ^{223}Ra, so you decide to use it to boil water for tea. You fill a well-insulated container with 100 mL of water at 18° C and drop in the cube of radium.
    a. How long will it take the water to boil?
    b. Will the water have been altered in any way by this method of boiling? If so, how?

56. A sample of 1.0×10^{10} atoms that decay by alpha emission has a half-life of 100 min. How many alpha particles are emitted between $t = 50$ min and $t = 200$ min?

57. A sample contains radioactive atoms of two types, A and B. Initially there are five times as many A atoms as there are B atoms. Two hours later, the numbers of the two atoms are equal. The half-life of A is 0.50 hours. What is the half-life of B?

58. Radioactive isotopes often occur together in mixtures. Suppose a 100 g sample contains ^{131}Ba, with a half-life of 12 days, and ^{47}Ca, with a half-life of 4.5 days. If there are initially twice as many calcium atoms as there are barium atoms, what will be the ratio of calcium atoms to barium atoms 2.5 weeks later?

59. The technique known as potassium-argon dating is used to date old lava flows. The potassium isotope ^{40}K has a 1.28 billion year half-life and is naturally present at very low levels. ^{40}K decays by beta emission into ^{40}Ar. Argon is a gas, and there is no argon in flowing lava because the gas escapes. Once the lava solidifies, any argon produced in the decay of ^{40}K is trapped inside and cannot escape. A geologist brings you a piece of solidified lava in which you find the ^{40}Ar/^{40}K ratio to be 0.12. What is the age of the rock?

60. The half-life of the uranium isotope ^{235}U is 700 million years. The earth is approximately 4.5 billion years old. How much more ^{235}U was there when the earth formed than there is today? Give your answer as the then-to-now ratio.

61. A 75 kg patient swallows a 30 μCi beta emitter that is to be used as a tracer. The isotope's half-life is 5.0 days. The average energy of the beta particles is 0.35 MeV with an RBE of 1.5. Ninety percent of the beta particles are absorbed within the patient's body and 10% escape. What total dose (in mrem) does the patient receive?

62. What dose in rads of gamma radiation must be absorbed by a block of ice at 0°C to transform the entire block to liquid water at 0°C?

63. A chest x ray uses 10 keV photons with an RBE of 0.85. A 60 kg person receives a 30 mrem dose from one x ray that exposes 25% of the patient's body. How many x ray photons are absorbed in the patient's body?

64. The rate at which a radioactive tracer is lost from a patient's body is the rate at which the isotope decays *plus* the rate at which the element is excreted from the body. Medical experiments have shown that stable isotopes of a particular element are excreted with a 6.0 day half-life. A radioactive isotope of

the same element has a half-life of 9.0 days. What is the effective half-life of the isotope in a patient's body?

65. The plutonium isotope ^{239}Pu has a half-life of 24,000 years and decays by the emission of a 5.2 MeV alpha particle. Plutonium is not especially dangerous if handled because the activity is low and the alpha radiation doesn't penetrate the skin. However, there are serious health concerns if even the tiniest speck of plutonium is inhaled and lodges deep in the lungs. This could happen following any kind of fire or explosion that disperses plutonium as dust. Let's determine the level of danger.
    a. Soot particles are roughly 1 μm in diameter, and it is known that these particles can go deep into the lungs. How many atoms are in a 1.0-μm-diameter particle of ^{239}Pu? The density of plutonium is 19,800 kg/m^3.
    b. What is the activity, in Bq, of a 1.0-μm-diameter particle?
    c. The activity of the particle is very small, but the penetrating power of alpha particles is also very small. The alpha particles are all stopped, and each deposits its energy in a 50-μm-diameter sphere around the particle. What is the dose, in rem/year, to this small sphere of tissue in the lungs? Use an average RBE of 15 and assume that the tissue density is that of water.
    d. Is this exposure likely to be significant? How does it compare to the natural background of radiation exposure?

Challenge Problems

66. The uranium isotope ^{238}U is naturally present at low levels in many soils. One of the nuclei in the decay series of ^{238}U is the radon isotope ^{222}Rn with $t_{1/2} = 3.82$ days. Radon is a gas, and it tends to seep from soil into basements. The Environmental Protection Agency recommends that homeowners take steps to remove radon, by pumping in fresh air, if the radon activity exceeds 4 pCi per liter of air.
    a. How many ^{222}Rn atoms are there in 1 m^3 of air if the activity is 4 pCi/L?
    b. The range of alpha particles in air is $\approx$3 cm. Suppose we model a person as a 180-cm-tall, 25-cm-diameter cylinder with a mass of 65 kg. Only decays within 3 cm of the cylinder can cause exposure, and only $\approx$50% of the decays direct the alpha particle toward the person. Determine the dose in mrem per year for a person who spends the entire year in a room where the activity is 4 pCi/L. Assume an average RBE of 15.
    c. Does the EPA recommendation seem appropriate? Why or why not?

67. Estimate the stopping distance in air of a 5.0 MeV alpha particle. Assume that the particle loses an average 30 eV per collision.

68. Beta-plus decay is $^{A}X_Z \rightarrow {}^{A}Y_{Z-1} + e^+ + \nu$.
    a. Determine the mass threshold for beta-plus decay. That is, what is the minimum atomic mass m_X for which this decay is energetically possible? Your answer will be in terms of the atomic mass m_Y and the electron mass m_e.
    Hint: Start with the nuclear masses, then add an equal number of electrons to both sides of the reaction to get atomic masses.
    b. Can ^{13}N undergo beta-plus decay into ^{13}C? If so, how much energy is released in the decay?

69. All the very heavy atoms found in the earth were created long ago by nuclear fusion reactions in a supernova, an exploding star. The debris spewed out by the supernova later coalesced into the gases from which the sun and the planets of our solar system were formed. Nuclear physics suggests that the uranium isotopes ^{235}U and ^{238}U should have been created in roughly equal numbers. Today, 99.28% of uranium is ^{238}U and only 0.72% is ^{235}U. How long ago did the supernova occur?

70. It might seem strange that in beta decay the positive proton, which is repelled by the positive nucleus, remains in the nucleus while the negative electron, which is attracted to the nucleus, is ejected. To understand beta decay, let's analyze the decay of a free neutron that is at rest in the laboratory. We'll ignore the antineutrino and consider the decay n → p$^+$ + e$^-$. The analysis requires the use of relativistic energy and momentum, from Chapter 36.
    a. What is the total kinetic energy, in MeV, of the proton and electron?
    b. Write the equation that expresses the conservation of relativistic energy for this decay. Your equation will be in terms of the three masses m_n, m_p, and m_e and the relativistic factors γ_p and γ_e. Then rearrange your equation to get the mass loss $\Delta m = m_n - m_p - m_e$ on one side.
    c. Write the equation that expresses the conservation of relativistic momentum for this decay. Let v represent speed, rather than velocity, then write any minus signs explicitly.
    d. You have two simultaneous equations in the two unknowns v_p and v_e. To help in solving these, first prove that $\gamma v = (\gamma^2 - 1)^{1/2} c$.
    e. Solve for v_p and v_e. (It's easiest to solve for γ_p and γ_e, then find v from γ.) First get an algebraic expression for each, in terms of the masses. Then evaluate each, giving v as a fraction of c.
    f. Calculate the kinetic energy in MeV of the proton and the electron. Verify that their sum matches your answer to part a.
    g. Now explain why the electron is ejected in beta decay while the proton remains in the nucleus.

71. Alpha decay occurs when an alpha particle tunnels through the Coulomb barrier. Figure CP42.71 shows a simple one-dimensional model of the potential-energy well of an alpha particle in a nucleus with $A \approx 235$. The 15 fm width of this one-dimensional potential-energy well is the *diameter* of the nucleus. Further, to keep the model simple, the Coulomb barrier has been modeled as a 20-fm-wide, 30-MeV-high rectangular potential-energy barrier. The goal of this problem is to calculate the half-life of an alpha particle in the energy level $E = 5.0$ MeV.
    a. What is the kinetic energy of the alpha particle while inside the nucleus? What is its kinetic energy after it escapes from the nucleus?
    b. Consider the alpha particle within the nucleus to be a point particle bouncing back and forth with the kinetic energy you stated in part a. What is the particle's *collision rate*, the number of times per second it collides with the wall of the potential-energy barrier?
    c. What is the tunneling probability P_{tunnel}?
    d. P_{tunnel} is the probability that on any one collision with the wall the alpha particle tunnels through instead of reflecting. The probability of *not* tunneling is $1 - P_{tunnel}$. Hence the probability that the alpha particle is still inside the nucleus after N collisions is $(1 - P_{tunnel})^N \approx 1 - NP_{tunnel}$, where we've used the binomial approximation because $P_{tunnel} \ll 1$. The half-life is the *time* at which half the nuclei have not yet decayed. Use this information to determine (in years) the half-life of this nucleus.

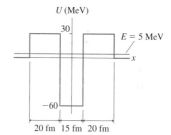

FIGURE CP42.71

STOP TO THINK ANSWERS

Stop to Think 42.1: 3. Different isotopes of an element have different numbers of neutrons but the same number of protons. The number of electrons in a neutral atom matches the number of protons.

Stop to Think 42.2: d. To keep A constant, increasing N by 1 requires decreasing Z by 1.

Stop to Think 42.3: No. A Geiger counter responds only to ionizing radiation. Visible light is not ionizing radiation.

Stop to Think 42.4: c. One-quarter of the atoms are left. This is one-half of one-half, or $(1/2)^2$.

Stop to Think 42.5: b. An increase of Z with no change in A occurs when a neutron changes to a proton and an electron, ejecting the electron.

Relativity and Quantum Physics

Niels Bohr was right on target with his remark, "Anyone who is not shocked by quantum theory has not understood it." Quantum mechanics *is* shocking. The predictability of Newtonian physics has been replaced by a mysterious world in which physical entities that by all rights should be waves sometimes act like particles. Electrons and neutrons somehow produce wave-like interference with themselves. These discoveries stood common sense on its head.

According to quantum mechanics, the wave function and its associated probabilities are *all we can know* about an atomic particle. This idea is so unsettling that many great scientists were reluctant to accept it. Einstein famously said, "God does not play dice with the universe." But Einstein was wrong. As strange as it seems, this is the way that nature really is.

As we conclude our journey into physics, the knowledge structure for Part VII summarizes the important ideas of relativity and quantum physics. Whether you're shocked or not, these are the scientific theories behind the emerging technologies of the 21st century.

KNOWLEDGE STRUCTURE VII Relative and Quantum Physics

ESSENTIAL CONCEPTS	Reference frame, event, atom, photon, quantization, wave function, probability density
BASIC GOALS	What are the properties and characteristics of space and time?
	How do we know about light and atoms?
	How are atomic and nuclear phenomena explained by energy levels, wave functions, and photons?

GENERAL PRINCIPLES	**Principle of relativity**	All the laws of physics are the same in all inertial reference frames.
	Schrödinger's equation	$\dfrac{d^2\psi}{dx^2} = -\dfrac{2m}{\hbar^2}[E - U(x)]\psi(x)$
	Pauli exclusion principle	No more than one electron can occupy the same quantum state.
	Uncertainty principle	$\Delta x \Delta p \geq h/2$

RELATIVITY It follows from the principle of relativity that:

- The speed of light c is the same in all inertial reference frames. No particle or causal influence can travel faster than c.

- Length contraction: The length of an object in a reference frame in which the object moves with speed v is

$$L = \sqrt{1 - \beta^2}\, \ell \leq \ell$$

where ℓ is the proper length and $\beta = v/c$.

- Time dilation: The proper time interval $\Delta\tau$ between two events is measured in a reference frame in which the two events occur at the same position. The time interval Δt in a frame moving with relative speed v is

$$\Delta t = \Delta\tau/\sqrt{1 - \beta^2} \geq \Delta\tau$$

- $E = mc^2$ is the energy equivalent of mass. Mass can be transformed into energy and energy into mass.

QUANTUM PHYSICS Quantum systems are described by a wave function $\psi(x)$.

- The probability that a particle will be found in the narrow interval δx at position x is Prob(in δx at x) = $P(x)\,\delta x$. The probability density is $P(x) = |\psi(x)|^2$.

- The wave function must be normalized

$$\int_{-\infty}^{\infty} |\psi(x)|^2\, dx = 1$$

- The wave function can penetrate into a classically forbidden region with penetration distance

$$\eta = \hbar/\sqrt{2m(U_0 - E)}$$

- A particle can tunnel through an energy barrier of height U_0 and width w with probability $P_{\text{tunnel}} = e^{-2w/\eta}$

Properties of light

- A photon of light of frequency f has energy $E_{\text{photon}} = hf$.

- Photons are emitted and absorbed on an all-or-nothing basis.

Properties of atoms

- Quantized energy levels, found by solving the Schrödinger equation, depend on quantum numbers n and l.

- An atom can jump from one state to another by emitting or absorbing a photon of energy $E_{\text{photon}} = \Delta E_{\text{atom}}$

- The ground-state electron configuration is the lowest-energy configuration consistent with the Pauli principle.

Properties of nuclei

- The nucleus is held together by the strong force, an attractive short-range force between any two nucleons.

- Nuclei are stable only if the proton and neutron number fall along the line of stability.

- Unstable nuclei decay by alpha, beta, or gamma decay. The number of nuclei decreases exponentially with time.

Quantum Computers

All the systems we studied in Part VII were in a single, well-defined quantum state. For example, a hydrogen atom was in the $1s$ state or, perhaps, the $2p$ state. But there's another possibility. Some quantum systems can exist in a *superposition* of two or more quantum states.

We hinted at the possibility of superposition when we re-examined the double-slit interference experiment in the light of quantum physics. We noted that a photon or electron must, in some sense, go through both slits and then interfere with itself to produce the dot-by-dot buildup of an interference pattern on the screen. Suppose we say that an electron that has passed through the top slit in the figure is in quantum state ψ_a. An electron that has passed through the bottom slit is in state ψ_b.

The electron through the top slit is in state ψ_a.

The electron behind the slits is in the superposition state $\psi = \alpha\psi_a + \beta\psi_b$.

Incident electron

The electron through the bottom slit is in state ψ_b.

FIGURE PSVII.1 The electron emerging from the double slit is in a superposition state.

To say that the electron goes through both slits is to say that the electron emerges from the double slit in the *superposition state* $\psi = \alpha\psi_a + \beta\psi_b$, where the coefficients α and β must satisfy $\alpha^2 + \beta^2 = 1$. (Notice that this is like finding the magnitude of a vector from its components.) If we were to detect the electron, α^2 and β^2 are the probabilities that we would find it to be in state ψ_a or state ψ_b, respectively. But until we detect it, the electron exists in the superposition of *both* state ψ_a and ψ_b. It is this superposition that allows the electron to interfere with itself to produce the interference pattern.

But what does this have to do with computers? As you know, everything a modern digital computer does, from surfing the Internet to crunching numbers, is accomplished by manipulating binary strings of 0s and 1s. The *concept* of computing with binary bits goes back to Charles Babbage in the mid-19th century, but it wasn't until the mid-20th century that scientists and engineers developed the technology that gives this concept a physical representation.

A binary bit is always a 1 or a 0; there's no in-between state. These are represented in a modern microprocessor by small capacitors that are either charged or uncharged. Suppose we wanted to represent information not with capacitors but with a quantum system that has two states. We could say that the system represents a 0 when it is in state ψ_a and a 1 when it is in ψ_b. Such a quantum system is an ordinary binary bit as long as the system is in one state or the other.

But the quantum system, unlike a classical bit, has the possibility of being in a superposition state. Using 0 and 1, rather than ψ_a and ψ_b, we could say that the system can be the state $\psi = \alpha \cdot 0 + \beta \cdot 1$. This basic unit of quantum computing is called a *qubit*. It may seem at first that we could do the same thing with a classical system by allowing the capacitor charge to vary, but a partially charged capacitor is still a single, well-defined state. In contrast, the qubit—like the electron that goes through both slits—is simultaneously in both state 0 *and* state 1.

To illustrate the possibilities, suppose you have three classical bits and three qubits. The three bits can represent eight different numbers (000 to 111), but only one at a time. The three qubits represent all eight numbers *simultaneously*. To perform a mathematical operation, you must do it eight times on the three bits to learn all the possible outcomes. But you would learn all eight outcomes simultaneously from *one* operation on the three qubits. In general, computing with n qubits provides a theoretical improvement of 2^n over computing with n bits.

We say "theoretical" because quantum computing is still mostly in the concept stage, much as digital computers were 150 years ago. What kind of quantum systems can actually be placed in an appropriate superposition state? How do you manipulate qubits? How do you read information in and out? What kinds of computations would be improved by quantum computing?

These are all questions that are being actively researched today. Quantum computing is in its infancy, and the technology for making a real quantum computer is largely unknown. Just as Charles Babbage couldn't possibly have imagined today's computers, the uses of tomorrow's quantum computers are still unforeseen. But, quite possibly, there are uses that some of you may help to invent.

FIGURE PSVII.2 This string of beryllium ions held in an ion trap is being studied as a possible quantum computer. The quantum states of the ions are manipulated with laser beams.

Mathematics Review

Algebra

Using exponents:

$$a^{-x} = \frac{1}{a^x} \qquad a^x a^y = a^{(x+y)} \qquad \frac{a^x}{a^y} = a^{(x-y)} \qquad (a^x)^y = a^{xy}$$

$$a^0 = 1 \qquad a^1 = a \qquad a^{1/n} = \sqrt[n]{a}$$

Fractions:

$$\left(\frac{a}{b}\right)\left(\frac{c}{d}\right) = \frac{ac}{bd} \qquad \frac{a/b}{c/d} = \frac{ad}{bc} \qquad \frac{1}{1/a} = a$$

Logarithms:

If $a = e^x$, then $\ln(a) = x$ $\qquad \ln(e^x) = x \qquad e^{\ln(x)} = x$

$$\ln(ab) = \ln(a) + \ln(b) \qquad \ln\left(\frac{a}{b}\right) = \ln(a) - \ln(b) \qquad \ln(a^n) = n\ln(a)$$

The expression $\ln(a + b)$ cannot be simplified.

Linear equations: The graph of the equation $y = ax + b$ is a straight line. a is the slope of the graph. b is the y-intercept.

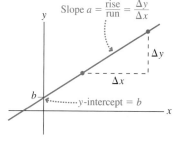

Proportionality: To say that y is proportional to x, written $y \propto x$, means that $y = ax$, where a is a constant. Proportionality is a special case of linearity. A graph of a proportional relationship is a straight line that passes through the origin. If $y \propto x$, then

$$\frac{y_1}{y_2} = \frac{x_1}{x_2}$$

Quadratic equation: The quadratic equation $ax^2 + bx + c = 0$ has the two solutions $x = \dfrac{-b \pm \sqrt{b^2 - 4ac}}{2a}$.

Geometry and Trigonometry

Area and volume:

Rectangle
$A = ab$

Rectangular box
$V = abc$

Triangle
$A = \frac{1}{2}ab$

Right circular cylinder
$V = \pi r^2 l$

Circle
$C = 2\pi r$
$A = \pi r^2$

Sphere
$A = 4\pi r^2$
$V = \frac{4}{3}\pi r^3$

Arc length and angle: The angle θ in radians is defined as $\theta = s/r$.

The arc length that spans angle θ is $s = r\theta$.

$2\pi \text{ rad} = 360°$

Right triangle: Pythagorean theorem $c = \sqrt{a^2 + b^2}$ or $a^2 + b^2 = c^2$

$$\sin\theta = \frac{b}{c} = \frac{\text{far side}}{\text{hypotenuse}} \qquad \theta = \sin^{-1}\!\left(\frac{b}{c}\right)$$

$$\cos\theta = \frac{a}{c} = \frac{\text{adjacent side}}{\text{hypotenuse}} \qquad \theta = \cos^{-1}\!\left(\frac{a}{c}\right)$$

$$\tan\theta = \frac{b}{a} = \frac{\text{far side}}{\text{adjacent side}} \qquad \theta = \tan^{-1}\!\left(\frac{b}{a}\right)$$

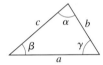

General triangle: $\alpha + \beta + \gamma = 180° = \pi \text{ rad}$

Law of cosines $c^2 = a^2 + b^2 - 2ab\cos\gamma$

Identities:

$$\tan\alpha = \frac{\sin\alpha}{\cos\alpha} \qquad\qquad \sin^2\alpha + \cos^2\alpha = 1$$

$$\sin(-\alpha) = -\sin\alpha \qquad\qquad \cos(-\alpha) = \cos\alpha$$

$$\sin(\alpha \pm \beta) = \sin\alpha\cos\beta \pm \cos\alpha\sin\beta \qquad \cos(\alpha \pm \beta) = \cos\alpha\cos\beta \mp \sin\alpha\sin\beta$$

$$\sin(2\alpha) = 2\sin\alpha\cos\alpha \qquad\qquad \cos(2\alpha) = \cos^2\alpha - \sin^2\alpha$$

$$\sin(\alpha \pm \pi/2) = \pm\cos\alpha \qquad\qquad \cos(\alpha \pm \pi/2) = \mp\sin\alpha$$

$$\sin(\alpha \pm \pi) = -\sin\alpha \qquad\qquad \cos(\alpha \pm \pi) = -\cos\alpha$$

Expansions and Approximations

Binomial expansion: $(1 + x)^n = 1 + nx + \dfrac{n(n-1)}{2}x^2 + \ldots$

Binomial approximation: $(1 + x)^n \approx 1 + nx$ if $x \ll 1$

Trigonometric expansions: $\sin\alpha = \alpha - \dfrac{\alpha^3}{3!} + \dfrac{\alpha^5}{5!} - \dfrac{\alpha^7}{7!} + \ldots$ for α in rad

$\cos\alpha = 1 - \dfrac{\alpha^2}{2!} + \dfrac{\alpha^4}{4!} - \dfrac{\alpha^6}{6!} + \ldots$ for α in rad

Small-angle approximation: If $\alpha \ll 1$ rad, then $\sin\alpha \approx \tan\alpha \approx \alpha$ and $\cos\alpha \approx 1$.

The small-angle approximation is excellent for $\alpha < 5°$ (≈ 0.1 rad) and generally acceptable up to $\alpha \approx 10°$.

Periodic Table of Elements

Key: Atomic number — 27 · Symbol — Co · Atomic mass — 58.9

Main Table

Period	1	2	3	4	5	6	7	8	9	10	11	12	13	14	15	16	17	18
1	1 H 1.0																	2 He 4.0
2	3 Li 6.9	4 Be 9.0											5 B 10.8	6 C 12.0	7 N 14.0	8 O 16.0	9 F 19.0	10 Ne 20.2
3	11 Na 23.0	12 Mg 24.3											13 Al 27.0	14 Si 28.1	15 P 31.0	16 S 32.1	17 Cl 35.5	18 Ar 39.9
4	19 K 39.1	20 Ca 40.1	21 Sc 45.0	22 Ti 47.9	23 V 50.9	24 Cr 52.0	25 Mn 54.9	26 Fe 55.8	27 Co 58.9	28 Ni 58.7	29 Cu 63.5	30 Zn 65.4	31 Ga 69.7	32 Ge 72.6	33 As 74.9	34 Se 79.0	35 Br 79.9	36 Kr 83.8
5	37 Rb 85.5	38 Sr 87.6	39 Y 88.9	40 Zr 91.2	41 Nb 92.9	42 Mo 95.9	43 Tc 96.9	44 Ru 101.1	45 Rh 102.9	46 Pd 106.4	47 Ag 107.9	48 Cd 112.4	49 In 114.8	50 Sn 118.7	51 Sb 121.8	52 Te 127.6	53 I 126.9	54 Xe 131.3
6	55 Cs 132.9	56 Ba 137.3	57 La 138.9	72 Hf 178.5	73 Ta 180.9	74 W 183.9	75 Re 186.2	76 Os 190.2	77 Ir 192.2	78 Pt 195.1	79 Au 197.0	80 Hg 200.6	81 Tl 204.4	82 Pb 207.2	83 Bi 209.0	84 Po 209.0	85 At 210.0	86 Rn 222.0
7	87 Fr 223.0	88 Ra 226.0	89 Ac 227.0	104 Rf 261	105 Db 262	106 Sg 263	107 Bh 264	108 Hs 269	109 Mt 268	110 Ds 271	111 272	112 285						

Transition elements

Inner Transition Elements

Lanthanides (Period 6):

58 Ce 140.1	59 Pr 140.9	60 Nd 144.2	61 Pm 144.9	62 Sm 150.4	63 Eu 152.0	64 Gd 157.3	65 Tb 158.9	66 Dy 162.5	67 Ho 164.9	68 Er 167.3	69 Tm 168.9	70 Yb 173.0	71 Lu 175.0

Actinides (Period 7):

90 Th 232.0	91 Pa 231.0	92 U 238.0	93 Np 237.0	94 Pu 239.1	95 Am 241.1	96 Cm 244.1	97 Bk 249.1	98 Cf 252.1	99 Es 257.1	100 Fm 257.1	101 Md 258.1	102 No 259.1	103 Lr 262.1

Atomic and Nuclear Data

Atomic Number (Z)	Element	Symbol	Mass Number (A)	Atomic Mass (u)	Percent Abundance	Decay Mode	Half-Life $t_{1/2}$
0	(Neutron)	n	1	1.008 665		β^-	10.4 min
1	Hydrogen	H	1	1.007 825	99.985	stable	
	Deuterium	D	2	2.014 102	0.015	stable	
	Tritium	T	3	3.016 049		β^-	12.33 yr
2	Helium	He	3	3.016 029	0.000 1	stable	
			4	4.002 602	99.999 9	stable	
			6	6.018 886		β^-	0.81 s
3	Lithium	Li	6	6.015 121	7.50	stable	
			7	7.016 003	92.50	stable	
			8	8.022 486		β^-	0.84 s
4	Beryllium	Be	7	7.016 928		EC	53.3 days
			9	9.012 174	100	stable	
			10	10.013 534		β^-	1.5×10^6 yr
5	Boron	B	10	10.012 936	19.90	stable	
			11	11.009 305	80.10	stable	
			12	12.014 352		β^-	0.020 2 s
6	Carbon	C	10	10.016 854		β^+	19.3 s
			11	11.011 433		β^+	20.4 min
			12	12.000 000	98.90	stable	
			13	13.003 355	1.10	stable	
			14	14.003 242		β^-	5 730 yr
			15	15.010 599		β^-	2.45 s
7	Nitrogen	N	12	12.018 613		β^+	0.011 0 s
			13	13.005 738		β^+	9.96 min
			14	14.003 074	99.63	stable	
			15	15.000 108	0.37	stable	
			16	16.006 100		β^-	7.13 s
			17	17.008 450		β^-	4.17 s
8	Oxygen	O	14	14.008 595		EC	70.6 s
			15	15.003 065		β^+	122 s
			16	15.994 915	99.76	stable	
			17	16.999 132	0.04	stable	
			18	17.999 160	0.20	stable	
			19	19.003 577		β^-	26.9 s
9	Fluorine	F	17	17.002 094		EC	64.5 s
			18	18.000 937		β^+	109.8 min
			19	18.998 404	100	stable	
			20	19.999 982		β^-	11.0 s
10	Neon	Ne	19	19.001 880		β^+	17.2 s
			20	19.992 435	90.48	stable	
			21	20.993 841	0.27	stable	
			22	21.991 383	9.25	stable	

Atomic Number (Z)	Element	Symbol	Mass Number (A)	Atomic Mass (u)	Percent Abundance	Decay Mode	Half-Life $t_{1/2}$
11	Sodium	Na	22	21.994 434		β^+	2.61 yr
			23	22.989 770	100	stable	
			24	23.990 961		β^-	14.96 hr
12	Magnesium	Mg	24	23.985 042	78.99	stable	
			25	24.985 838	10.00	stable	
			26	25.982 594	11.01	stable	
13	Aluminum	Al	27	26.981 538	100	stable	
			28	27.981 910		β^-	2.24 min
14	Silicon	Si	28	27.976 927	92.23	stable	
			29	28.976 495	4.67	stable	
			30	29.973 770	3.10	stable	
			31	30.975 362		β^-	2.62 hr
15	Phosphorus	P	30	29.978 307		β^+	2.50 min
			31	30.973 762	100	stable	
			32	31.973 908		β^-	14.26 days
16	Sulfur	S	32	31.972 071	95.02	stable	
			33	32.971 459	0.75	stable	
			34	33.967 867	4.21	stable	
			35	34.969 033		β^-	87.5 days
17	Chlorine	Cl	35	34.968 853	75.77	stable	
			36	35.968 307		β^-	3.0×10^5 yr
			37	36.965 903	24.23	stable	
18	Argon	Ar	36	35.967 547	0.34	stable	
			38	37.962 732	0.06	stable	
			39	38.964 314		β^-	269 yr
			40	39.962 384	99.60	stable	
			42	41.963 049		β^-	33 yr
19	Potassium	K	39	38.963 708	93.26	stable	
			40	39.964 000	0.01	β^+	1.28×10^9 yr
			41	40.961 827	6.73	stable	
20	Calcium	Ca	40	39.962 591	96.94	stable	
			42	41.958 618	0.64	stable	
			43	42.958 767	0.13	stable	
			44	43.955 481	2.08	stable	
			47	46.954 547		β^-	4.5 days
			48	47.952 534	0.18	stable	
24	Chromium	Cr	50	49.946 047	4.34	stable	
			51	50.944 772		EC	27 days
			52	51.940 511	83.79	stable	
			53	52.940 652	9.50	stable	
			54	53.938 883	2.36	stable	
25	Manganese	Mn	55	54.938 048	100	stable	
26	Iron	Fe	54	54.939 613	5.9	stable	
			55	54.938 297		EC	2.7 yr
			56	55.934 940	91.72	stable	

Atomic Number (Z)	Element	Symbol	Mass Number (A)	Atomic Mass (u)	Percent Abundance	Decay Mode	Half-Life $t_{1/2}$
			57	56.935 396	2.1	stable	
			58	57.933 278	0.28	stable	
27	Cobalt	Co	59	58.933 198	100	stable	
			60	59.933 820		β^-	5.27 yr
28	Nickel	Ni	58	57.935 346	68.08	stable	
			60	59.930 789	26.22	stable	
			61	60.931 058	1.14	stable	
			62	61.928 346	3.63	stable	
			64	63.927 967	0.92	stable	
29	Copper	Cu	63	62.929 599	69.17	stable	
			65	64.927 791	30.83	stable	
30	Zinc	Zn	64	63.929 144	48.6	stable	
			66	65.926 035	27.9	stable	
			67	66.927 129	4.1	stable	
			68	67.924 845	18.8	stable	
47	Silver	Ag	107	106.905 091	51.84	stable	
			109	108.904 754	48.16	stable	
48	Cadmium	Cd	106	105.906 457	1.25	stable	
			109	108.904 984		EC	462 days
			110	109.903 004	12.49	stable	
			111	110.904 182	12.80	stable	
			112	111.902 760	24.13	stable	
			113	112.904 401	12.22	stable	
			114	113.903 359	28.73	stable	
			116	115.904 755	7.49	stable	
53	Iodine	I	127	126.904 474	100	stable	
			129	128.904 984		β^-	1.6×10^7 yr
			131	130.906 124		β^-	8 days
54	Xenon	Xe	128	127.903 531	1.9	stable	
			129	128.904 779	26.4	stable	
			130	129.903 509	4.1	stable	
			131	130.905 069	21.2	stable	
			132	131.904 141	26.9	stable	
			133	132.905 906		β^-	5.4 days
			134	133.905 394	10.4	stable	
			136	135.907 215	8.9	stable	
55	Cesium	Cs	133	132.905 436	100	stable	
			137	136.907 078		β^-	30 yr
56	Barium	Ba	131	130.906 931		EC	12 days
			133	132.905 990		EC	10.5 yr
			134	133.904 492	2.42	stable	
			135	134.905 671	6.59	stable	
			136	135.904 559	7.85	stable	
			137	136.905 816	11.23	stable	
			138	137.905 236	71.70	stable	

Atomic Number (Z)	Element	Symbol	Mass Number (A)	Atomic Mass (u)	Percent Abundance	Decay Mode	Half-Life $t_{1/2}$
79	Gold	Au	197	196.966 543	100	stable	
80	Mercury	Hg	198	197.966 743	9.97	stable	
			199	198.968 253	16.87	stable	
			200	199.968 299	23.10	stable	
			201	200.970 276	13.10	stable	
			202	201.970 617	29.86	stable	
			204	203.973 466	6.87	stable	
81	Thallium	Tl	203	202.972 320	29.524	stable	
			205	204.974 400	70.476	stable	
			207	206.977 403		β^-	4.77 min
82	Lead	Pb	204	203.973 020	1.4	stable	
			205	204.974 457		EC	1.5×10^7 yr
			206	205.974 440	24.1	stable	
			207	206.975 871	22.1	stable	
			208	207.976 627	52.4	stable	
			210	209.984 163		α, β^-	22.3 yr
			211	210.988 734		β^-	36.1 min
83	Bismuth	Bi	208	207.979 717		EC	3.7×10^5 yr
			209	208.980 374	100	stable	
			211	210.987 254		α	2.14 min
			215	215.001 836		β^-	7.4 min
84	Polonium	Po	209	208.982 405		α	102 yr
			210	209.982 848		α	138.38 days
			215	214.999 418		α	0.001 8 s
			218	218.008 965		α, β^-	3.10 min
85	Astatine	At	218	218.008 685		α, β^-	1.6 s
			219	219.011 294		α, β^-	0.9 min
86	Radon	Rn	219	219.009 477		α	3.96 s
			220	220.011 369		α	55.6 s
			222	222.017 571		α, β^-	3.823 days
87	Francium	Fr	223	223.019 733		α, β^-	22 min
88	Radium	Ra	223	223.018 499		α	11.43 days
			224	224.020 187		α	3.66 days
			226	226.025 402		α	1 600 yr
			228	228.031 064		β^-	5.75 yr
89	Actinium	Ac	227	227.027 749		α, β^-	21.77 yr
			228	228.031 015		β^-	6.15 hr
90	Thorium	Th	227	227.027 701		α	18.72 days
			228	228.028 716		α	1.913 yr
			229	229.031 757		α	7 300 yr
			230	230.033 127		α	75.000 yr
			231	231.036 299		α, β^-	25.52 hr
			232	232.038 051	100	α	1.40×10^{10} yr
			234	234.043 593		β^-	24.1 days

Atomic Number (Z)	Element	Symbol	Mass Number (A)	Atomic Mass (u)	Percent Abundance	Decay Mode	Half-Life $t_{1/2}$
91	Protactinium	Pa	231	231.035 880		α	32.760 yr
			234	234.043 300		β^-	6.7 hr
92	Uranium	U	233	233.039 630		α	1.59×10^5 yr
			234	234.040 946		α	2.45×10^5 yr
			235	235.043 924	0.72	α	7.04×10^8 yr
			236	236.045 562		α	2.34×10^7 yr
			238	238.050 784	99.28	α	4.47×10^9 yr
93	Neptunium	Np	236	236.046 560		EC	1.15×10^5 yr
			237	237.048 168		α	2.14×10^6 yr
94	Plutonium	Pu	238	238.049 555		α	87.7 yr
			239	239.052 157		α	2.412×10^4 yr
			240	240.053 808		α	6 560 yr
			242	242.058 737		α	3.73×10^6 yr
			244	244.064 200		α	8.1×10^7 yr

Answers

Answers to Odd-Numbered Exercises and Problems

Solutions to questions posed in the Part Overview captions can be found at the end of this answer list.

Chapter 36

1. 5 m, 1 s; −5 m, 5 s
3. 345 m/s, 15 m/s
5. a. 15 m/s b. −5 m/s c. 11.2 m/s
7. 3.00×10^8 m/s
9. 167 ns
11. 2 μs
13. Bolt 2 first, by 20 μs
15. Simultaneously
17. 0.866c
19. a. 0.8c b. 16 y
21. 4.8 ns
23. Yes
25. 0.78 m
27. 9.5×10^4 m/s
29. (8.25×10^{10} m, 325 s)
31. 0.36c
33. 0.71c
35. 0.8c
37. 0.71c
39. a. 1.8×10^{16} J b. 9.0×10^9
41. 0.943c
43. 50 g ball: 1.33 m/s to the right; 100 g ball: 3.33 m/s to the right
45. 1st ball: 4.0 m/s to the right; 2nd ball: 2.0 m/s to the left
47. 11.2 hr
49. a. No b. 67.1 y
51. a. 0.9965c b. 59.8 ly
53. 4600 kg/m^3
55. a. 8.50 ly, 17 y b. 7.36 ly, 14.7 y
57. 0.96c
59. 0.9997c
61. a. 0.98c b. 8.49×10^{-11} J
63. b. Lengths perpendicular to the motion are not affected.
65. a. $u'_y = u_y/\gamma(1 - u_x v/c^2)$ b. 0.36c
67. 3.87mc
69. 0.786c
71. a. 7.56×10^{16} J b. 0.84 kg
73. 7.5×10^{-13} J
75. 1.06×10^{-12} m
77. 22 m
79. 0.846c
81. Yes

Chapter 37

3. 6.25×10^{10}
5. (5.0×10^{-3} T, out of page)
7. 0.521 μm
9. a. 71.2 eV b. −14.4 eV c. 5.0 keV
11. a. 5.93×10^6 m/s b. 3.10×10^7 m/s c. Alpha particle
15. a. 3, 4, and 5 b. 6, 6, and 6 c. 4, 7, and 8

17. a. ^{2}H b. ^{14}N^{++}
19. a. 82 protons, 82 electrons, 125 neutrons
    b. 1.66×10^7 V, 2.34×10^{21} V/m
21. a. 2 and 3; 2 and 4; 2 and 5; 2 and 6 b. 397.1 nm
23. 121.6 nm, 102.6 nm, 97.3 nm, 95.0 nm
25. a. 6.66 GeV b. 3.63 MeV
27. 0.512 MeV and 939 MeV
29. 173 MeV
31. (0.0457 T, into page)
33. a. 7.2×10^{13} b. 1.16 μA
35. 0.000000000058% occupied, 99.999999999942% empty
37. a. 50,000 kg/m^3 b. 3.2×10^{-10} m c. 1.67×10^{17} kg/m^3
39. a. 57.6 N b. 4.65×10^{-35} N
    c. Very strong, very short range, independent of charge
41. 1.76×10^7 V
43. a. 3.43×10^7 m/s b. 6.14×10^6 V
45. 2.53×10^5 m/s, 65° below the +x-axis
47. a. mg/E_0 b. $-mg/b$ d. 2.40×10^{-18} C e. 15

Chapter 38

3.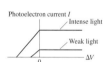

5. 6.25×10^{13}
7. 3.20 eV
9. 3.11 eV
11. a. Aluminum b. 2.0 V
13. a. 4140 nm; infrared b. 414 nm; visible c. 41.4 nm; ultraviolet
15. 497 nm
17. 6.0×10^{-6} V
19. 0.427 nm
21. 0.354 nm
23. a. Yes b. 0.5 eV
25. Yes to $n = 2$, no to $n = 3$
27. a. 69 b. 3.2×10^4 m/s, −0.0029 eV
29. 3.40 eV
33. 91.12 nm
35. a.

n	r_n(nm)	v_n(m/s)	E_n(eV)
1	0.026	4.38×10^6	−54.4
2	0.106	2.19×10^6	−13.6
3	0.238	1.46×10^6	−6.0

37. 1.44
39. a. 1.73×10^{18} b. 1.73×10^{26} photons/s
41. Potassium: a. 5.56×10^{14} Hz b. 540 nm c. 1.08×10^6 m/s d. 3.35 V;
    Gold: a. 1.23×10^{15} Hz b. 244 nm c. 4.4×10^5 m/s d. 0.55 V

43. Sodium
45.

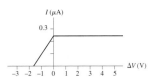

47. a. $p = E/c$ b. $\lambda = h/p$ c. $\lambda = h/mv$
49. a. 2.09×10^{-4} eV b. 1.985 nm c. 3.6 m
51. 0.427 nm
53.

55. a. 6.5 eV b. 355 nm, 276 nm c. Both ultraviolet d. 6.16×10^5 m/s
57. -0.284 eV
59. 1876 nm
61. a. $n = 99$: 518 nm, 2.21×10^4 m/s; $n = 100$: 529 nm, 2.19×10^4 m/s
    b. 6.79×10^9 Hz, 6.59×10^9 Hz c. 6.68×10^9 Hz d. 0.15%
63. 10.28 nm, 7.62 nm, 6.80 nm; ultraviolet
65. 4.16 eV
67. a. ϵ^N b. 2.4 mA c. 4.5×10^6 d. 3.0
69. a. 1.518×10^{-16} s b. 1.32×10^6
71. a. 4.26×10^{-5} nm, 1.31×10^7 m/s b. 0.0164 nm c. X ray
    d. 7.3×10^{13} orbits

Chapter 39

1. 20%, 10%
3. a. 7.7% b. 25%
5. a. 1/6 b. 1/6 c. 5/18
7. 2.0×10^7
11. a. 3333 b. 1111
13. a. 5.0×10^{-3} b. 2.5×10^{-3} c. 0 d. 2.5×10^{-3}
15. a. 0.25 mm^{-1} b.

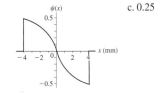

c. 0.25

17. a. 0.354 mm$^{-1/2}$ b.

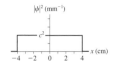

c. 0.25

19. 25 ns
21. 100,000
23. 18 μm
25. 0 m/s $\leq v \leq 2.5 \times 10^7$ m/s
27. 42.5 m
29. 1.0×10^5 pulses/s
31. a.

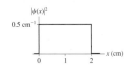

b. 1.0% c. 10^4 d. 0.50 cm^{-1}

33. a. Yes b.

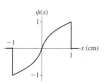

c. 0.000, 0.0050, 0.0010 d. 900
35. a. $\sqrt{3/8}$ mm$^{-1/2}$
    b.

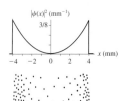

c.

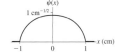

d. 0.125
37. a. 0.27% b. 31.8%
39. a. 0.866 cm$^{-1/2}$
    b.

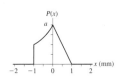

c.

d. 3440
41. a. $a = b$ b.

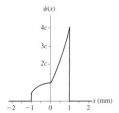

c. Both 0.838 d. 58.1%
43. No; 1.4×10^{-27} m
45. a. $0 \leq v \leq 1.8 \times 10^{10}$ m/s b. Not possible
47. a. 1.5×10^{-13} m; no b. 6.4×10^{13} m
49. a. $b = c$
    b.

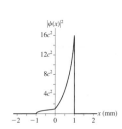

c. 90%

Chapter 40

1. 0.74 nm
3. 0.752 nm
7. 0.135
9. 0.038 eV

11.

13. a.

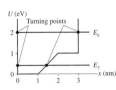

b.

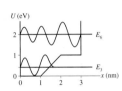

15. 150 nm
17. a. 0.49 eV, 1.46 eV, 2.43 eV b. 640 nm
19. 1.34 N/m
21. 1.25%
25. a. $\lambda = 8mcL^2/3h$ b. 0.795 nm
29. a.

n	b. Most	c. Least	d. Probability	e. Probability
1	$\frac{1}{2}L$	0 and L	$< \frac{1}{3}$	0.195
2	$\frac{1}{4}L, \frac{3}{4}L$	$0, \frac{1}{2}L, L$	$< \frac{1}{3}$	0.402
3	$\frac{1}{6}L, \frac{3}{6}L, \frac{5}{6}L$	$0, \frac{1}{3}L, \frac{2}{3}L, L$	$= \frac{1}{3}$	0.333

31. 4.77×10^7 m/s
35. a. $(\pi b^2)^{-1/4}$ b. $2(\pi b^2)^{-1/2} \int_b^\infty \exp(-x^2/b^2)\, dx$ c. 15.7%
37. a. $\dfrac{1}{2h\sqrt{1-(y/h)}}$ b.

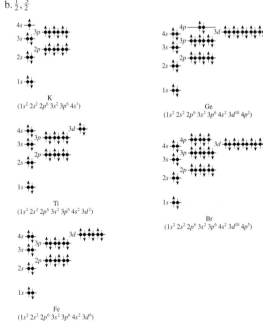

39. a. 4.95 eV b. 4.80 eV c. 4.55 eV
43. b. 0.0091 eV, 0.273 eV, 0.455 eV, 0.0637 eV c. 68 μm; infrared
45. $10^{-1.17\times10^{-32}}$ or $10^{-117000000000000000000000000000000}$

Chapter 41

1. a. $\sqrt{2}\hbar$ b. $\sqrt{12}\hbar$
3. a. f b. -0.85 eV
5. -0.378 eV; $\sqrt{12}\hbar$
7. a. 2 b. 1
11. Si: $1s^2 2s^2 2p^6 3s^2 3p^2$; Ge: $1s^2 2s^2 2p^6 3s^2 3p^6 4s^2 3d^{10} 4p^2$;
    Pb: $1s^2 2s^2 2p^6 3s^2 3p^6 4s^2 3d^{10} 4p^6 5s^2 4d^{10} 5p^6 6s^2 4f^{14} 5d^{10} 6p^2$.
13. a. Fluorine; excited state b. Nickel; ground state
15. $1s^2 3s^1$
19. a. Yes; 2.21 μm b. No; $\Delta l \neq 1$
21. 0.020
23. a. 9.0×10^5 b. 8.7 ns
25. 5.3×10^{22}
27. a. 1.48×10^{-34} Js b. $-1, 0, 1$
    c.

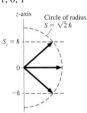

29. $\sqrt{6}\hbar$
31. a. 3.68×10^{-3} b. 5.41×10^{-3} c. 2.93×10^{-3}
37. a.

L_z	S_z	J_z	m_j
$\hbar$	$\frac{1}{2}\hbar$	$\frac{3}{2}\hbar$	$\frac{3}{2}$
$\hbar$	$-\frac{1}{2}\hbar$	$\frac{1}{2}\hbar$	$\frac{1}{2}$
0	$\frac{1}{2}\hbar$	$\frac{1}{2}\hbar$	$\frac{1}{2}$
0	$-\frac{1}{2}\hbar$	$-\frac{1}{2}\hbar$	$-\frac{1}{2}$
$-\hbar$	$\frac{1}{2}\hbar$	$-\frac{1}{2}\hbar$	$-\frac{1}{2}$
$-\hbar$	$-\frac{1}{2}\hbar$	$-\frac{3}{2}\hbar$	$-\frac{3}{2}$

b. $\frac{1}{2}, \frac{3}{2}$

39.

K $(1s^2\, 2s^2\, 2p^6\, 3s^2\, 3p^6\, 4s^1)$

Ge $(1s^2\, 2s^2\, 2p^6\, 3s^2\, 3p^6\, 4s^2\, 3d^{10}\, 4p^2)$

Ti $(1s^2\, 2s^2\, 2p^6\, 3s^2\, 3p^6\, 4s^2\, 3d^2)$

Br $(1s^2\, 2s^2\, 2p^6\, 3s^2\, 3p^6\, 4s^2\, 3d^{10}\, 4p^5)$

Fe $(1s^2\, 2s^2\, 2p^6\, 3s^2\, 3p^6\, 4s^2\, 3d^6)$

41. a. $6s \rightarrow 5p$, $6s \rightarrow 4p$, and $6s \rightarrow 3p$ b. 7300 nm; 1630 nm; 515 nm
43. a.

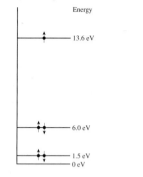

b. 21.68 eV
45. 1.13×10^6 m/s
47.

Transition	a. Wavelength	b. Type	c. Absorption
$2p \rightarrow 2s$	670 nm	VIS	Yes
$3s \rightarrow 2p$	816 nm	IR	No
$3p \rightarrow 2s$	324 nm	UV	Yes
$3p \rightarrow 3s$	2170 nm	IR	No
$3d \rightarrow 2p$	611 nm	VIS	No
$3d \rightarrow 3p$	25 μm	IR	No
$4s \rightarrow 2p$	498 nm	VIS	No
$4s \rightarrow 3p$	2430 nm	IR	No

49. a. 6.25×10^8 s^{-1} b. 0.17 ns
51. 5.72 ns
53. 5.0×10^{16} s^{-1}
55. a. 1.06 μm b. 1.87 W
57. 0.677
61. $1.5a_B$; $5a_B$

Chapter 42

1. a. 1 proton; 2 neutrons b. 18 protons; 22 neutrons
    c. 20 protons; 20 neutrons d. 94 protons; 145 neutrons
3. a. 3.8 fm b. 8.2 fm c. 14.5 fm
5. 3.6×10^{26} protons; 3.6×10^{26} neutrons
7. 1.2×10^{11} kg
9. a. ^{36}S and ^{36}Ar b. 5, 8
11. ^{40}Ar: 344 MeV, 8.59 MeV/nucleon; ^{40}Ca: 342 MeV,
     8.55 MeV/nucleon
13. ^{12}C: 7.68 MeV/nucleon; ^{13}C: 7.47 MeV/nucleon; ^{12}C
15. 20.179 u
17. 8000 N

19. 2.3×10^{-38}
21. a.

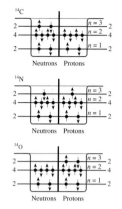

b. ^{14}N stable; ^{14}C beta-minus decay; ^{14}O beta-plus decay
23. a. 9.3×10^{11} b. 4.7×10^{11} c. 5.5×10^8
25. a. 3.32 b. 6.64
27. 80.4 days
29. a. ^{228}Th b. ^{207}Tl c. ^{7}Li d. ^{60}Ni
31. 5.24 MeV
33. 0.4.82 MeV
35. 60 mrem
37. 0.225 J
39. a. 3.50×10^7 m/s b. 25.6 MeV
41. a. 12.7 km b. 779 μs
43. a. 1.46×10^{-8} u; 1.45×10^{-6}% b. 0.0304 u; 0.76%
45. 6.0 MeV
47. a. ^{17}N, ^{17}O, ^{17}F b. ^{17}O
    c. ^{17}F $\rightarrow$ ^{17}O + e$^+$ + ν; ^{17}N $\rightarrow$ ^{17}O + e$^-$ + $\bar{\nu}$
51. 7.16×10^{11} Bq or 19.4 Ci
53. 2.73×10^{17}
55. a. 18.9 s b. No
57. 1.19 hr
59. 200 million years
61. 69.7 mrem
63. 3.31×10^{12}
65. a. 2.61×10^{10} b. 0.0239 Bq c. 1.436×10^7 rem/year
    d. Yes; $\approx$50 million times the natural background.
67. 15 cm
69. $\approx$6 billion years
71. a. 65.0 MeV; 5.0 MeV b. 3.7×10^{21} c. 8.01×10^{-39}
    d. 540 million years

Index

For users of the five-volume edition: pages 1–481 are in Volume 1; pages 482–607 are in Volume 2; pages 608–779 are in Volume 3; pages 780–1194 are in Volume 4; and pages 1148–1383 are in Volume 5.
Pages 1195–1383 are not in the Standard Edition of this textbook.

electric dipole, 795–96
  in electrostatic equilibrium, 870–72, 939–41
  isolated, 792
Conservation laws, 237–38, 338–39
Conservation of angular momentum, 259–60, 401–2
Conservation of charge, 790–91
Conservation of current, 883–84, 892–93
Conservation of energy, 238, 270, 323–28, 339.
    See also Thermodynamics: first law of
  inside capacitor, 903–4
  in charge interactions, 912
  in fluid flow, 465–67
  Kirchhoff's loop law, 938–39
  in motional emf, 1047
  relativity and, 1187–88
  in simple harmonic motion, 423
Conservation of mass, 237–38
Conservation of mechanical energy, 278–79, 318
Conservation of momentum, 246–52, 1180
Conservative forces, 316–18, 901, 902, 906
Constant-pressure (isobaric) processes, 503, 518
Constant-temperature (isothermal) processes, 503–5, 518–19, 525–26
Constant-volume gas thermometer, 492
Constant-volume (isochoric) process, 502–3, 518, 525, 533
Constructive interference, 649, 661, 667, 668–69, 696
Contact forces, 98, 104, 130
Continuity, equation of, 463–65
Continuous spectrum, 758
Contour map, 670–71, 915–17, 918
Converging lens, 732–33, 736
Coordinate axes, 85
Coordinate systems, 7–8, 36, 84–88
  acceleration and, 164
  axes of, 85
  Cartesian, 84
  inertial reference frame, 112–13
  origin of, 7, 10
  with tilted axes, 91–92
Copernicus, Nicholas, 344, 481
Corpuscles, 685, 1197
Correspondence principle, 1291–93
Coulomb electric field, 1061, 1085, 1091
Coulomb's law, 796–802, 1146
  electrostatic forces and, 799
  Gauss's law vs., 861–62
  units of charge, 797–98
  using, 798–802
Couples, 380, 840
Covalent bond, 1305–7
Crests, 626
Critical angle, 725
Critical point, 494
Critical speed, 195
Crookes tubes, 1199–1200
Crossed-field experiment, 1201–2
Crossed polarizers, 1113
Crossover frequency, 1128
Cross product, 397–99, 1003–4

Crystal lattice, 761, 1316
Crystals, 486, 712
Current, 781, 782, 791, 890. *See also*
    Induced current
  batteries, 889–90
  conservation of, 883–84
  conventional, 891
  creating, 884–89
  direction of, 891
  displacement, 1099–1102
  eddy, 1048–49
  effect on compass, 999
  electric field in a wire, 885–87
  electric potential and, 943–46
  electron, 880–84
  of inductor, 1131
  magnetic field of, 1004–8
  model of conduction, 887–89
  peak, 1125
  root-mean-square (rms), 1135–36
  in series *RLC* circuit, 1133
  tunneling, 1310
Current balance, 1026
Current density, 890–93
Current loop, 1006, 1008–9
  forces and torques on, 1026–28
  as magnetic dipole, 1009
  magnetic field of, 1006–9
  parallel, 1026–27
Curve of binding energy, 1358–59
Cyclotron, 1021–22
Cyclotron frequency, 1019–21
Cyclotron motion, 1019–21
Cylindrical symmetry, 850, 852

D

Damped oscillation, 432–34
Damped systems, energy in, 433–34
Damping constant, 432
Daughter nucleus, 1369
Davisson, Clinton, 766, 768
Davisson-Germer experiment, 766
DC circuits, 961, 1122
De Broglie, Louis-Victor, 767
De Broglie wavelength, 767–68, 773, 1231
  Schrödinger equation and, 1278–79
Decay. *See also* Radioactivity
  exponential, 433–34
  nuclear, 1364–66
  rate of, 1341, 1364
Decay equation, 1340–42
Decay mechanisms, 1369–73. *See also*
    Radioactivity
  alpha decay, 929, 1193, 1369–70
  beta decay, 235, 265, 929, 1203, 1362, 1370–71
  decay series, 1372–73
  gamma decay, 1372
  weak interaction, 1371–72
Defibrillator, 953
Degrees, 179
Degrees of freedom, 557, 558–59

Density(ies), 487, 824
  average, 460
  of fluids, 445–47
Destructive interference, 650, 661, 662, 667, 668, 690, 697
Deuterium, 488, 1355
Diatomic gases, 489, 584
Diatomic molecules, thermal energy of, 558–61
Dichroic glass, 646, 665
Diesel cycle, 586, 605
Diffraction, 685, 695
  circular-aperture, 700–702
  of electrons, 768–69
  of matter, 768–70
  of neutron, 769
  order of, 693
  single-slit, 695–99
  x-ray, 760–63
Diffraction grating, 693–95
  reflection gratings, 695, 713
  resolution of, 712
  transmission grating, 695
Diffraction-limited lens, 747
Diffuse reflection, 719
Diode, 963
  resonant tunneling, 1310–11
Dipole(s)
  acceleration of, 840
  electric, 795–96
  electric field of, 821–22, 823
  electric potential of, 920
  induced, 847
  induced magnetic, 1030–31
  magnetic, 998, 1008–11
  in nonuniform electric field, 840–41
  potential energy of, 909–10
  torque on, 839–40
  in uniform electric field, 839–40
Dipole moment, 822
Direct current (DC), 1121
Direct current (DC) circuits, 961, 1122
Discharging, 787, 793–94
Discrete spectrum, 758, 1212
Disk of charge
  electric fields of, 830–32
  electric potential of, 923, 934–35
Disorder, 565–66
Disordered systems, 566–67
Displaced fluid, 459
Displacement, 9–11, 13, 80–83
  algebraic addition to find, 90
  angular, 179–80
  from equilibrium, 281
  graphical addition to find, 81
  net, 80–81, 82, 83
  of sinusoidal wave, 622–23
  velocity and, 49–53, 83
  of wave, 618–20
Displacement current, 1099–1102
  Ampère's law and, 1100–1101
  induced magnetic field and, 1101–2
Displacement vectors, 12, 80, 153